MECOPTICS

WHERE ENERGY COMES FROM AND HOW MUCH

MECOPTICS: Where Energy Comes from and How Much

Also by Larry Bell:

Scared Witless: Prophets and Profits of Climate Doom
Cosmic Musings: Contemplating Life beyond Self
Climate of Corruption: Politics and Power behind the Global Warming Hoax
Reflections on Oceans and Puddles
Reinventing Ourselves: How Technology Is Rapidly and Radically Transforming Humanity
What Makes Humans Truly Exceptional: Including Us
Cyberwarfare: Targeting America
Thinking Whole: Rejecting Half-Witted Left & Right Brain Limitations
The Weaponization of AI and the Internet
Beyond Footprints and Flagpoles: Pioneering Pathways to the Future (with Buzz Aldrin)
Imaginings and Amusements: Carving Doorways to Dreams

Print ISBN 978-1-960405-53-1

eBook ISBN 978-1-960405-54-8

Cover Art Work by Surabhi Sharma

Cover Design by Guy D. Corp—www.GrafixCorp.com

STAIRWAY PRESS

www.StairwayPress.com
1000 West Apache Trail
Suite 126
Apache Junction, AZ 85120 USA

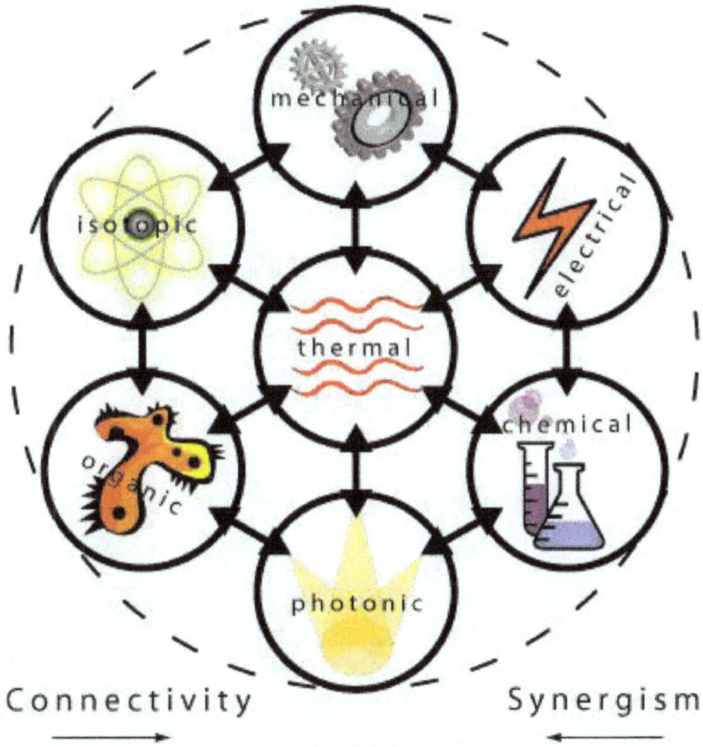

What Does MECOPTICS Mean?

The term "MECOPTICS," incorporates initials of seven general categories of processes that create and convert natural resources. For convenience and comprehension, MECOPTICS processes and devices are grouped into general categories that characterize important features and uses. Each type typically works in combination with others to accomplish desired outcomes.

- Mechanical processes provide kinetic energy to power vehicles and electric generators.
- Electrical processes produce electromagnetic energy for motors, lighting, heating, and a wide range of other uses.
- Chemical processes convert organic and inorganic feedstocks to gaseous and liquid fuels, commercial materials and other products.
- Organic processes use microorganisms to transform plant and animal biomass wastes into methane and fertilizers.
- Photonic processes convert sunlight into leafy plant materials, electricity and energy for water and space heating.
- Thermal processes harness and utilize heat energy for power generation and biomaterial processing.
- Isotopic processes create heat energy through atomic decay to provide power for terrestrial, and outer space uses.
- Connectivity ties multiple processes together that combine capabilities and create new ones.
- Synergy causes processes to work in concert with one another as integrated systems that increase capacities and efficiencies.

A Comprehensive Guide to Ingredients for a Sustainable Energy Future

The global demand for energy is currently growing at an unprecedented rate and is projected to increase globally by as much as 50-100% depending upon what forecasts one sees. There is uncertainty for sure – but the certainty is that growth is in our future.

Today less than 35% of our energy is consumed as electricity with the balance as fuels, petrochemicals, and products – nearly all of which are hydrocarbons. As the world's population expands in the developing world and the basic needs for energy as well, energy must be provided to support societies that cannot be left in poverty. The basic needs of civilization today require energy that is truly sustainable. Sustainable being available and reliable, cost effective and affordable, and environmentally responsible. It is our most challenging aspect of the future we create.

Also, today it is a fact that two-thirds of the energy we produce is wasted before any consumption by people or industry. It is directly related to the types of energy we produce and the manner in which it is translated to consumers. Efficiencies of transmission for electricity, combustion, conversion to useful propulsion in autos, and transportation of all types – it all has massive waste in the translation from production to consumption.

And as if the current state and future we envision is not enough with population and the basic needs of society – here comes data centers, artificial intelligence, etc. along with a host of societal electrification that is massive.

We are not in an energy "transition." That would indicate we are moving away from something to replace it with something else. We cannot grow by subtraction. We instead require a "transformation" of our energy system, and it is not with just what we know, but with what we must create. We cannot succeed without such a transformation and the creativity we unleash must come from all areas of science and study.

This book is committed to thinking in conventionally unconventional ways…opening our eyes and minds to see problems and possibilities through far-sighted lenses while applying technologically and geographically targeted practical solutions that address today's realities as well.

MECOPTICS is not a menu of solutions, but rather, offers a comprehensive and objective discussion of available ingredients to make the future sustainable.

Charles McConnell
Executive Director, University of Houston Center for Carbon Management in Energy

Larry Bell is a Professor of Architecture and Endowed Professor of Space Architecture at the University of Houston where he founded and is Director Emeritus of the Sasakawa International Center for Space Architecture (SICSA) and the Space Architecture Graduate Program. Professor Bell and his organization's work have appeared in numerous popular magazine and media broadcast features in the US and abroad, including PBS, the History Channel and the Discovery Channel.

Larry has authored dozens of technical conference papers and professional journal articles addressing a broad variety of space and terrestrial design and technology topics and has co-founded several successful high-tech companies. In addition to NASA Headquarters achievement and recognition certificates, he has received two of the highest honors awarded by the Federation of Astronautics and Cosmonautics of the former Soviet Union for his contributions to international space development, the Konstantin Tsiolkovsky Gold Medal and the Yuri Gagarin Diploma. His name was placed on the Russian rocket that launched the first astronaut crew to the International Space Station. He is an Associate Fellow of the American Institute of Aeronautics and Astronautics (AIAA), and a Fellow emeritus of the Explorers Club.

About the Author

World population growth, expansion of commerce, competition for energy, environmental concerns, and dwindling fossil resources are prompting general recognition of a need to use available resources more wisely. Diverse types of organizations are working to do this. Many are exploring new operating policies, conducting technology R&D programs, putting responsive plans into action, and are demonstrating successes. Such initiatives are the right things to do and often make good practical and economic sense as well.

1. Bernd Untiedt
2. Osvaldo Gago
3. NATO
4. The Green Car Co., NZ

5. NASA
6. Michael Jastremski
7. Walter Siegmund
8. J. Stein Carter

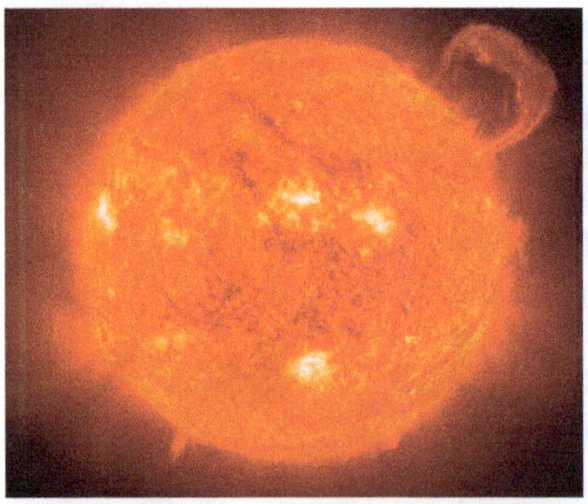

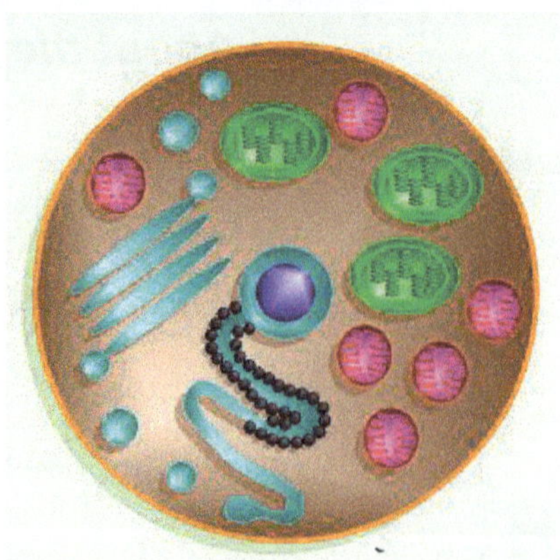

MECOPTICS is intended as a “model” to help support and frame broader and clearer thinking about ways that natural and human-assisted processes transform substances we often ignore into resources we care about. It illustrates that sources and products of life connect together in remarkable ways and in an enormous range of sizes. It directs attention to ways that unseen phenomena operate, interact and converge with marvelous synergistic outcomes. It addresses ways that Nature works, and ways we can more effectively apply those lessons.

9. Kyle R. Burton
11. Turbo Squid

10. Filip Maljkovic, Papeline
12. Stan Shebs, IV Design

The MECOPTICS model presents no methodologies, and only a few basic formulas which you either already know or don't really need to. It offers no prescriptions, proposes no devices, and comes with no operating instructions.

MECOPTICS simply looks at Nature's resources and possible ways we can use them resourcefully. In doing so, we are encouraged to expand our perspectives to consider relationships between living systems on both much larger and smaller scales. Ultimately, we may realize that everything is part of everything else, including ourselves.

There can be no doubt that the short era of abundant fossil fuel energy is over and that products of human activities are impacting Earth's fragile heat balance and environments.

- It is not the purpose of MECOPTICS to focus upon what we are doing wrong, but rather, to emphasize what we are beginning to do right. It reveals ways to do even more and better, including:

- Transitioning from fossil fuels and feedstocks to renewable sources and appropriate end-use technologies.

- Applying cost - and energy-efficient production methods, including parallel processing to recycle output wastes.

- Developing practical means and incentives to beneficially collect and convert municipal and agricultural wastes.

- Optimizing source-to-production-to-consumer transportation economies.

13. Larry Bell

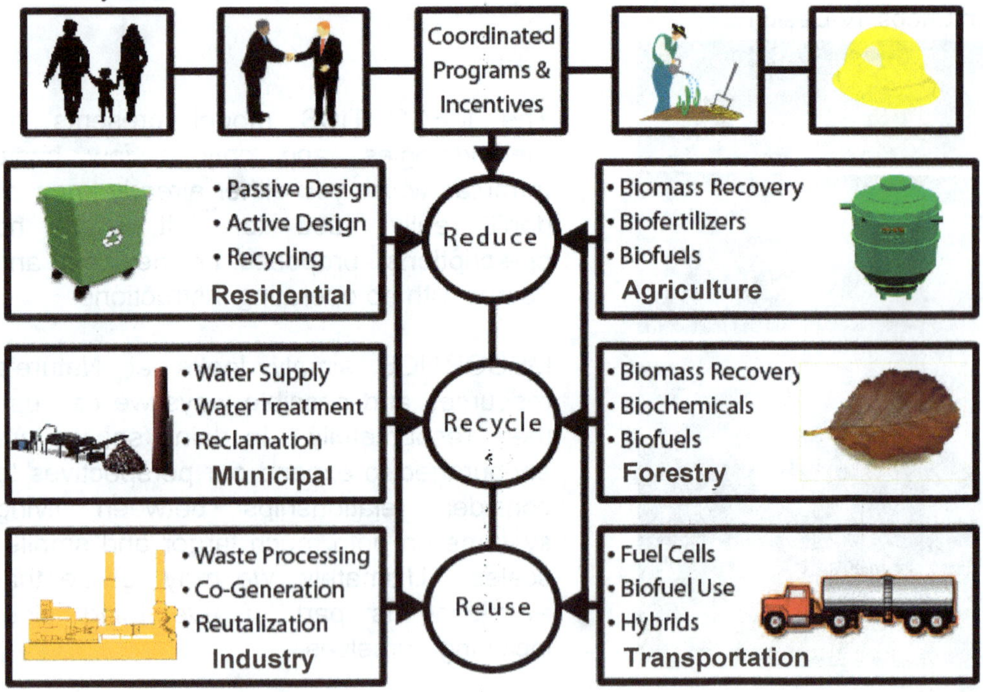

MECOPTICS applies broadly recognized conservation fundamentals.

- Many municipalities, industries, businesses and households recognize and apply three basic principles.
  - Reduce consumption through sensible economies.
  - Recycle rather than discard useful leftover products.
  - Reuse salvaged products, including biomass and heat.

Shared Principles

MECOPTICS emphasizes how natural systems operate and connect together in intricate ways and over extreme ranges in scale as lessons we can apply through practical approaches.

- Recognition that these systems are seamless, interactive and interdependent can reveal ways our personal and community connections to Nature and to our small planet home can be more resourceful:
  - We can more fully appreciate that human systems are not exempt from natural laws that govern survival of all life.
  - We can also become more aware of ways that natural principles and processes can be assisted through human devices to extend resources we currently enjoy, and to preserve them for future generations.

Part I: Natural Foundations

Observations and information about ways natural systems work and connect together, available resources, demand trends, and asset considerations.

Part II: MECOPTICS Frameworks

Overview of natural models and means to conserve, produce, and recycle energy and materials to achieve more with less.

Part III: Working Applications

An organizing approach for mapping efficient resource processing, transport and utilization opportunities across industry sectors and at national, regional, local and domestic scales.

Part IV: Challenges/Possibilities

Realities and uncertainties concerning resource depletion, climate change, their causes and consequences, and intervention priorities to protect our global future.

Contents

Table of Contents

15. Eindhoven Tech. U. 16. The Cavalier Daily 17. Simon Ho 18. Isreal EICI
19. Filipe Fortes 20. Trek Earth 21. Trek Earth
22. Langfel Designs 23. Space Daily

PART I: NATURAL FOUNDATIONS

We humans are integral parts of marvelous, complex, intricate and dynamic natural "systems" that are constantly in a precarious state of balance.

- Unlike categories of fixed and bounded systems that we often define to divide Nature into more understandable parts and phenomena; true Nature is connected and synergetic.
  - It is important to realize that when we focus exclusively upon any of the parts in order to isolate and examine aspects that we are most interested in, we risk missing a much larger picture that more fully reveals Nature's true nature.
  - We should also recognize that we and our systems of living are parts of that picture, and that consequences of those systems are inseparable from the other parts which sustain us.

24. NASA

25. NASA
26. NASA
27. NASA
28. NASA
29. Wash. Native Plant Soc.
30. Associated Content

Common use of the term "system" to identify certain interrelationships between natural objects and phenomena separate from others can sometimes imply boundaries of scale or function that obfuscates connectivity within larger frameworks.

- The way we view any system depends upon our perspective and focal range:
  - From a "big picture" perspective, we can see the Sun both as part of a much larger Cosmic System, and also as the central energy source in our Solar System powered by fusion.
  - Earth viewed from space (our Geo System) reveals weather systems on closer inspection that are driven by the Sun and Moon.
  - From close up we witness living ecosystems influenced by weather and by tiny bio-systems that provide nutrients that support them.

Systems of Systems

31. Spirit Alembic 32. Micropoy UK
33. J. W. Adams Comb. School 34. Soc. For Archeology

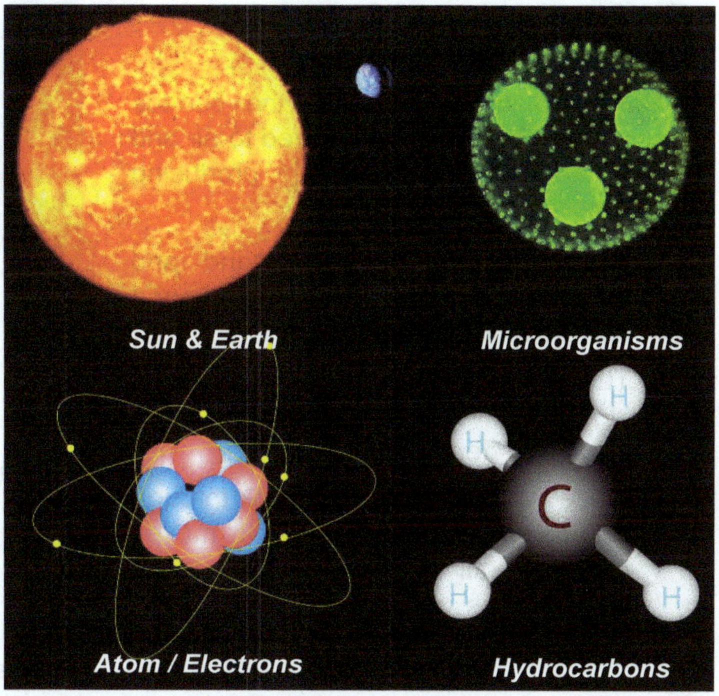

In Nature's grand model, all parts are connected and synergistic, with complex forces and processes acting in concert.

- Even the largest celestial bodies and phenomena derive their essential substance from the tiniest elements:
  - Processes that occur on the Sun drive Earth's weather and enable life to exist.
  - All plants and creatures depend upon nutrients produced by microorganisms.
  - Carbon and hydrogen constitute the building blocks of all organic life.
  - Interactions between electrons that bind atoms to form molecules create the energy fuel of all chemistry.

35. NASA

The Sun is composed entirely of gas and contains more than 99.8% of the total mass in our Solar System (about 70% hydrogen and 28% helium).

The Sun supports nearly all life on Earth and drives our planet's weather.

- Fusion processes convert hydrogen to helium and produce photons that travel at the speed of light, reaching Earth (a distance of about 94 million miles) in approximately 8.5 minutes:
  - Photons are mass-less, have no electric charge, and do not spontaneously decay in empty space.
  - Although some of the photons are absorbed in Earth's atmosphere, about 1,000 watts per square meter are deposited on surfaces exposed to direct sunlight.
  - Leafy plants use photons to convert CO_2 to oxygen and reduced carbon compounds that support living organisms.

Sun-Earth Relationships

The Sun is a primary influence on Earth's climates and weather conditions.

- Solar radiation that heats the surface and water bodies is transferred to air and circulated by convection, producing winds:
  - As heated air layers expand and become less dense, they are pushed upwards by buoyancy.
  - The rising air cools as it reaches higher altitudes in the atmosphere, becoming denser.
  - Since the denser air can't sink through the rising air beneath it, it moves laterally until it reaches the surface again and is drawn back into a rising column.

36. Video Weather & Weatherstreet

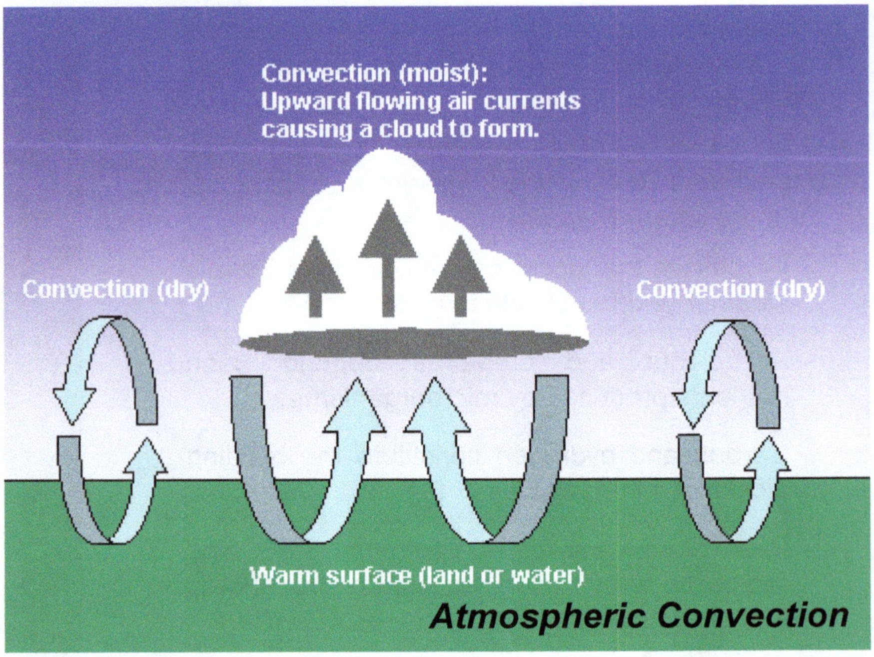

Convection currents produced by solar heat cause winds, thermals, cyclones, thunderstorms, and at a larger scale, atmospheric circulation.

37. Carleton College Science Research Center

Solar heating and salinity differences produce convection effects in oceans that influence regional climates.

Oceanic convection driven by solar radiation and by differences in water salinity have important influences upon continental climates, particularly in coastal regions.

- Examples are the Gulf Stream which makes northwest Europe much more temperate than other regions at the same latitude, and the Hawaiian Islands, where the climate is cooler due to the California Current:
  - Equatorial water heated by solar radiation tends to circulate towards the poles, while polar water heads towards the equator.
  - "Thermoline convection" driven by differences in salinity causes relatively warm saline water to sink, while colder fresh water tends to rise and reverse heat transport.

38. By Kaidor (NASA)

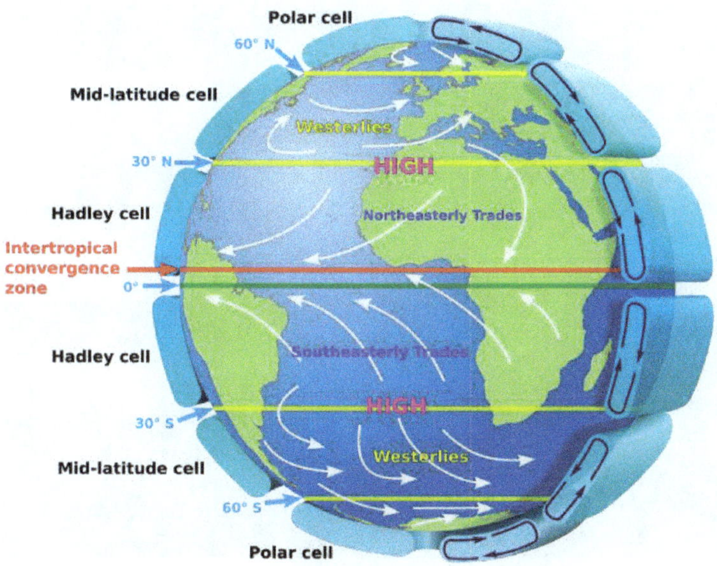

Coriolis Influences Upon Weather

Coriolis forces produced by the Earth's rotation influence wind patterns that balance atmospheric heat, and transfer moisture to arid regions that need it.

Earth's rotation also influences wind and weather.

- Winds are comprised of air molecules that are subject to Coriolis forces which cause air to turn as it flows:
  - Coriolis forces deflect wind flows to the right in the Northern Hemisphere, and to the left in the Southern Hemisphere.
  - The Coriolis deflections contribute to formations of tropical cyclones that produce extremely strong winds, tornados, torrential rain and huge waves.
  - Cyclones that often cause great human and ecological disasters also provide important mechanisms to transport heat away from the tropics and maintain equilibrium in the Earth's troposphere.
  - Hurricanes in the northeastern Pacific supply moisture to the southwestern US.

Sun/Moon-Driven Tides

The gravitational pull of the Sun and Moon produce daily sea level changes that are essential to support marine and coastal ecological systems.

- Although these gravitational forces act upon all Earth surfaces, solid land areas remain intact, while water is lifted:
  - These forces produce two bulges in oceans, one in the direction of the Moon, and one on the opposite side of the Earth.
  - Even though the Moon's mass is 27 million times smaller than the Sun's, its closer proximity to Earth gives it a tide-generating force that is more than twice greater.
  - High tides are produced in areas where water flows between the bulges, and low tides (areas of depression) occur between bulges due to withdrawal of surging water.

39. Case Western University

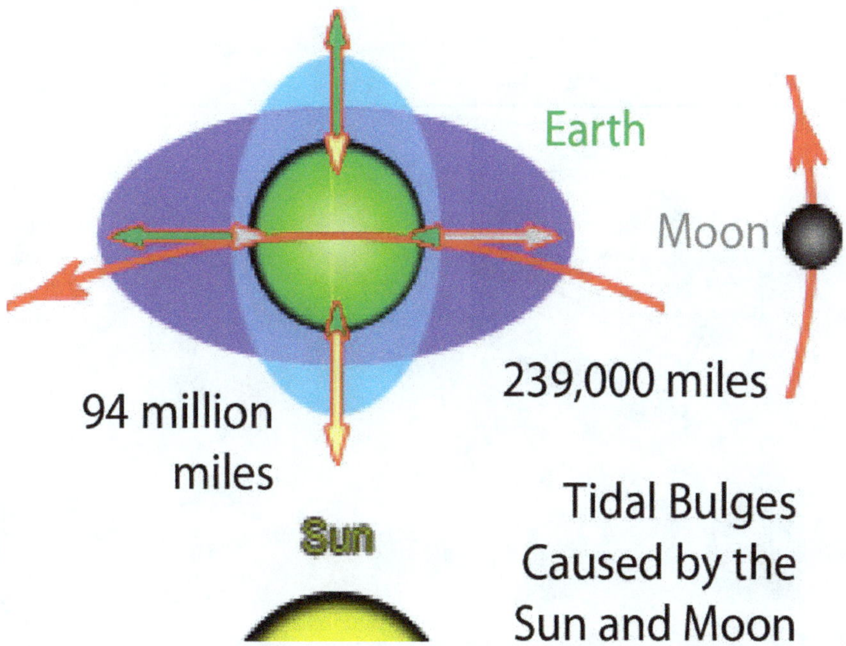

Since distances between the Earth, Sun and Moon vary considerably throughout a year due to elliptical orbits, tide generation forces are greatest when the Earth is closest in solar orbit (perihelion), and when the Moon is closest in its Earth orbit (perigee).

Tidal effects of the Sun and Moon provide critical mechanisms to support marine life and numerous types of plants and animals that reside in marshes and estuaries.

- Tidal circulation serves a variety of essential functions:
  - Tides deliver suspended sediments necessary to replace surface soils removed by runoff land erosion.
  - Flooding tides enhance plant growth by delivering oxygen-enriched water to marsh soils and removing toxic materials.
  - Tidal waters are vehicles that carry plant nutrients onto marshlands during periods of abundance, and transport excess nutrients back to estuaries at other times.
  - Exported nutrients support growth of single-celled phytoplankton that are the foundation of marine food chains.

40. National Oceanic and Atmospheric Administration

Tides deliver vital nutrients to coastal ecosystems and remove accumulated toxins that endanger living organisms that support food chains.

Tidal Benefits

41. Science TREK

Most ecosystems on Earth are "solar powered", depending upon the food energy created by photosynthesis.

42. Purves, et al, *Life: the Science of Biology*

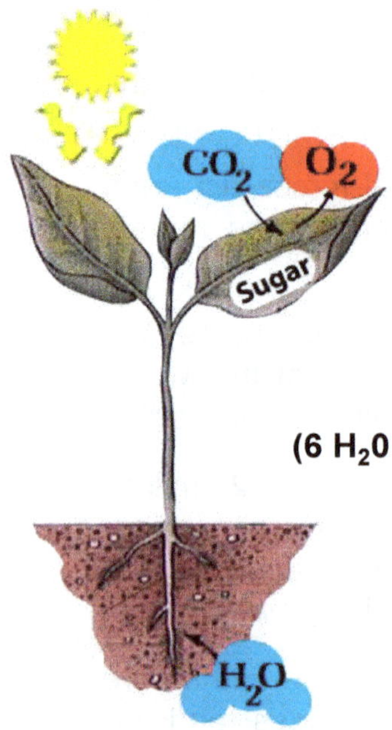

Six molecules of water taken up through roots and leaves plus six molecules of CO_2 produce one molecule of sugar plus six molecules of O_2.

($6\ H_20 + 6\ CO_2 \rightarrow C_6H_{12}O_6 + 6\ O_2$).

Photosynthesis

Photosynthesis uses sunlight to remove CO_2 (a greenhouse gas) from the air and convert it into oxygen, sugars, starches and carbohydrates that support life.

Solar-Powered Life Systems

43. Oregon State University

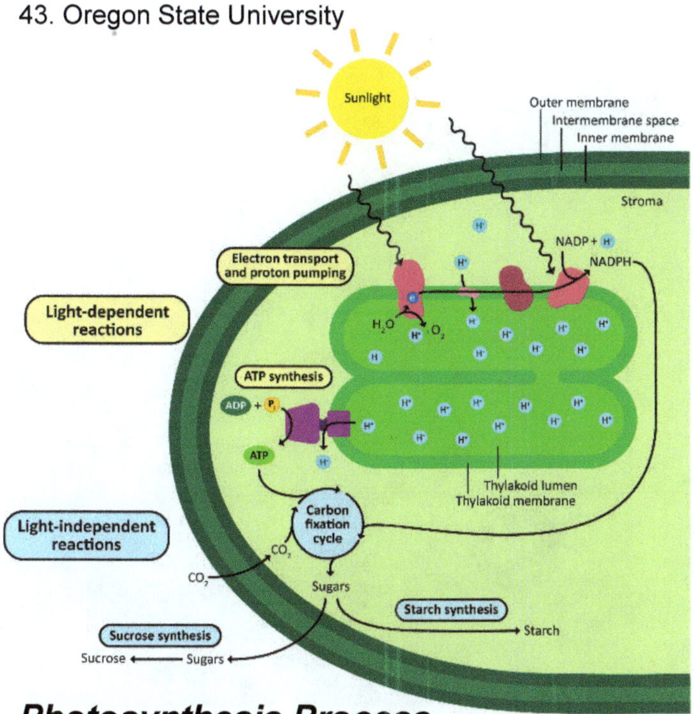

Photosynthesis Process

Photosynthesis occurs in higher plants, algae, some bacteria and some single-celled protists use light energy to produce glucose with oxygen as a byproduct.

- The process occurs through two complex phases:
  - In a first phase "light-dependent reaction" a pigment chlorophyll molecule absorbs light, and loses an electron in the process.
  - The released electron moves down a transport chain to produce high-energy NADPH and ATP molecules, and the chlorophyll molecule regains its electron by splitting water, releasing oxygen.
  - In a second phase "dark reaction" an enzyme RuBisCO captures CO_2 from the atmosphere and through a complex Calvin-Benson cycle releases 3-carbon sugars that combine to form glucose.

44. Larry Bell

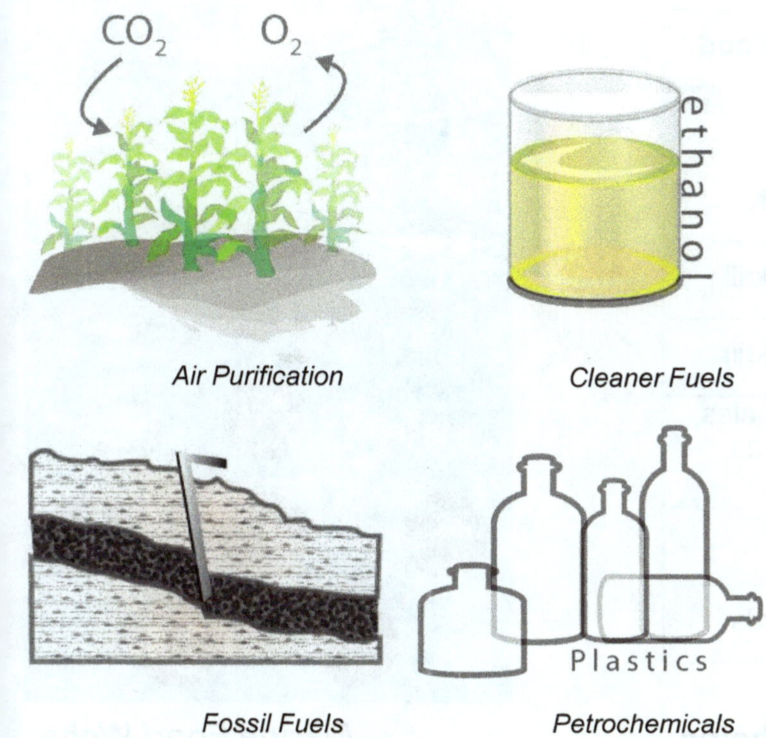

Photosynthesis Products

Solar-powered photosynthesis is arguably the most important biological process on Earth.

- Photosynthesis converts CO_2 and water into breathable oxygen to help ameliorate greenhouse gas produced by combustion, creating carbohydrates that we use for many other purposes as additional by-products:
  - One of the carbohydrates is cellulose that can be converted to sugar and fermented into ethanol (grain alcohol) for direct use as automotive fuel and as a gasoline additive to reduce pollutant emissions.
  - Ethanol can be readily converted to ethylene, an important feedstock for petrochemicals.
  - Carbon "fixed" through photosynthesis in ancient times created the biomass that is the source of fossil fuels we use today.

Solar-Powered Resources

45. Bigelow Laboratory for Ocean Sciences.

In addition to higher plants, many types of microscopic organisms ("phototrophs") use photosynthesis to create essential food energy in marine environments.

- Despite their small size, these algae and cyanobacteria are vital to our planet's ecological survival because they provide the base of the marine food web:
  - Each level of a food chain is called a "tropic" or "feeding level" that classifies organisms as either "primary producers" or "consumers".
  - Phototrophs are primary producers because they convert inorganic materials to organic forms that "heterotrophs" depend upon.
  - Heterotrophs include carnivores, omnivores, vegetarians and decomposers.

Marine Phototrophs

46. Surabhi Sharma, SICSA

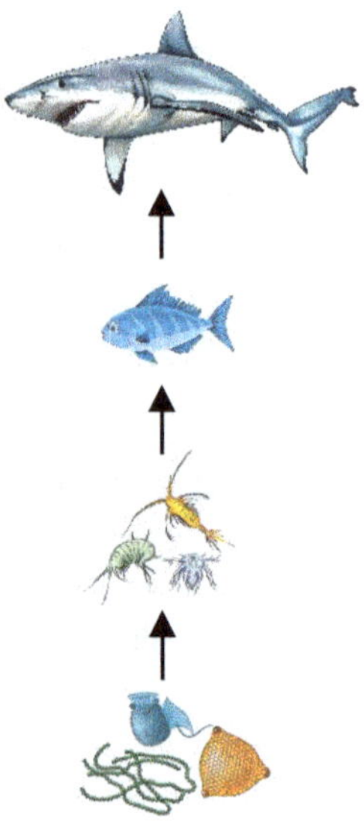

Organism	Trophic Type	Prey/Food
algae	primary producer	---
birds	carnivorous consumer	krill, fish
blue whales	planktivorous consumer	algae, krill
fish	omnivorous consumer	algae, krill
killer whales	top consumer	blue whales, fish, birds, seals
krill	herbivorous consumer	algae
seals	carnivorous consumer	fish, birds

Marine Food Chains

47. Britannica

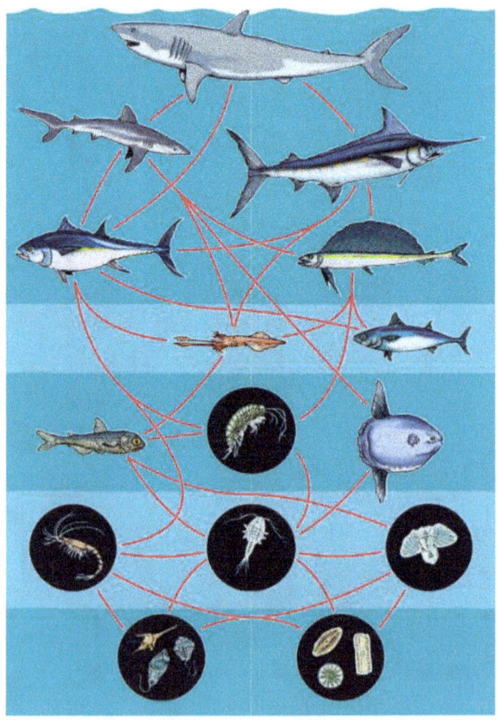

Marine Food Webs

The Earth is often characterized as being comprised of four general types of connected zones or "spheres".

- The atmosphere provides a protective gaseous blanket:
  - It contains oxygen and nitrogen essential for life, and provides weather.
- The geosphere consists of Earth's solid material and inner core:
  - It contains minerals and compounds that comprise organic and inorganic materials.
  - About 97% is contained in oceans.
- The biosphere contains all living systems:
  - This includes organisms ranging from viruses to humans that exist in the geosphere and hydrosphere.

48. Surabhi Sharma, SICSA

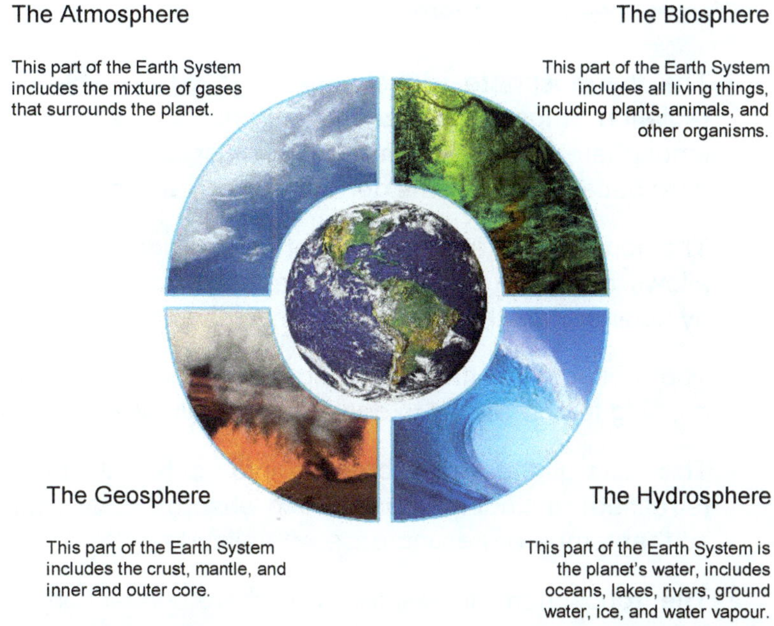

The Earth Viewed as Systems

Viewed as systems within a larger system, the spheres of influence interact in dynamic and complex ways to maintain the equilibrium necessary to sustain life.

- The atmosphere serves as a large global transport system:
  - It circulates oxygen, water, nitrogen, carbon dioxide and heat among regions.
- The geosphere affords a foundation for life:
  - It provides inorganic substances to support organic chemistry, geothermal heat, and a CO_2 containment service.
- The hydrosphere supports marine life and land creatures that depend upon it:
  - It also influences weather, provides water for precipitation and cooling, and absorbs CO_2 from the atmosphere.
- The biosphere exchanges life support products and biowastes with other spheres:
  - It is the source of biomass fuels and nutrients, and converts CO_2 into O_2 and back again.

49. Larry Bell

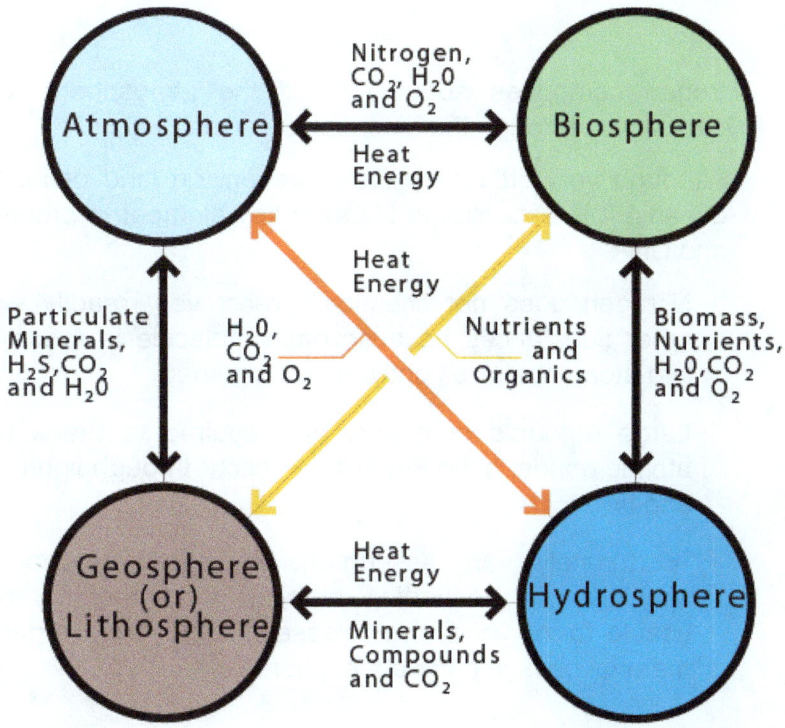

Connections and Synergies

Our Dynamic Planet

SPHERES OF INFLUENCE

50. Larry Bell

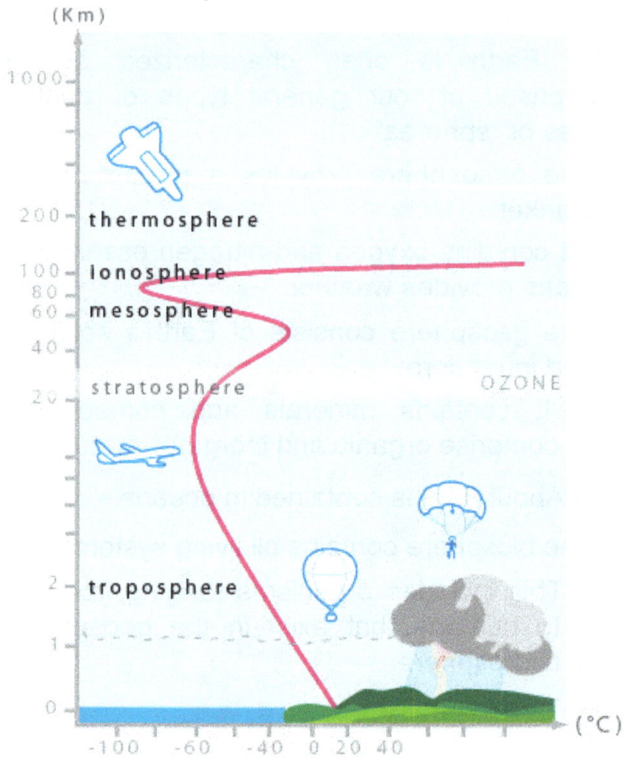

Atmospheric Zones

The atmosphere is also characterized as being comprised of different zones, or "spheres".

- The thermosphere is the uppermost region where solar radiation strips off (ionizes) electrons from atoms in the ionosphere below. (Temperatures increase with altitude here because the atoms don't absorb solar radiation).
- The ionosphere is a layer that traps radio signals and allows them to travel around the world (unless disturbed by solar storms that disrupt communications).
- The mesosphere has no strong solar radiation absorbers, causing temperatures to decrease with altitude.
- The stratosphere is broken into different materials (stratified), including ozone which absorbs solar radiation so that temperatures increase with altitude.
- The troposphere (lowest level) is where convection occurs to produce weather, and temperatures increase with altitude.

51. Windows to the Universe, UCAR

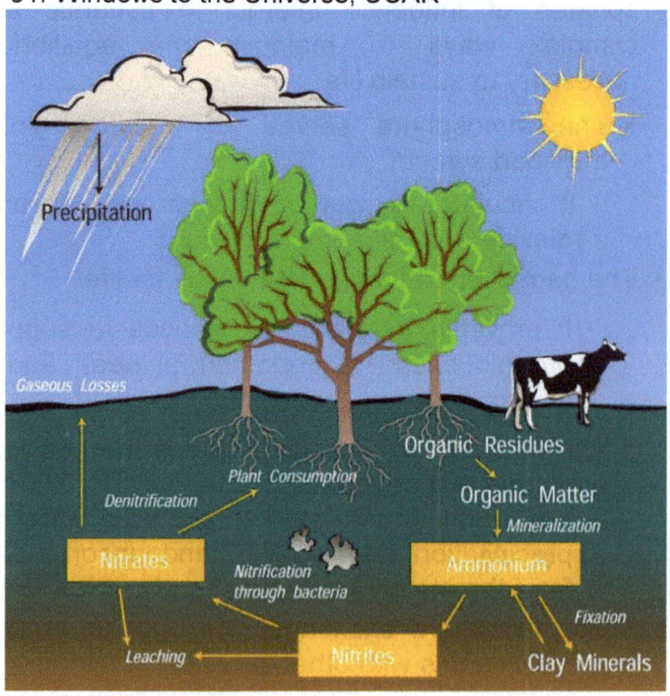

The Nitrogen Cycle

Nitrogen comprises about 78% of the atmosphere and cycles slowly through the biosphere.

- Although very little nitrogen is present on land or in the oceans, it is essential to life as a key element in proteins and DNA:
  - Nitrogen does not chemically react very readily with other substances because each molecule is made of two atoms that are tightly bound together.
  - Large amounts of energy are required to break the atomic bonds, although this can occur through lightning and fires.
  - Fortunately, an assortment of bacterial species specialize in converting nitrogen from the air into usable forms, and also release nitrogen from organic material back into the atmosphere.

52. US Geological Survey

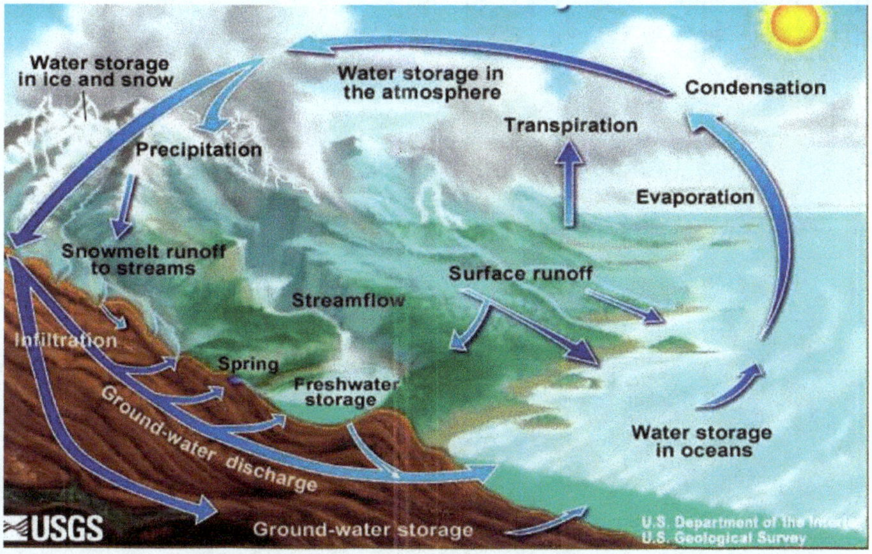

The Water Cycle

Water in liquid and vapor form cycles continuously through the geosphere, hydrosphere, atmosphere and biosphere.

- Water molecules comprise the primary substance of all higher plants and animals:
  - It is transferred between the hydrosphere and biosphere by evaporation and precipitation, releasing heat energy in the process.
  - In the air, water vapor circulates and condenses to form clouds, and rain that falls back to the surface.
  - Plants and animals use and reuse water, and release the vapor back into the air.
  - Some water leaches into underground aquifers, releasing suspended particles in the filtering process.

53. Daniel Alan Vallero

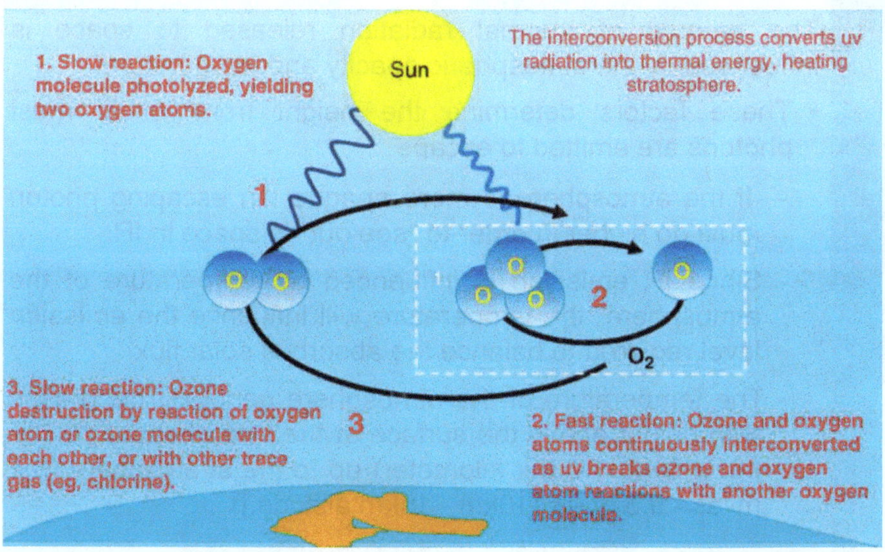

Radiation-powered reactions in the stratosphere continuously convert oxygen molecules to ozone, which eventually reforms back to oxygen.

Oxygen-Ozone Cycles

Oxygen comprises about 21% of the atmosphere.

- Although oxygen is very reactive and rapidly combines with other elements and “disappears”, relatively stable concentrations are replaced through photosynthesis and plant respiration:
  - Some oxygen is lofted into upper reaches of the atmosphere (the stratosphere) where a series of reactions powered by solar radiation convert it into ozone (O_3).
  - The ozone beneficially serves to absorb biologically-damaging ultraviolet (UV) radiation, reducing the amount that reaches the Earth’s surface to a tiny fraction.
  - Ozone molecules eventually break apart to reform into O_2, which circulates again through the atmosphere to repeat the cycle.

Atmospheric Cycles

The Earth's thermal balance is in a steady state when the energy stored in the atmosphere and oceans equals the incident solar energy radiated back to space.

- This radiation includes reflected solar radiation and emitted thermal infrared radiation (IR):
  - The Earth receives about 30% of the incident solar radiation flux, and the remaining 70% is absorbed, warms the land, atmosphere and oceans and powers life on the planet.
  - Eventually the absorbed incident energy is radiated to space as IR photons.
  - This thermal (IR) radiation increases with rising temperatures.

54. Online Biology Book. Estrella Mountain Community College.

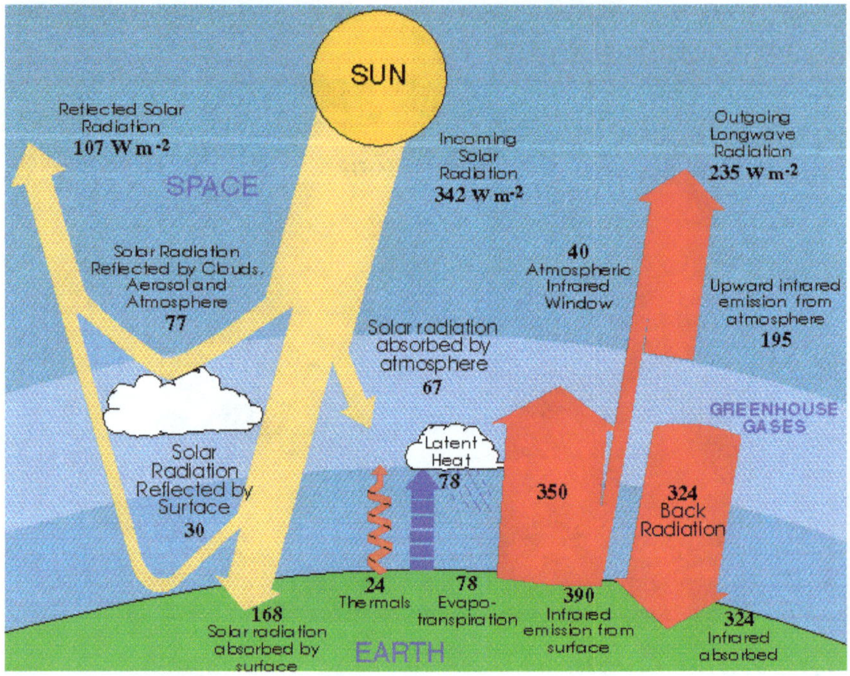

Steady State Thermal Balance

55. Northern Arizona University.

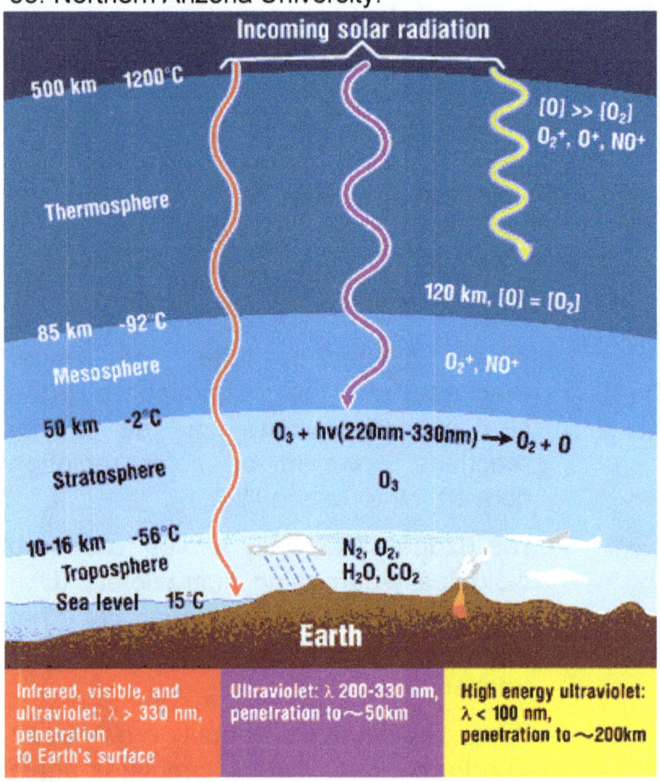

Atmosphere Transparency vs. Altitude and Incoming Radiation Wavelengths

The amount of thermal radiation released to space is influenced by the atmospheric opacity and temperature.

- These factors determine the height from which most photons are emitted to escape:
  - If the atmosphere is more opaque, an escaping photon must go higher in order to "see out" to space in IR.
  - Since IR emission is influenced by temperature of the atmosphere, the temperature will influence the emission level required to balance the absorbed solar flux.
  - The temperature of the atmosphere generally decreases with height above the surface at the rate of about 15.7°F per mile (6.5°C per kilometer) up to the stratosphere (6.2 miles - 9.3 miles [10km - 15km altitude.])

Since the atmosphere emits IR both upwards and downwards, the temperature balance between energy radiated to space must be equal to the direct incident radiation absorbed plus the downward IR flux.

- The atmosphere is relatively transparent to the direct incident radiation in the visible range, but strongly absorbs thermal IR wavelengths emitted by the surface and atmosphere:
  - Most IR escaping to space is emitted from the upper atmosphere (not the surface).
  - Accordingly, most IR photons emitted by the surface are absorbed by the atmosphere and do not escape.
  - A large part of the thermal energy transfer to the atmosphere occurs through convection (sensible heat transport) and evaporative condensation of water vapor (latent heat transfer).*

Atmospheric greenhouse gas concentrations influence the amount of IR energy absorbed by the surface or radiated back to space:

- Dominant infrared-absorbing gases in Earth's atmosphere are water vapor and carbon dioxide:
  - Clouds are very important IR absorbers, and become more prominent through evaporation as surface water temperatures increase.
  - CO_2 is a linear molecule which becomes "floppy" and easily excited by IR so that it bends with carbon in the middle, moving one way, and oxygen on the ends moving the other to vibrate and absorb IR energy.
  - CO_2 absorption decreases logarithmically with atmospheric "saturation" concentrations causing warming influences to diminish.

56. Wikimedia

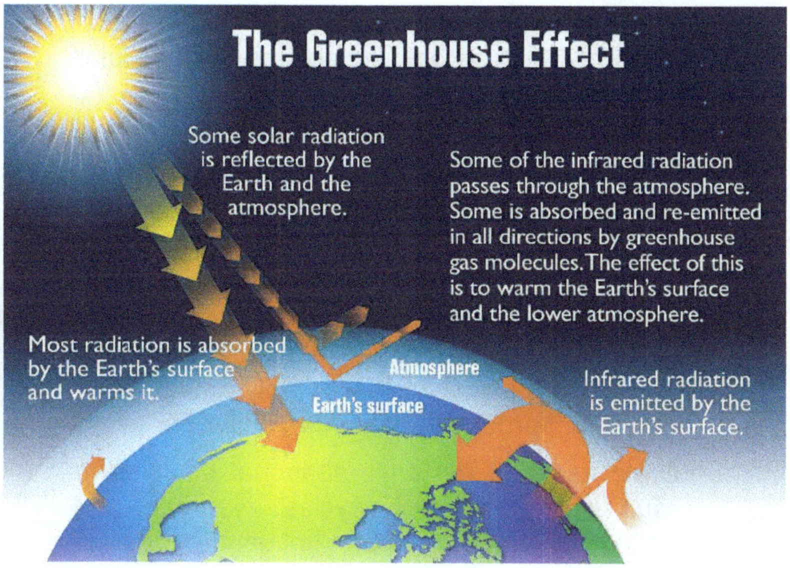

The "Greenhouse Effect"

57. Global Warming Art.

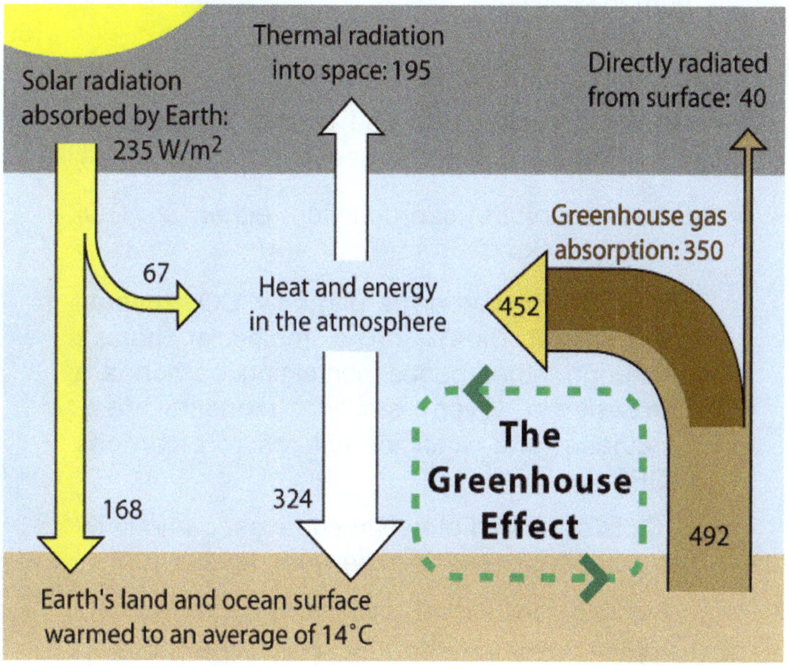

Greenhouse Mechanisms

*See Appendix: 1

Atmospheric CO_2 derives from many natural sources, including volcanoes, ocean sediments, combustion of organic matter and fossil fuels, respiration of living aerobic organisms, and cellular respiration of microorganisms.

- Records from polar ice cores and ocean sediments indicate that atmospheric CO_2 concentrations have naturally fluctuated greatly throughout the Earth's history, reaching levels higher than those observed in recent decades:
  - Volcanic activity was essential to create a warm, stable climate essential for life when the Earth was young, but now releases only about 1% of the amount produced by human activities.
  - Levels rise and fall on a seasonal basis as a result of plant growth cycles, warming and cooling changes that influence ocean absorption rates and other factors.
  - The US National Oceanographic and Atmospheric Administration estimates that about 97% of annual atmospheric CO_2 comes from natural sources, and about 3% from human activities.

Carbon, typically in the form of CO_2, cycles between "reservoirs" provided by the atmosphere, biosphere, hydrosphere and geosphere.

- Natural reservoirs can function either as CO_2 sources or sinks:
  - Atmospheric carbon is primarily CO_2 (overall only about 0.04% on a molecular basis), although other gases containing carbon are present in even smaller amounts (e.g. methane and artificial chlorofluroro-carbons [CFCs]).
  - CO_2 is dissolved at ocean surfaces, particularly in polar regions where the water is cold.
  - Aerobic respiration by plants and animals breaks down glucose into CO_2 and water, and photosynthesis reverses the process.
  - Limestone, marble and chalk mainly comprised of calcium carbonate are eroded by ocean water to produce CO_2 and carbonic acid.

58. Hanues Grobe, et al. Inst.for Polar and Marine Research

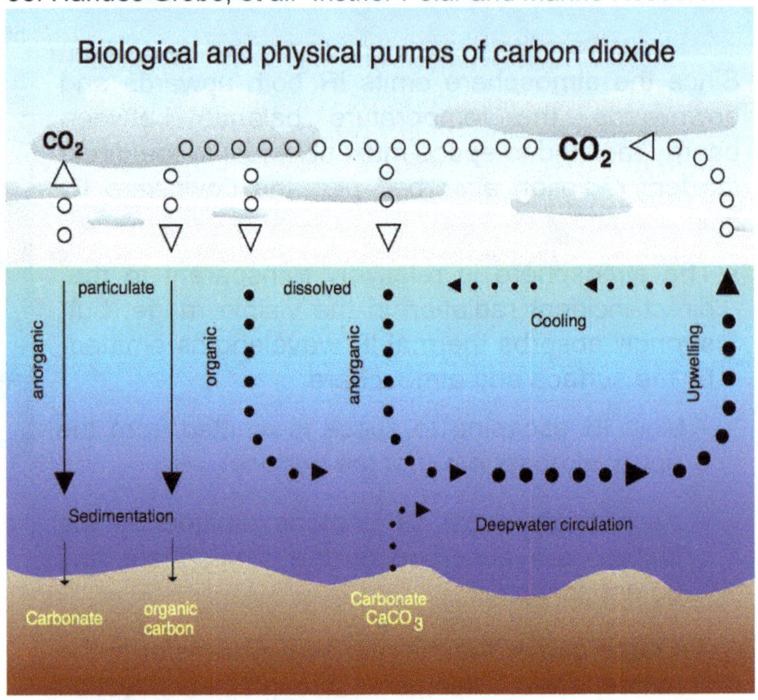

CO_2 from Ocean Carbonate

59. NASA

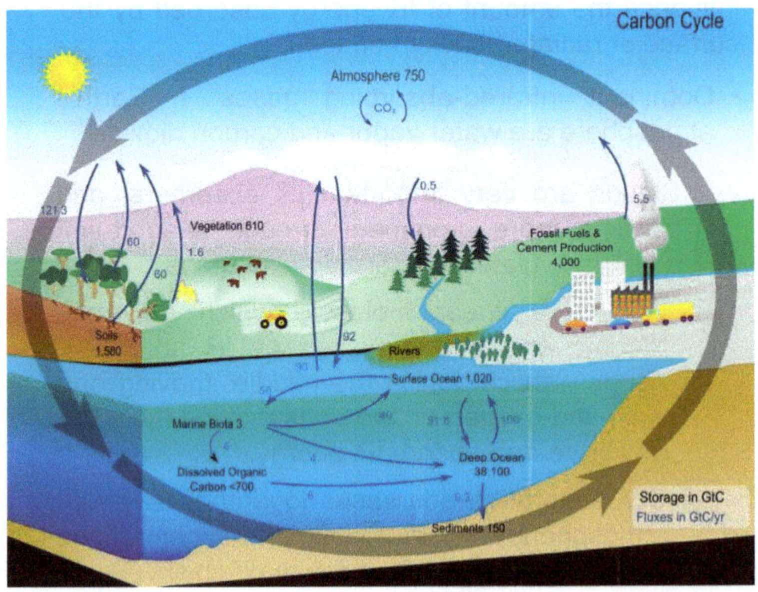

CO_2 is constantly being released into the atmosphere and hydrosphere from hydraulic erosion of limestone deposits, biomass decomposition, fossil fuel combustion and other sources.

60. NASA

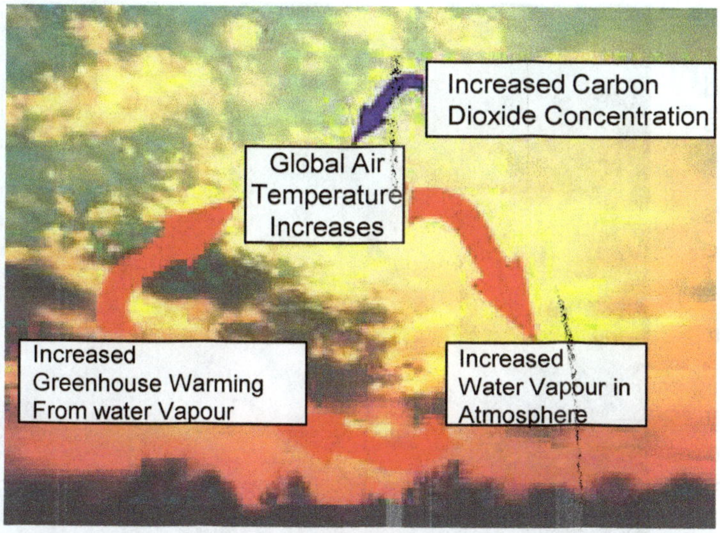

Water vapor in the atmosphere is known to act as a greenhouse gas that contributes to Earth warming.

Water Vapor as a Greenhouse Gas

Water vapor in combination with CO_2 and other greenhouse gases are believed to have feedback influences that compound warming effects.

- Since the concentration of a greenhouse gas in the atmosphere is a function of temperature, a feedback produced by another greenhouse gas can increase the temperature, to in turn increase the concentration of the original gas.
  - As surface temperatures rise, more water from oceans evaporates, adding to atmospheric vapor levels .
  - Research based upon data from NASA's Upper Atmosphere Research Satellite (UARS) has confirmed this effect but suggests that the impacts may not be as large as previous climate change models have assumed.*

Greenhouse Gas Feedbacks

61. Larry Bell

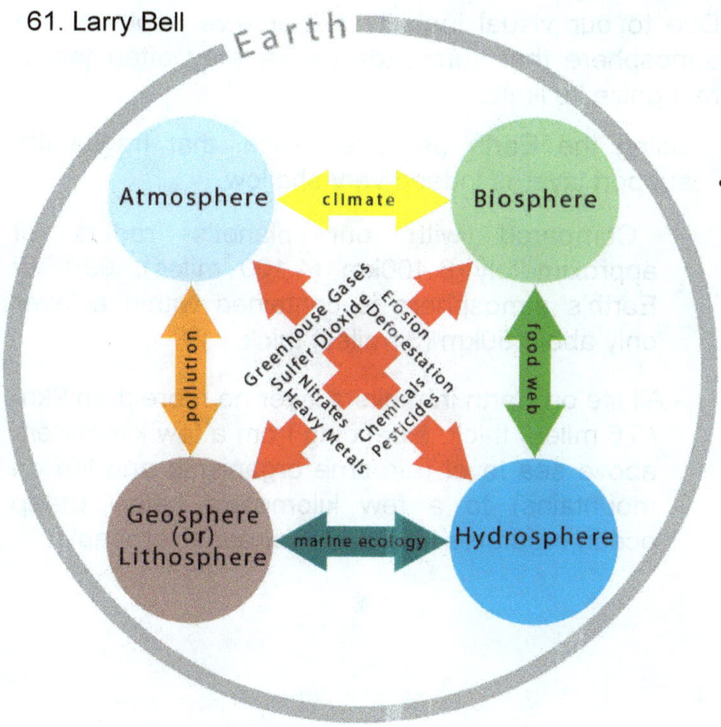

Contributions to Global Changes

The spheres of influence that impact life on our planet are always in a dynamic, complex and delicate state of balance.

- Human activities clearly impact the equilibrium of these influences, and are doing to at an accelerating rate:
  - Poor agricultural and forestry practices have influenced surface "albedo" reflection of solar IR back to space and leafy plant conversion of carbon dioxide to oxygen we breathe.
  - Toxic pesticides created to increase food production needed to support growing world populations has entered sensitive marine ecosystems.
  - Nitrates and particulates released from combustion are polluting the atmosphere.
  - Short-sighted agricultural, mining and deforestation practices have sacrificed topsoils, leached chemicals into land and water, and effected the oxygen-CO_2 balance.

The Human Factor

***See Appendix: 2**

Gazing upwards towards the nighttime sky, none of us alive today in "developed" regions, other than astronauts who have orbited above our small globe, have witnessed the bright and expansive starfields that our ancestors observed a few generations past.

- Lacking such a reference of change, and with views obscured by atmospheric pollutants, we may often lose sight of important realities:
  - We may be unaware of the extent and rate our activities are effecting essential life support systems.
  - We may tend not to recognize that what every community on Earth puts into the atmosphere (and oceans) is rapidly distributed through natural transport to global destinations.
  - We may often forget that nature mixes all environmental influences together with combined and interactive consequences.

62. NASA

63. Larry Bell/ Candy Feuer

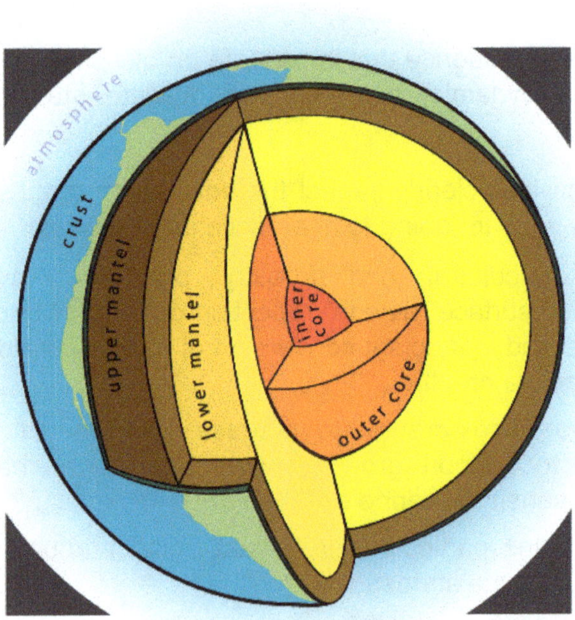

Earth Layers	Depth	Thickness
1 inner core	-6378 km	1216 km
2 outer core	-5150 km	2270 km
3 lower mantel 4 upper mantel	-2981 km	2900 km
5 crust	0 km	5-70 km
6 atmosphere	50 km	50 km

Due to our visual inability to perceive depth of the atmosphere that surrounds us we may often fail to recognize its limited dimension.

- Using the Earth as a reference, that fragile life support layer is indeed very shallow.
  - Compared with our planet's radius of approximately 6,400km (4,100 miles), 99% of Earth's atmosphere is contained within a layer only about 50km (31miles) thick.
  - All life on Earth inhabits a layer no more than 9km (5.6 miles) thick, extending from a few kilometers above sea level (airborne organisms and life on mountains) to a few kilometers below (deep ocean creatures and subterranean microbes).

64. Larry Bell/ Candy Feuer

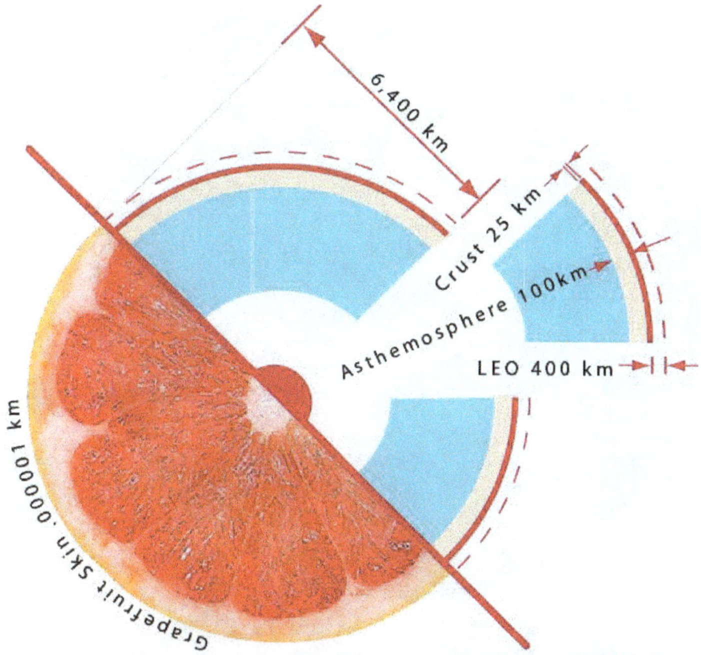

(Distance Reference: Earth/LEO vs. Grapefruit)

A Thin Slice of Life

Moving farther out into "space" at an altitude where space stations orbit, we continue to find ourselves well within Earth's immediate territorial border.

- Using the proportions of a grapefruit to put dimensions into perspective we are now only moving about the outer skin surface:
  - Low-Earth Orbit (LEO) at a distance of about 250 miles (400km) is less than 5% of Earth's radius away.
  - Earth's atmosphere estimated to extend outward from the surface to about 62 miles (100km) represents about 1.5% of the radius (the inner grapefruit skin).
  - Earth's crust, which constitutes all solid structures, extends inwards about 15 miles (25km) (about 0.4% of the radius) barely penetrating into the grapefruit's juicy part.

65. Larry Bell/ Candy Feuer

Living Only In the Here and Now

Viewed from vantage points of our everyday lives it is convenient to perceive the world as an open and unbounded place with limitless self-healing ability and resource abundance.

- From this perspective we may be even more inclined to view ourselves and our communities as small, isolated parts of overwhelmingly large systems and events where our actions are inconsequential or futile:
  - Everything, including the weather, comes from "somewhere else" and we have no effect upon it.
  - If we lack adequate supplies of something we can find them "somewhere else", or eventually substitute other things that will work as well or even better.
  - We might imagine that adverse changes in our environments will occur slowly and reversibly so that we can afford to worry about them later.

A Short-Sighted View

66. NASA

Looking Back

Moving outward to LEO we observe a dynamic Earth where forces of nature and consequences of human activities connect, interact and often collide.

- This view reveals the contiguous surface of our planet, where the only real boundaries between human nations and jurisdictions are defined only by land features and water bodies.
  - Weather forces form windswept patterns that rapidly circumnavigate the globe, distributing water and dispersing contaminants along their paths.
  - Storm runoffs and river currents transport topsoils and chemicals to marshlands and estuaries for deposit as sediments, or transfer to more distant locations by ocean tides.

A realistic and resourceful world view recognizes that each and all of us are integral parts of nature and that it constitutes what we are made of.

- This perspective makes no distinction or excuse regarding size since elements and actions at all levels of scale are connected:
  - Being parts of an unimaginably large and marvelous Universe empowers rather than diminishes us (and everything is tiny compared to the Universe).
  - Small actions and examples can bring about large effects by establishing precedents for others to participate.
  - Almost nothing in nature can be truly isolated from an influence upon everything else.
  - Prevention of a problem now is always less expensive and hazardous than trying to fix it later.

67 Candy Feuer

Everything is Small Compared to the Universe, and Everything is Important

Looking Forward

68. Monash University Australia 69. Tony Boom
70. Looking Glass Productions Ltd 71. Chapel Electrics

Freeway Traffic

Electrical Power

Industry

Heating

Fossil Fuel Consumption

Although it took only about 150 years to create our current fossil fuel-energized society where oil is the most widely-traded world commodity, this dependence upon finite Earth resources cannot continue much longer.

- Many leading geologists believe that most oil and natural gas reservoirs will be exhausted well before the end of this century, and coal will be depleted by the middle of the next one:
  - It is broadly predicted that oil production capacities are already peaking, and that new reservoirs will not be sufficient to offset declines in old fields, much less sustain continued consumption growth.
  - While oil will continue to flow for another 50-75 years, price increases and supply shortages will drive essential transitions to alternative, renewable energy sources at an accelerating rate.

Coal is the world's most abundant fossil fuel, providing much of the total primary heat and electrical energy.

- The International Energy Agency (IEA) projects a Steady demand for coal through 2027:
  - This trend is causing escalating public concerns about CO_2 emissions into the atmosphere that are contributors to global warming.
  - In response, the US and many other countries are investing heavily to develop "clean coal" technologies that dramatically reduce CO_2 and other pollutant emissions.
  - Coal gasification is a leading approach to extract hydrogen and other useful energy forms from coal in the process of removing CO_2.

72. Oakridge National Laboratory

Gasification and other "clean coal" technologies are being developed and advanced to remove CO_2 and other pollutants such as sulfur and nitrogen that form droplets of weak sulfuric and nitric acid ("acid rain").

"Clean Coal" Gasification Plant

Coal

Efforts to reduce emissions from coal power plants have led to the development of advanced technologies aimed at minimizing environmental impact.

- Coal gasification appears to offer one of the most versatile ways to do this, producing usable products in the process:
  - Gasification uses hot steam and controlled amounts of air or oxygen under high temperatures and pressures to break down coal (or any other carbon-based feedstock) into a mixture of carbon monoxide, hydrogen and other gaseous compounds.
  - The hydrogen and other coal gases can, in turn, be used for fuel cells and to power turbines in power-generating plants.
  - Gasification of coal, biomass and organic municipal/industrial wastes can also produce gaseous materials used to create plastic photographic film and methanol fuel for automobiles.

73. World Energy Council

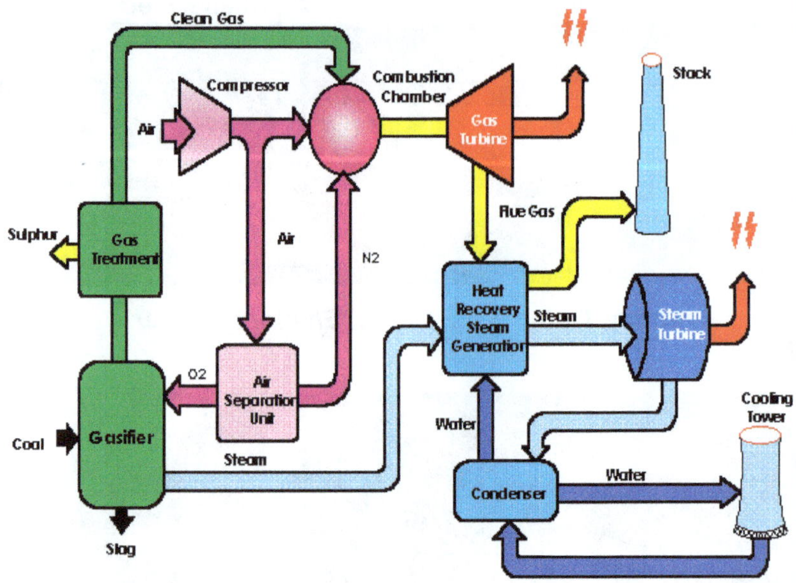

Coal Gasification Elements

74 BBC Co., UK

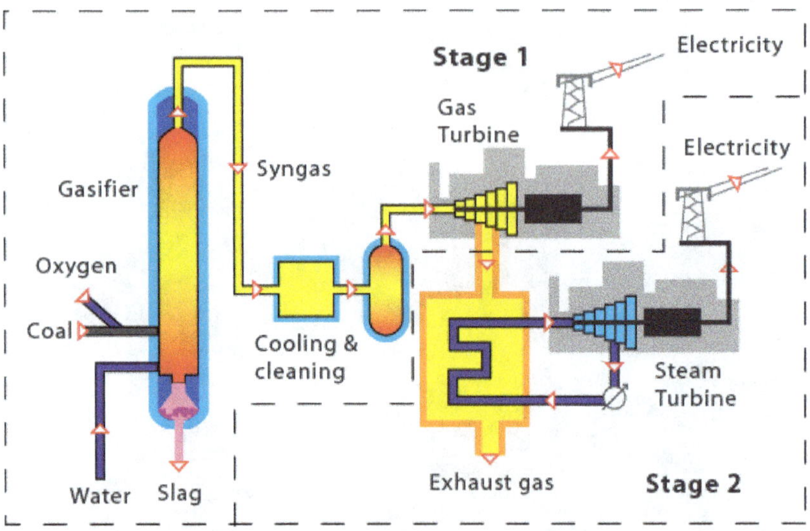

1) The coal gas (syngas) comprised principally of hydrogen and carbon monoxide is first burned to drive a gas turbine to create electricity.

2) The exhaust gases are then heat-exchanged with water/steam to drive a steam turbine.

The second stage steam can also be exported for use by nearby factories and district heating plants.

IGCC Coal Power Production Stages

One of the most pressing problems associated with use of coal as an energy source is to remove and potentially utilize CO_2 before it is combusted.

- A recent processing approach called "Integrated Gasification Combined Cycle" (IGCC) enables use of low-quality solid and liquid fuels, while also meeting stringent emission requirements:
  - IGCC plants use oxygen as a gasifier instead of air so that the CO_2 is emitted in a concentrated gas stream that can be more easily captured and removed.
  - Power generation efficiencies are boosted 15-20% or more over conventional plants through a two-stage process which first uses the gaseous product to drive gas turbines and then uses exhaust gases to provide steam for steam turbines.

CO_2 produced by coal processing must be either stored (sequestered) or used.

- While the scale of need for CO_2 disposed far exceeds today's uses, the product does offer practical value:
  - Millions of tons of CO_2 are injected into oil fields for enhanced oil recovery each year.
  - CO_2 is also injected into deep, unmineable coal seams where it is absorbed to displace methane. (While this is currently not as economical as enhanced oil recovery, the future potential is large.)
  - CO_2 is used for a variety of commercial applications, including carbonation of soft drinks and soda water, as a leavening agent in baking, for fire extinguishers, and as an inexpensive atmosphere for welding.

75. US Government

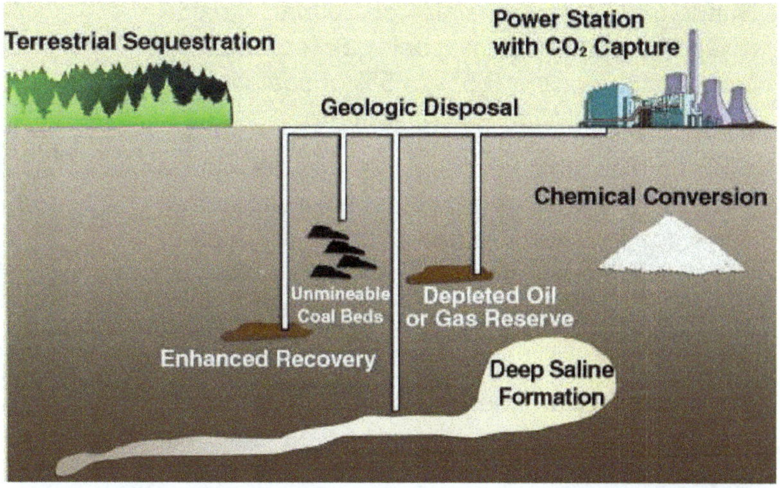

Much of the CO_2 collected from current coal processing is sequestered in deep geological strata such as aquifers and depleted oil fields. This gas can be of increasing value for enhanced oil and natural gas (methane) recovery as these resources are depleted.

Carbon Sequestration Options

76. Balanced Energy

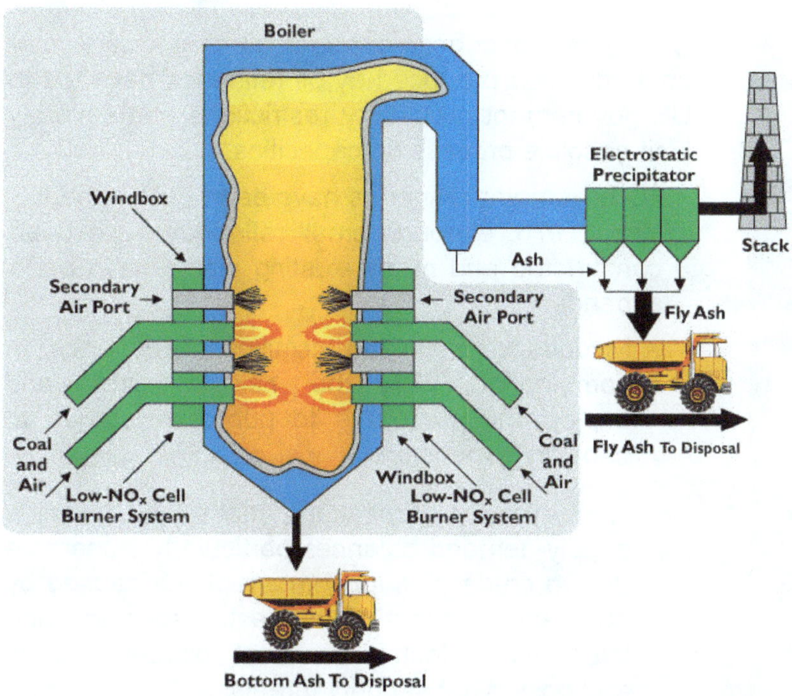

Staged combustion introduces a rich gas mixture into "low-NOx burners" that causes oxygen to combine with the fuel rather than the nitrogen and then sends the exhaust to a second combustion chamber to be burned again.

NOx Removal from Coal Gas

Nitrogen oxides (NOx) released into the atmosphere from combustion of coal and other fuels are the pollutants that cause brown smog haze and ground-level ozone seen around large cities, and contribute to acid rains that kill vegetation and erode building structures.

- Nitrogen can be removed from coal gases by a variety of processes, including use of chemical "scrubbers" and "staged combustions":
  - The scrubber approach uses chemical catalysts to remove NOx from coal flue gases.
  - Nitrogen obtained from scrubbers can potentially be sold as fertilizers and other ammonia-based products.

77. Power Technology

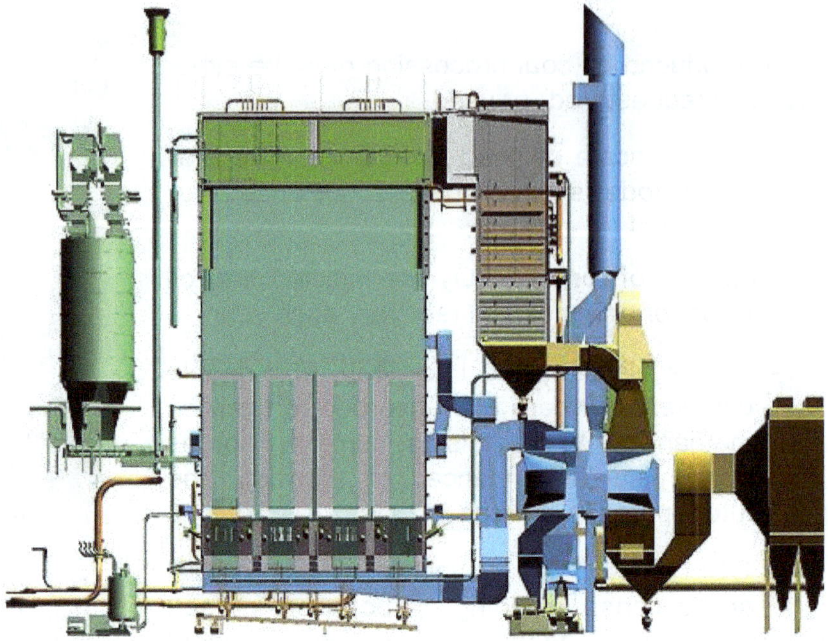

Sulfur that is responsible for sulfuric dioxide (SO_2) components of coal gas combustion can constitute from 0.5% to 5% of coal weight, depending upon its regional source.

- Some of this sulfur that exists in a particulate form can be washed away before the coal arrives at a power plant, while other contents may be chemically bonded to carbon molecules and removed using scrubbers:

- "Fluidized bed boilers" that apply a tumbling action, typically use limestone as a "sulfur sponge" that chemically reacts to produce dry powder calcium sulfate that can be used to create wallboards for building construction.

- These scrubbers can also remove nearly all of the NOx pollutants that produce smog and acid rain.

Advanced pressurized fluid bed boiler technology uses steam to remove sulfur and nitrogen pollutants and produces electricity-generating efficiencies that can be more than 50% higher than conventional plants.

Coal Gas

SO_2 and NOx Removal from Coal Gas

78. NASA

Vulnerable US Production

Strong public concerns about air and wastewater contamination produced by oil refineries have led to US government regulatory restrictions that severely limit gasoline processing capacities.

- No new major refineries have been built in the US since 1976, although small refineries have been constructed and many existing plants have been expanded:
  - Noxious odors in the vicinities of refineries, in combination with the risk of fires and explosions, contribute to public resistance to capacity expansion.
  - Such process limitations make US gasoline supply-demand balances particularly vulnerable due to crude oil supply interruptions caused by hurricanes impacting domestic sources, and international instabilities that disrupt imports and pose plant security threats.

Petroleum

79. Chemistry Daily

Processing of gasoline is but one of many products that compete for crude oil use.

- Refineries separate the raw oil into a variety of solid, liquid, and gaseous parts using fractional distillation furnaces:
  - Oil contains hydrocarbon molecules of varying masses and lengths, each with special chemical properties and uses.
  - Fractional distillation uses different temperatures in fractionating columns to separate the molecules, where lighter ones rise to cooler regions on top, and heavier fractions emerge at lower levels.
  - The heavier molecule fractions can then be broken up ("cracked") into lighter short chains to create other products.

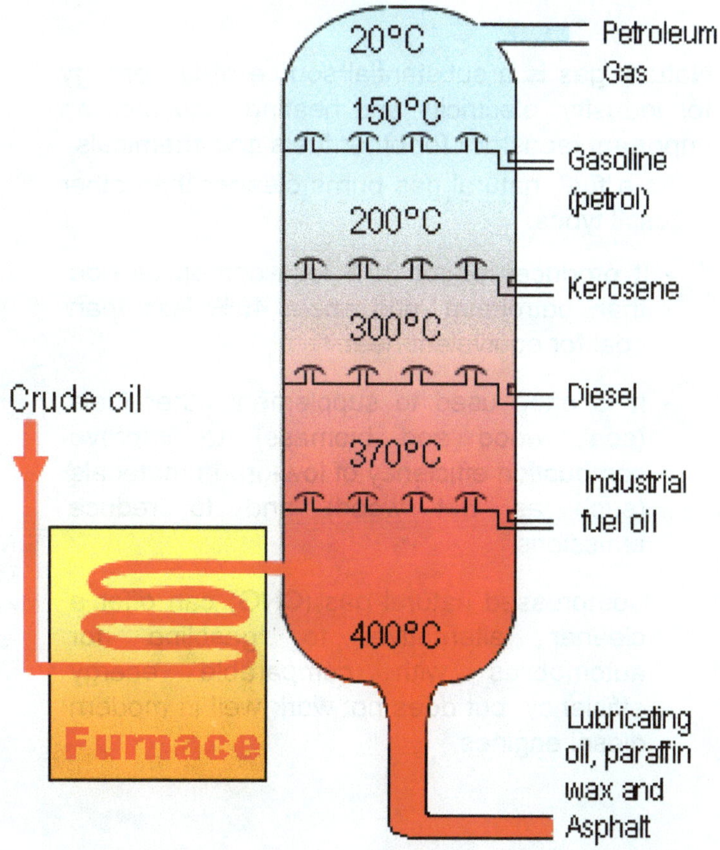

Oil Refining and Products

Octane ratings determine how resistant a particular gasoline mixture is to premature detonation in an engine, which causes knocking.

- Following fractional distillation and contaminant removal, different product streams can be recombined to meet specific octane requirements:
  - Octane can also be improved by catalytic reforming which strips hydrogen out of hydrocarbons to produce "aromatics" which have much higher ratings.
  - The final step in gasoline production is to blend fuels of different octane ratings, vapor pressures, and other properties.
  - Exact ratios depend upon the particular refinery (not all have the same processing units), the grade of crude oil available at a given time, and distributor demands.

80. *EIA*

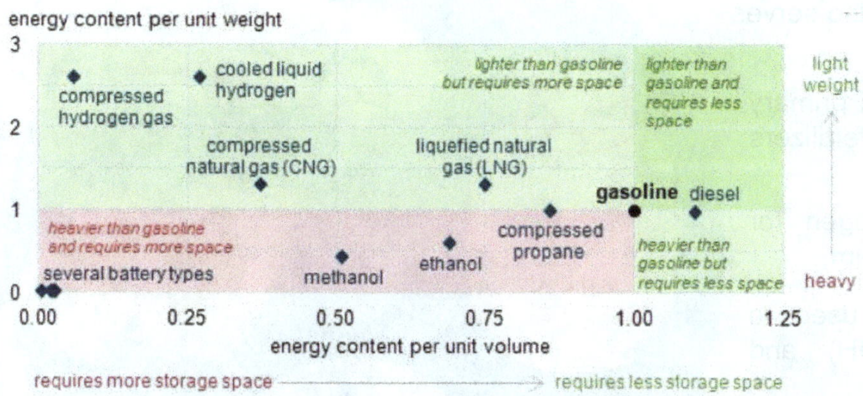

Energy Density Content of Different Fuels

Natural gas is a substantial source of US energy for industry, electricity and heating, and also an important feedstock for other fuels and chemicals.

- As a fuel, natural gas burns cleaner than other fossil types:
  - It produces about 30% less carbon dioxide than petroleum, and about 45% less than coal for equivalent heat.
  - It is often used to supplement other fuels (coal, wood and biomass) to improve combustion efficiency of low-grade materials (such as wet wood) and to reduce emissions.
  - Compressed natural gas (CNG) can offer a cleaner alternative to gasoline for automobiles with comparable energy efficiency, but does not work well in modern diesel engines.

81. US IEA

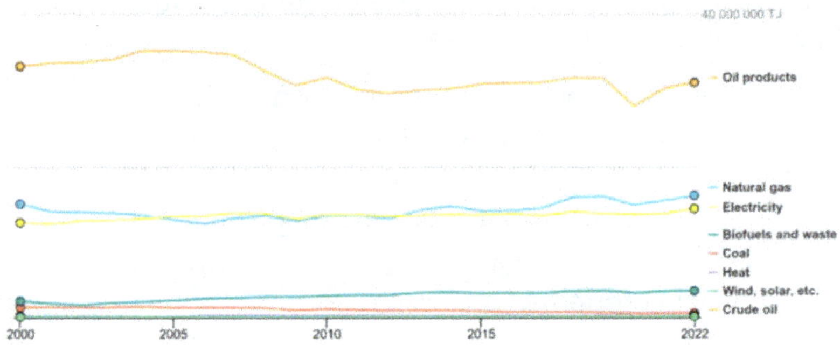

Natural gas supplies about 43% of all US energy. This includes nearly 41% of all electrical power, about 24% of all energy for industry, 5% of all energy for commercial facilities, and heating and cooling for more than 60 million households.

US Natural Gas Consumption

In addition to direct energy use, natural gas also serves a variety of other valuable purposes.

- Consisting mostly of methane (CH_4) it is a primary feedstock for producing secondary fuels, fertilizers and other chemicals:
  - Methane is a major source of hydrogen for industrial use and fuel cells, and also helium.
  - Steam methane reforming (SMR) is used to commercially produce syngas (CH_2OH) and methanol (CH_3OH).
  - Natural gas methane is an important feedstock for manufacturing butane, ethane, propane, and ammonia fertilizer.

82. Larry Bell

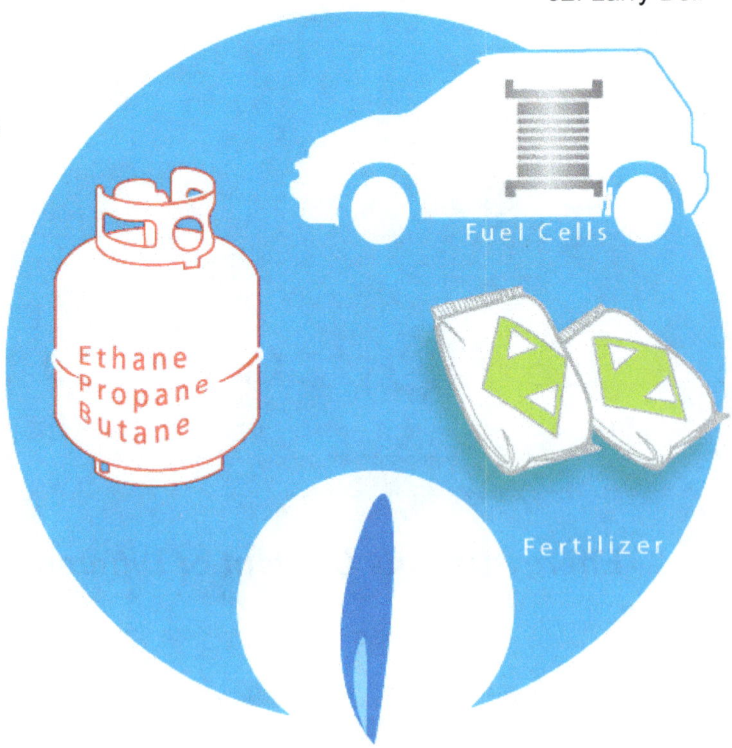

Natural Gas Products

Natural gas is collected from oil wells as “casinghead gas”, and is also obtained from gas fields and coal beds.

- Prior to use, the gas is cleansed of sulfur impurities such as hydrogen sulfide (H_2S):
  - The gas is a major source of electricity production through gas and steam turbines (often operated in combined cycles to optimize efficiency).
  - Compressed Natural Gas (CNG) is used as an alternative to liquid automotive fuel.
  - Although natural gas can be explosive when confined at concentrations of 5% - 15% of air, this is not a hazard with CNG used in vehicles.
  - Since processed gas is odorless, small amounts of thiol odorants are added to assist in leak detection.

83. Larry Bell

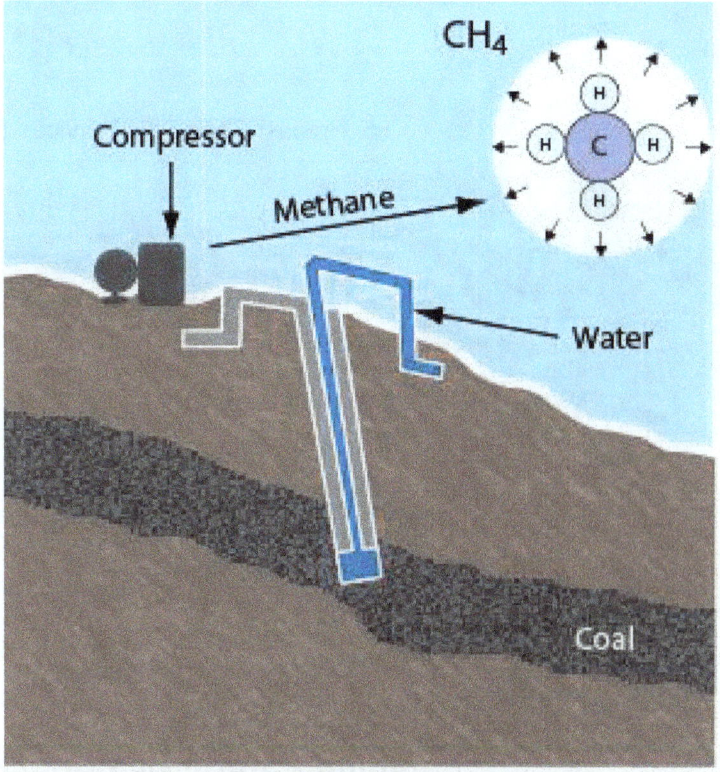

Natural Gas Collection and Processing

Although natural gas collected from oil field operations in recent years was not a commercially profitable product and was simply burned (“flared”), this circumstance no longer holds true.

- This gas is now recognized to be a very valuable, non-renewable resource:
  - Flaring is no longer practiced in the US because it pollutes the atmosphere.
  - Oil field gas is now re-injected back into a formation, stored in underground caverns formed inside depleted gas wells, in salt domes, or in LNG tanks for later recovery during high demand periods.
  - Storage near ultimate end-users is most convenient to address highly volatile demands but isn’t always possible due to land use restrictions associated with limited urban space and public safety concerns.

84. Ontario Ministry of Natural Resources

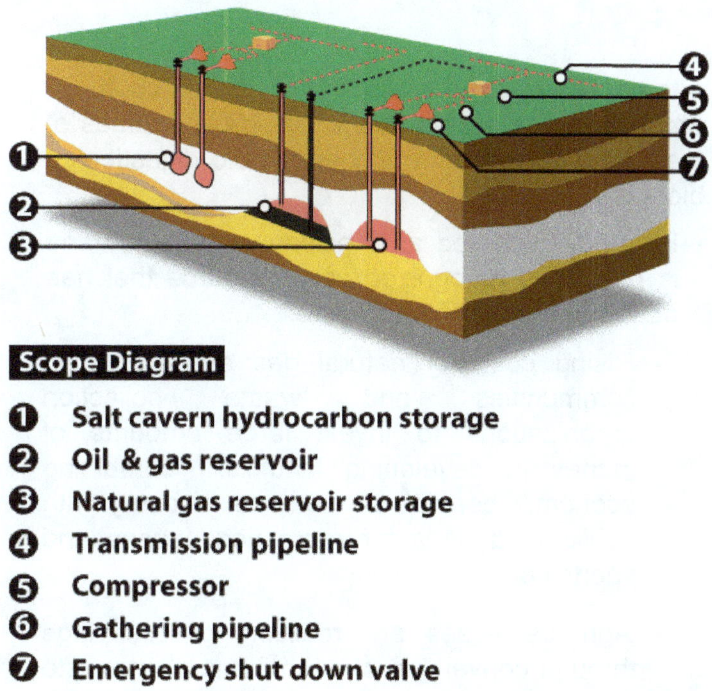

Natural Gas Storage Reservoirs

85. Precast Consulting Services

Looming Natural Gas Crisis

Many experts predict that known natural gas resources that provide about one-fourth of US energy and 20% worldwide will be exhausted by the end of this century.

- Continued dependence upon foreign imports to supplement domestic sources is unrealistic:
  - Global competition for all fossil fuels combined with geo-political uncertainties and conflicts have volatile influences upon supplies and costs.
  - Imports of LNG to the US are constrained by limited port terminal location options with adequately deep harbors sheltered from wind and waves.
  - New or expanded LNG terminal developments are inhibited by seaport congestion, right-of-way conflicts for pipelines, and problems in locating large LNG and end-use plants in high population density areas near ports.

Alternatives to fossil fuels and derivative chemical products are available using renewable sources of methane released through anaerobic digestion of biomass.

- Formerly regarded as a nuisance gas, methane is becoming recognized as a resource that has become too valuable to waste:
  - Rising costs of natural gas are motivating communities and waste collection organizations to invest large amounts of money in developing landfills, combining economic benefits of methane recovery with public land uses for recreation spaces and sports fields.
  - Agri-businesses are realizing cost savings through conversion of animal and plant waste into fuels for space heating, electricity, equipment and fertilizers.

86. Kanagawa Pref. Ed. Cntr. 87. Helsinki University of Tech
88. Kanagawa Pref. Ed. Cntr. 89. Loughborough University

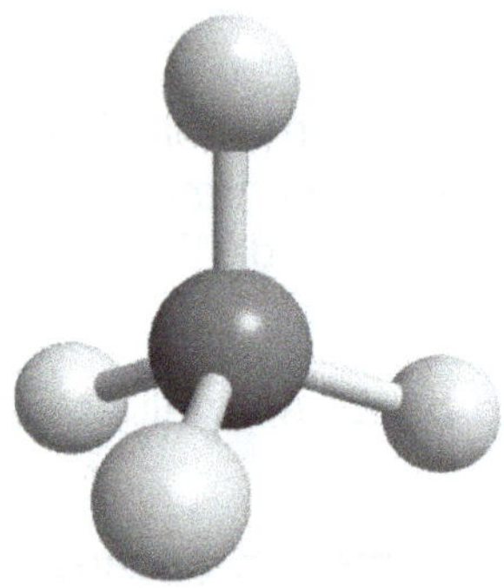

Methane

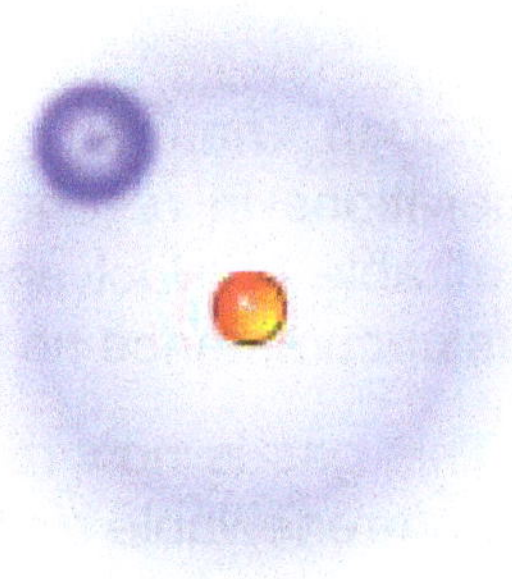

Hydrogen

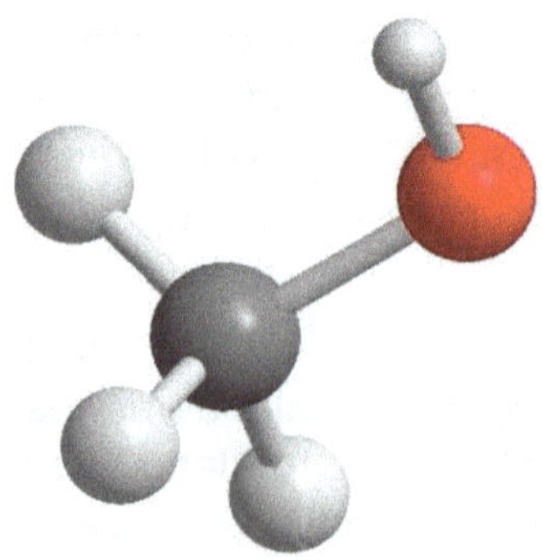

Methanol

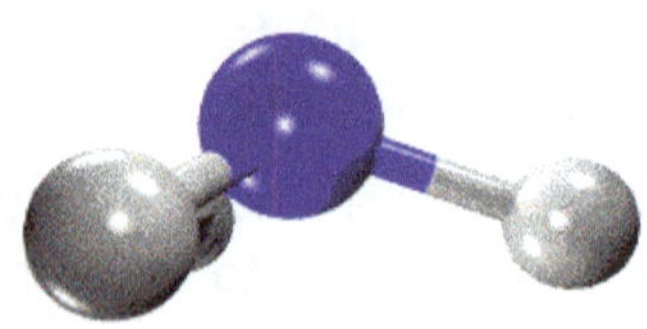

Ammonia

90. General Sources

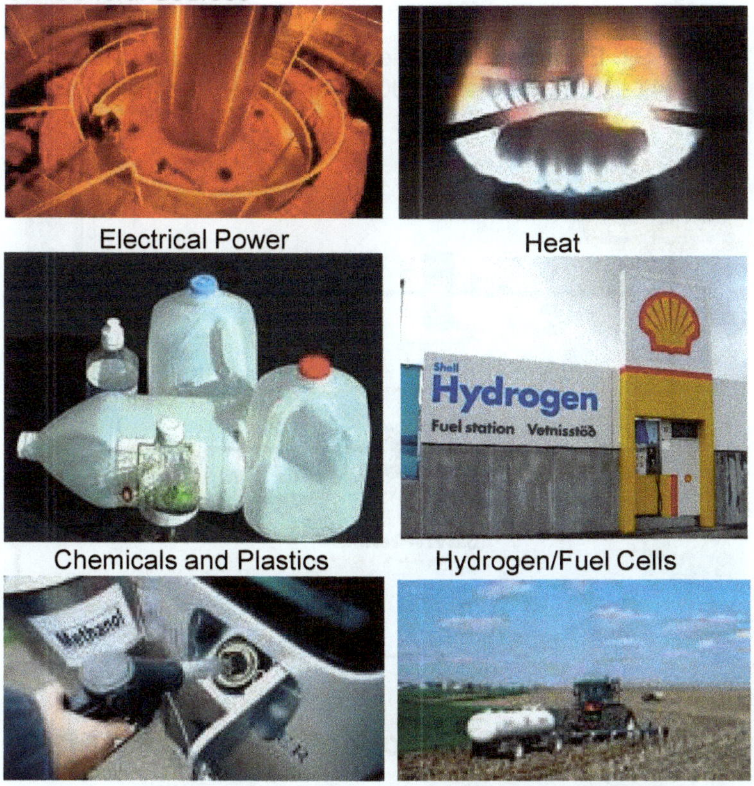

Electrical Power | Heat | Chemicals and Plastics | Hydrogen/Fuel Cells | Methanol | Ammonia Fertilizer

Methane/Natural Gas Uses and Derivatives

Many organizations recognize the value of municipal and agricultural wastes as renewable resources.

- Methane recovery is an important example:
  - This gas (CH_4) is the same material as natural gas from fossil sources that are rapidly being depleted.
  - Methane can be burned directly as fuel for power and heating, or can serve as a versatile feedstock for plastics and other substances.
  - Methane is processed as the leading commercial source of hydrogen that can power fuel cells.
  - Methane is also used to produce methanol, and can be reacted with methane-derived hydrogen to create ammonia and urea fertilizers.
  - Bio-oils obtained from the fertilized crops can be used directly for fuel, or can be combined with methanol to create bio-diesel fuel to power engines and drive turbines.

Living matter is made of organic compounds containing carbon and hydrogen.

- Methane is the simplest of all organic compounds, and is an important feedstock for other compounds and commercial hydrogen production.
  - Methane is produced through digestion of carbonaceous biomass by microscopic methanogenic organisms in the stomachs of mammals, in swamp sediments, in landfills, and in other anaerobic environments.
  - Much of the biomass comes from vegetation sources, including the plants that livestock and other animals feed upon.
  - Methane-derived fertilizers are used to provide nutrients to support plant growth.
  - Plant oils, in turn, combine with methane-derived methanol to create biodiesel fuel for tractors and power generation.

91. Larry Bell

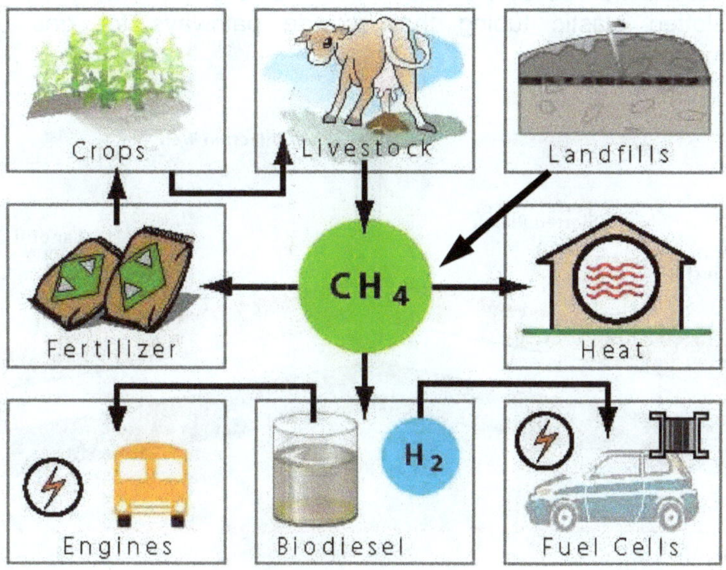

Life Cycles

Many large municipalities and waste management organizations are converting garbage to revenues.

- Landfills produce substantial amounts of methane that can be captured to help meet growing demands:
  - Landfill gas is created by natural anaerobic decomposition of discarded organic wastes that must often be removed to prevent high pressures from damaging the landfill structure, produce unpleasant odors, cause vegetation die-off and present explosion hazards.
  - When water vapor is removed, about half of the gas is methane and half is CO_2, along with small amounts of nitrogen, oxygen, hydrogen, and sulfides.
  - Landfill gas can only be distributed through natural gas pipelines after it is cleaned of impurities to the same quality standards.

92. Environmental Science and Engineering Magazine, March 2004, Vancouver Landfill Operation

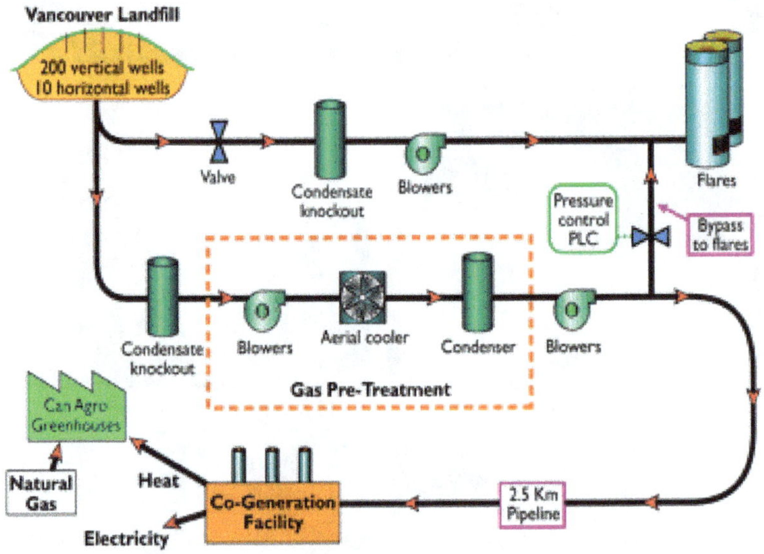

Landfill Gas Collection and Utilization

Landfill gas collectors use "wells" made of perforated or slotted plastic tubing that provide pathways for gas migration.

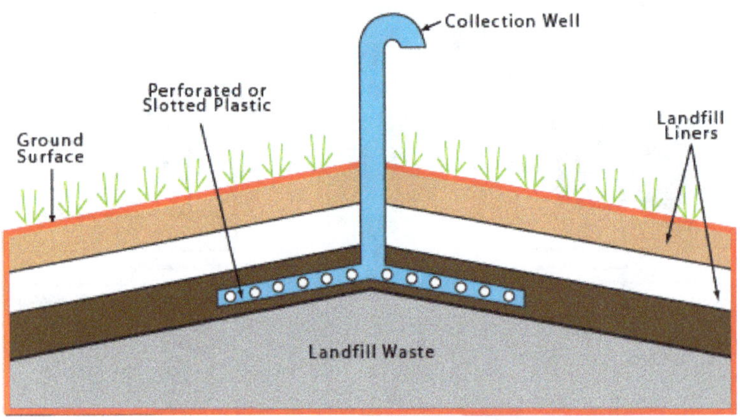

Passive systems are used in small recovery operations, and active systems provide for managed commercial capabilities.

Passive Landfill Gas Collector

93. ATSDR

Passive systems collect gas using impermeable liners on top, sides and bottom to limit venting into the atmosphere.

Active systems incorporate vacuum pumps and control valves to monitor and regulate gas flow. Sampling ports measure pressure and content.

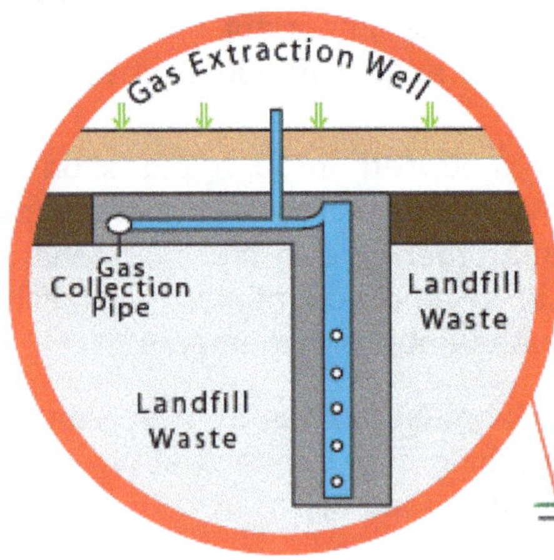

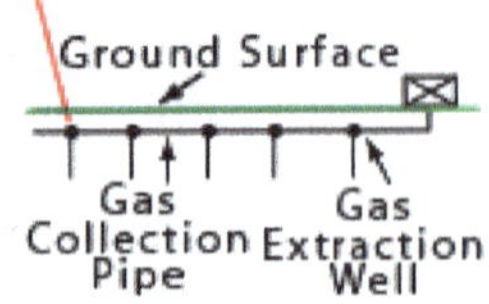

Active Landfill Gas Collector

Methane

Methane obtained from both fossil fuel sources and anaerobic digestion has a variety of direct and derivative uses.

- The gas can be burned directly, and can also provide feed components for important industrial products:
  - Methane combustion is a very efficient means to transfer heat for direct contact industrial water heaters.
  - Combined heat and power (CHP) cogeneration systems can use methane for "cogeneration" that produces electricity and recycles excess heat and steam for space heaters, water heaters and industrial boilers.
  - Synthetic gas (syngas) produced from methane through SMR and other processes are used to produce hydrogen and methanol, along with ethane (C_2H_6), propane (C_3H_8), and butane (C_4H_{10}).

94. Larry Bell

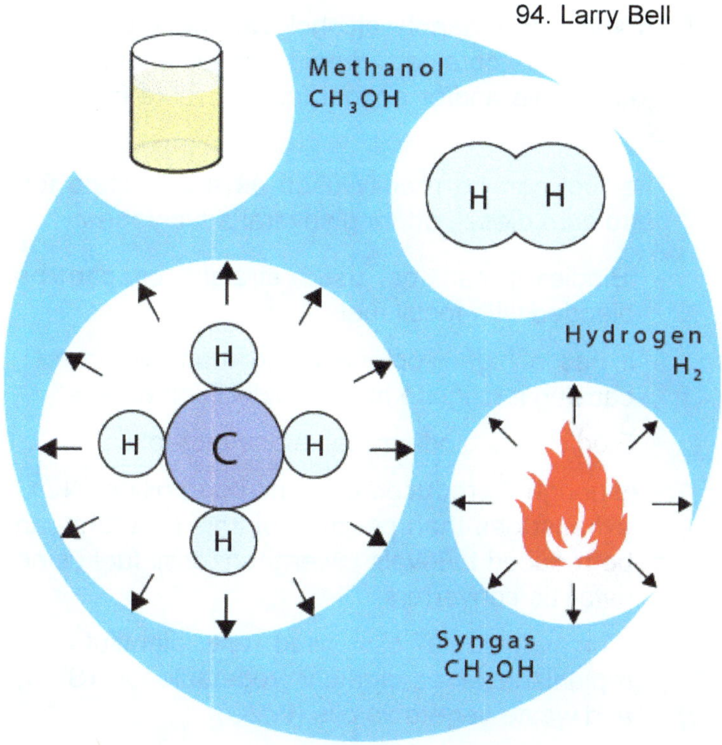

Methane Gas Products

Methanol (or methyl alcohol) is most commonly produced from the methane component of natural gas or syngas, but is also a product of coal gasification.

- The chemical compound (CH_3OH) is the simplest and cheapest alcohol created by anaerobic metabolism:
  - Methanol is used on a limited basis to fuel internal combustion engines, but is not nearly as flammable as gasoline.
  - It can be reacted with vegetable oils to produce biodiesel.
  - Methanol is often mixed with gasoline on an 85% methanol, 15% gasoline basis to boost energy performance ("methanol 85").
  - The drawback to its use as a fuel is its corrosive nature that degrades aluminum and other metals.
  - Its largest use by far is to obtain acetic acid to produce plastics, wood glue, paints, explosives and permanent press textiles.

95. Larry Bell

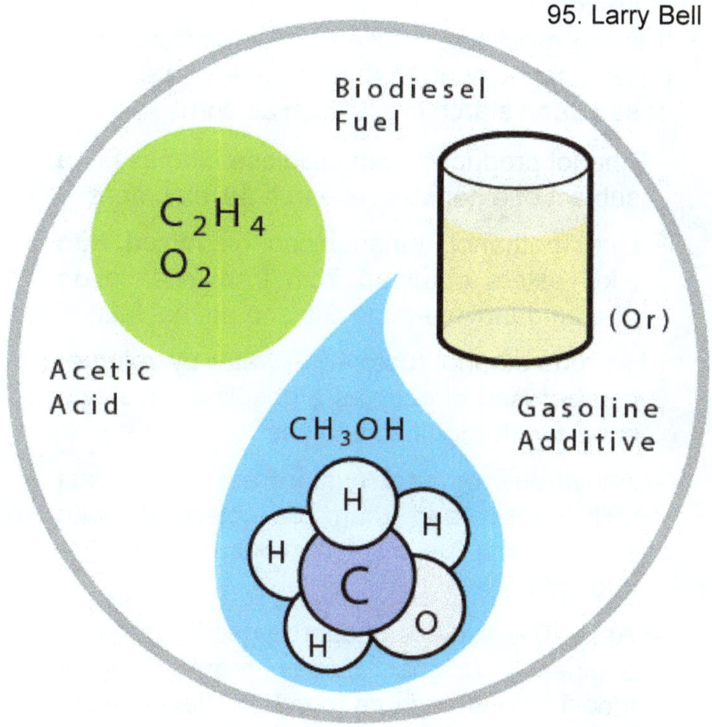

Methanol

Methanol (or ethanol) alcohol can be mixed with alkyl ethers obtained from transesterication of vegetable oils and/or animal fats to create biodiesel fuel.

- The fuel can be readily used as a substitute for petroleum diesel, offering several advantages:
  - Biodiesel can be used straight or can be blended with fossil fuels.
  - It has a higher-octane rating than petrodiesel, causing it to ignite more rapidly in an engine.
  - Biodiesel is biodegradable and non-toxic.
  - Although it produces more nitrogen oxide (NOx) exhaust gas than petrodiesel, this pollutant can be reduced to levels lower than fossil fuel using catalytic converters.
  - The vegetable oils used are plentiful and replenishable as straight vegetable oil (SVO) and waste vegetable oils (WVO).

96. National Renewable Energy Laboratory

Organic biodiesel can be produced by combining methanol with methyl esters obtained from three varieties of sources:

- *Rapeseed methyl ester (RME)*
- *Vegetable methyl ester (VME)*
- *Fat methyl ester (FME)*

It burns cleanly and fully in modern diesel engines, and works with catalysts, particulate traps and exhaust recirculation to eliminate most pollutants.

Biodiesel Production and Benefits

Ethanol

Ethanol is made by converting the carbohydrate portion of biomass into sugar, which is then fermented into alcohol in a process similar to brewing beer.

- It is the most widely used biofuel today with a current capacity of 15.4 billion gallons per year based upon starch crops such as corn:
  - Ethanol produced from cellulose biomass is a subject of extensive research development
  - Like methanol, ethanol can be mixed with alkyl esters obtained from transesterication of abundant vegetable oils and animal fats.
  - Hydrous ethanol (upto 4.9% water by volume) can be used directly as a gasoline substitute in cars with modified engines.
  - Anhydrous ethanol (water-free) is at least 99% pure and can be blended with conventional fuel in proportions between 5% and 85%.
  - As a 10% additive, it can be used in modern engines. (Higher blends require engine modifications such as found on "flexible-fuel" vehicles).

97. US DOE-EIA

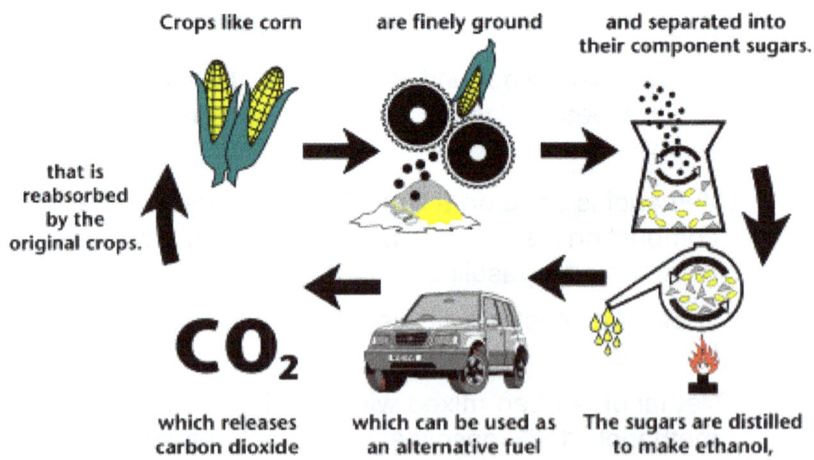

Ethanol is a readily available clean fuel that can be used in combustion engines as a petroleum substitute, additive or blend. It is also used to manufacture ethyl-tertiary-butyl-ether (ETBE), a fuel enhancer for conventional petroleum.

Ethanol Production and Benefits

Bio-Oil

Bio-Oil Production

98. US DOE-EERE

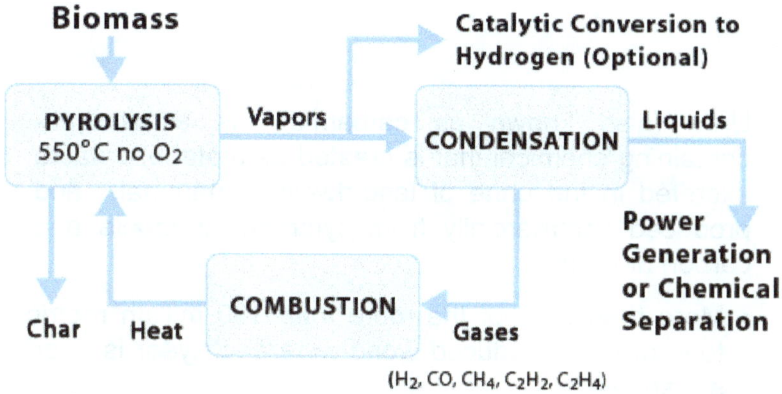

Fast pyrolysis of biomass is a thermal decomposition process that produces a liquid bio-oil that can be readily stored and transported. Pyrolysis oil has been successfully tested in engines, turbines and boilers, and has been upgraded to high- quality fuels, although presently, at uncompetitive energetic and financial costs.

Bio-oil is created using fast pyrolysis to process biomass obtained from agricultural and municipal wastes, forestry residue, oil crops, and short-rotation herbaceous and woody plants.

- Up to 40% bio-oil has been used as a mixture with standard diesel fuel, and continuing developments may enable direct use in diesel engines for trucks, tractors and other equipment:
  - Bio-oil is considered to be "CO_2 neutral", and emits no SOx, and only low NOx, when combusted.
  - The fuel can be combusted directly in boilers, gas turbines and slow-medium speed diesels for heat and power applications.
  - Bio-oil has only about half of the heating value of light or heavy oil per volume, but is superior to heavy fuel oil for viscosity and cold weather properties.

Nutrients

Ammonia Fertilizer

99. Larry Bell

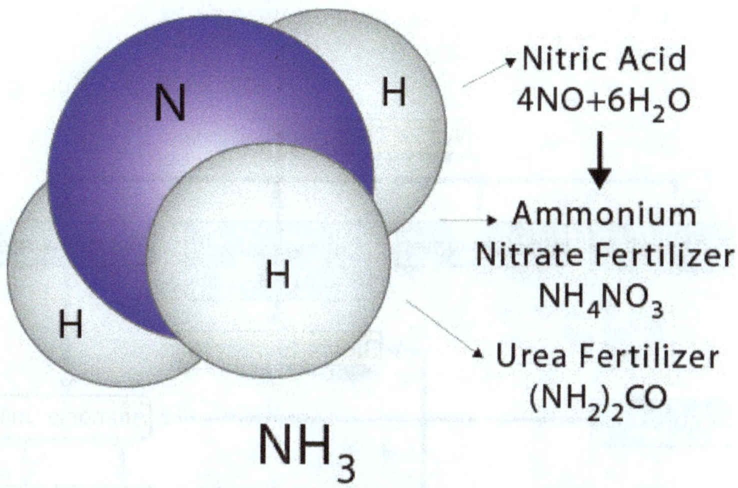

Ammonia is a colorless, pungent lighter-than-air gas.

- *Ammonia gas does not sustain combustion unless mixed with oxygen, but can catch fire and become explosive when combined with chlorine..*
- *Anhydrous ammonia corrodes copper and zinc alloys, and liquefies under pressure at ambient temperatures and when refrigerated below -28°F for transport and storage.*

Ammonia (NH_3) is produced naturally by the putrefaction of nitrogenous animal and plant matter, and since it readily dissolves in water, small quantities are found in rainwater.

- This nitrogen and hydrogen compound is one of the most commonly manufactured inorganic chemicals (more than one million metric tones world-wide annual production), primarily for fertilizers, explosives and polymers:
  - Ammonia used commercially is anhydrous (not the ammonia hydroxide solution used for household purposes).
  - The most important use is for production of ammonium nitrate (NH_4NO_3), a white powder at room temperature used as a high-nitrogen fertilizer.
  - Ammonia can also be used to create urea fertilizer, or can be used directly as a fertilizer when mixed with irrigation water for growing nitrogen-dependent crops such as corn.

Urea (also known as carbamide) is a nitrogen-containing chemical that is created as metabolic waste excreted in the urine of land-dwelling mammals, and produced commercially from synthetic ammonia and carbon dioxide.

- More than 90% of the more than 180 million metric tons of urea produced world-wide each year is used as fertilizer:
  - The chemical has the highest nitrogen content of all nitrogenous fertilizers (46.4%), and therefore, the lowest transportation cost per unit of nitrogen content.
  - Urea is produced using two raw materials that are released in large quantities during ammonia manufacturing – ammonia and carbon dioxide.
  - The commercial product is available in the form of granules, flakes, pellets, prill, crystals and solutions.

100. University of Minnesota

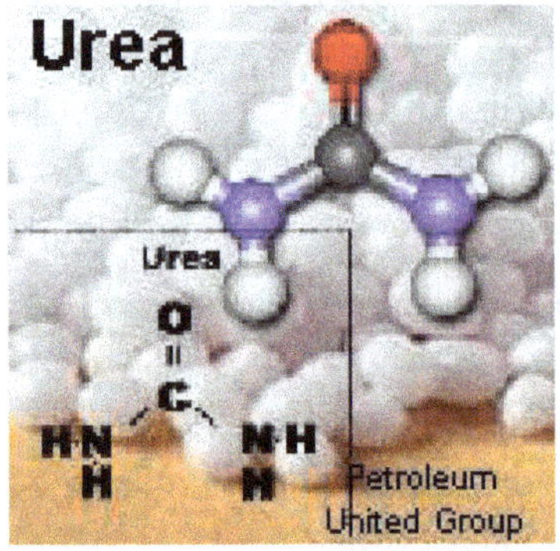

101. SOS-Arsenic

Urea is formed in nature as a waste product in the livers of mammals during the process of converting toxic ammonia through the "urea cycle". It is present in abundant quantities in livestock excretions.

The commercial product is available in the form of granules, flakes, pellets, prill, crystals and solutions.

Urea Fertilizer

102. PT Pupuk Kujang

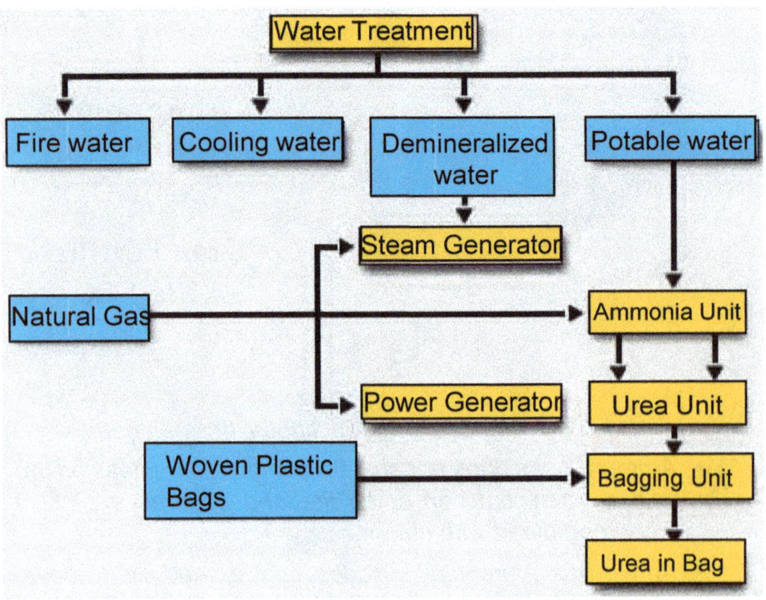

Urea Fertilizer Production

Urea is produced in connection with standard ammonia manufacturing processes.

- The main feedstock in these production processes are natural gas/methane, water and air:
  - The natural gas / methane is processed with water and air to produce nitrogen (N_2), hydrogen (H_2) and carbon dioxide (CO_2).
  - Nitrogen and hydrogen gases are then processed to produce ammonia (NH_3).
  - Ammonia and carbon dioxide are finally processed to produce urea "prill" with a diameter of 0.04in - 0.08in (1mm - 2 mm), which is then bagged for distribution.
  - In addition to fertilizer, other commercial uses include treatment of diesel engine exhaust, and as a raw material for certain plastic resins and glues.

Composting offers a low-technology means to recover nitrogen and other nutrients from organic wastes to provide fertilizer while also reducing disposal requirements.

- Although decomposition of organic matter by aerobic organisms takes place naturally in all but the most hostile of environments, composting under controlled conditions can expedite the process and optimize nutrient product yields:
  - Active (or hot) composting techniques are most generally used for commercial and industrial applications, and involve regular turning of the pile, monitoring temperatures, and adjusting ingredients of the mix to speed the process.
  - Passive (or cold) methods let Nature take a slower course, and are more often used for home and farm applications.

103. KNOWASTE

Active composting can produce finished product with useful fertilizer constituents for plant growth.

Water . . . 10-15%
Organic matter . . . 10-20%
Nitrogen . . 0.8%
(Highest yields-manure)

Phosphorus . . . 0.45%
Potassium . . . 1.45%
Lime . . . 1.25%
Magnesium . . . 0.3%

Compost Fertilizer

Active Composting

- Is used by a few urban centers around the world for large-scale co-processing of solid wastes and de-watered bio-solids.
- Is also used in smaller-scale agricultural operations to recover nutrients from plant residues and animal manure.
- Provides the most effective composting approach to enable decomposing bacteria to thrive, killing most pathogens and seeds.
- Produces higher temperatures (reaching 110° - 160°F) to achieve rapid decomposition with high-quality fertilizer products
- Is most effective when high-nitrogen organics (such as corn) are mixed with high-carbon materials.

Passive Composting

- Is simpler and more often used in rural/farm applications (throw everything in a pile and leave it for a year or two).
- Often used as a means to eliminate and treat organic kitchen and garden wastes to produce fertilizer mulch.
- Lets Nature act more slowly, leaving pathogens and seeds alive and/or dormant in the pile.
- Longer curing periods tend to increase the amount of useful nutrients (particularly nitrogen) to leach out, reducing fertilizer value.
- Requires that non-biodegradable materials (including colored paper, foils and plastics) be avoided. (Also applies to active systems).

Active and Passive Composting Techniques

Ammonia fertilizer is a source of nitrogen required for the synthesis of amino acids, the building blocks of protein in most living creatures. The hydrogen component of this hydrogen-nitrogen compound is often obtained through the "Haber" process which purifies the product of steam methane reforming or char gasification and reacts the H_2 with the nitrogen gas. Although H_2 recovery using biomass pyrolysis is possible, this method is not presently practiced for commercial fertilizer applications.

The Haber process begins after the hydrogen has been removed from natural gas/methane by SMR, or from coal/char through gasification:

1. Sulfur compounds are removed through catalytic hydrogenation by converting them to gaseous hydrogen sulfide (H_2S).
2. The H_2S is absorbed and removed by passing it through zinc oxide beds that convert it to zinc sulfide (ZnS).
3. Catalytic steam reforming of the sulfur-free feedstock is used to produce carbon monoxide and hydrogen ($CO + 3H_2$).
4. A catalytic shift conversion coverts the carbon monoxide to carbon dioxide and more hydrogen ($CO_2 + H_2$).
5. Carbon dioxide is removed by absorption, and catalytic methanation removes residual carbon monoxide from the hydrogen.
6. The hydrogen is then catalytically reacted with nitrogen from processed air to form anhydrous liquid ammonia.

The Haber Process for Ammonia Synthesis from Hydrogen

Hydrogen comprises about 75% of all matter in the Universe.

- It exists in three different forms:
  - The term "hydrogen" sometimes refers to its elemental form (H) with an atomic number 1, the lightest of all elements.
  - Nuclear fusion within the Sun's core compresses hydrogen into helium, releasing hydrogen plasma ("H dot") comprised of an electron and proton which are not bound together with high electrical conductivity and emissivity ("sunlight" photons).
  - Hydrogen is also properly defined to refer to a diatomic molecule (H_2) that can exist in a gaseous or liquid form as an "energy carrier".
  - About 95% of all H_2 is produced commercially from natural gas (the lowest-cost method presently available).

104. NASA

Hydrogen-sourced nuclear fusion in the Sun's core converts about four million tons of matter into sunlight every second.

The Nature of Hydrogen

Hydrogen

Hydrogen is commonly used as an additive to upgrade performance of conventional hydrocarbon fuels and to power fuel cells.

- H_2 is characterized as an energy carrier rather than an energy source because more energy is required to produce it than can be extracted.
  - Steam methane reforming which removes H_2 from natural gas is the most widely used extraction method, and requires energy to produce the steam.
  - A new, potentially more efficient method to extract H_2 from natural gas passes the methane along with oxygen-depleted air through non-porous ceramic membrane wafers under high-energy pressure and temperature (producing H_2 + CO).
  - H_2 production using biological anaerobic methane reduction and all other processes requires energy to pressurize it into a contained gaseous form, or to refrigerate it into a cryogenic liquid (LH_2).
  - Electrolysis generally involves very expensive materials (platinum and palladium) which don't scale up well for economic production.

105. US DOE, Office of Fossil Energy

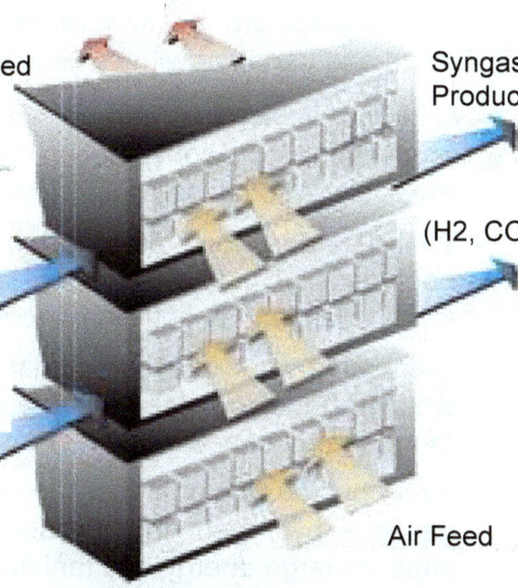

Hydrogen Extraction Through Partial Combustion of Natural Gas

Hydrogen Gas (H_2)

- Highly combustible, will burn in concentrations as low as 4% H_2 in air.
- Hydrogen explodes upon ignition when mixed with oxygen.
- Reacts violently in contact with chlorine and fluorine.
- Readily leaks due to small molecular size through porous materials, cracks or bad joints.
- Has good energy density per weight, but poor density per volume (compared with gasoline).

Liquid Hydrogen (H_2)

- Has higher volumetric energy density than gaseous H_2, but requires low temperature storage.
- Has much worse energy density per volume than gasoline.
- Expensive tank insulation is required to prevent boil-off (LH_2 boils at about -423°F [-253°C]).
- LH_2 is cold enough to freeze air, and can cause valves to plug up in automotive fuel applications.
- Production and transportation of LH_2 or H_2 can require more than twice the energy recovered.

Hydrogen Form Characteristics as Energy Carriers

Although hydrogen has been mass produced for more than 50 years, it may be some time before it has a major place in the automotive energy economy.

- Key problems are associated with high costs of storage and transport with current technology:
  - Although gaseous H_2 can be pumped directly from production to consumption areas, the tiny molecular size causes costly leakage.
  - While H_2 has good energy density/weight, it has poor density/volume, requiring larger and heavier storage tanks than gasoline or diesel fuels.
  - Increasing gas pressure enables tanks to be smaller, but also makes them heavier and requires more energy to power the compressors.
  - Liquid hydrogen has higher volumetric energy than gaseous hydrogen, though cryogenic liquefaction imposes large energy and tank insulation costs.

106. Larry Bell

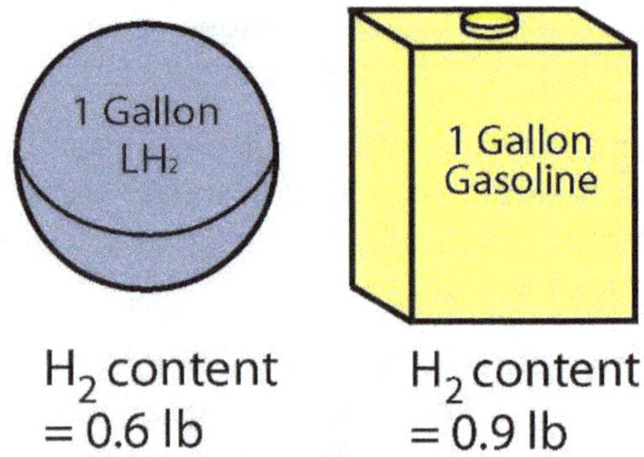

LH_2 has a worse energy density/volume than gasoline by a factor of about four. There is about 50% more H_2 in a gallon of gasoline weighing 0.9lb than in a gallon hydrogen weighing 0.6lb.

Hydrogen Storage Considerations

Hydrogen might be provided inexpensively using wind and hydropower that offer "free" sources of renewable energy applying mechanical forces supplied by Nature.

- Both of these sources are highly geographic-dependent, requiring that the kinetic power that is captured be used locally, stored or transferred in the form of electricity, or converted into a hydrogen carrier through water electrolysis with energy penalties:
  - Wind power development currently represents a very small but rapidly expanding segment of total renewable energy production.
  - Hydropower presently constitutes nearly half of all US renewable energy production, but is less likely to grow as dramatically due to limited major site options and environmental impacts upon water ecosystems.

107. National Renewable Energy Laboratories
108. National Renewable Energy Laboratories

Forces of Nature

109. Sandia National Laboratories

Offshore Wind Farm

The popularity of wind energy has always fluctuated with the prices of fossil fuels.

- Interest in wind turbines waned in the US and abroad when fuel prices fell after WW II:
  - Improvements in wind turbine technology along with oil embargoes of the 1970s promoted new interest and investment.
  - Wind farms connected to public utility grids have now become substantial alternative energy sources in the US and Europe.
  - Continuing R&D, along with investment growth, has made wind-generated electricity competitive with fossil fuel generators in some locations.
  - Wind is now the world's fastest-growing alternative energy source.

Although wind energy represents only a small percent of US renewable energy capacity and total production, it is gaining rapid expansion.

- Passage of a Production Tax Credit (PTC) by the US Congress has helped to boost corporate investment in this area:
  - US wind energy installations grew significantly, adding about 7.7 gigawatts (GW) of new capacity in 2025. (Each MW is enough to power approximately 270-300 homes.)
  - Growth is occurring at a time when US customers are facing electricity and natural gas hikes due to fossil fuel shortages.
  - About 156 GW of wind energy capacity (as of 2024) can potentially reduce natural gas consumption significantly.
  - Unlike large-scale drilling of natural gas fields and LNG terminal construction, wind plants can be rapidly built and permitted (requiring 1-2 years).

110. EIA

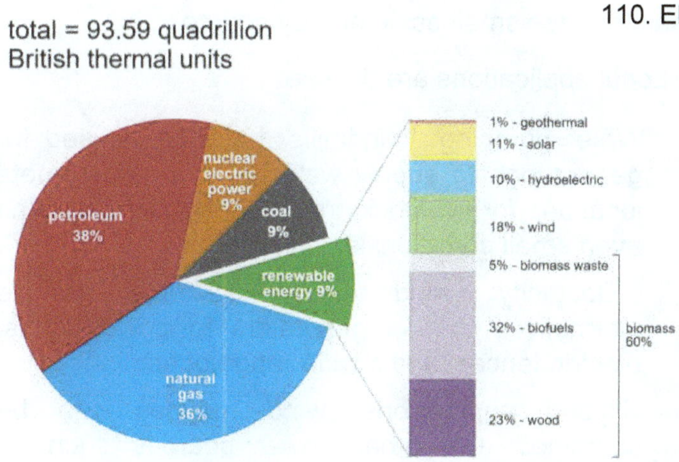

Wind Power as a Percent of US Renewable/Total Energy Consumption

Many large urban centers located along coastlines present major electricity demands.

- These areas are excellent candidates for offshore wind resources:
  - The DOE estimates that wind farms placed between 5 - 60 miles off coasts could provide as much as 4,000 GW of energy capacity.
  - There are currently 322 operating offshore wind projects globally, with a total installed capacity of 68,258 MW.
  - All installations are located in shallow coastal waters (less than 60 ft [18m] deep).
  - More than 80 GW of offshore wind projects are presently planned globally, with 30 GW in the US undergoing permitting processes.

111. US DOE-EERE

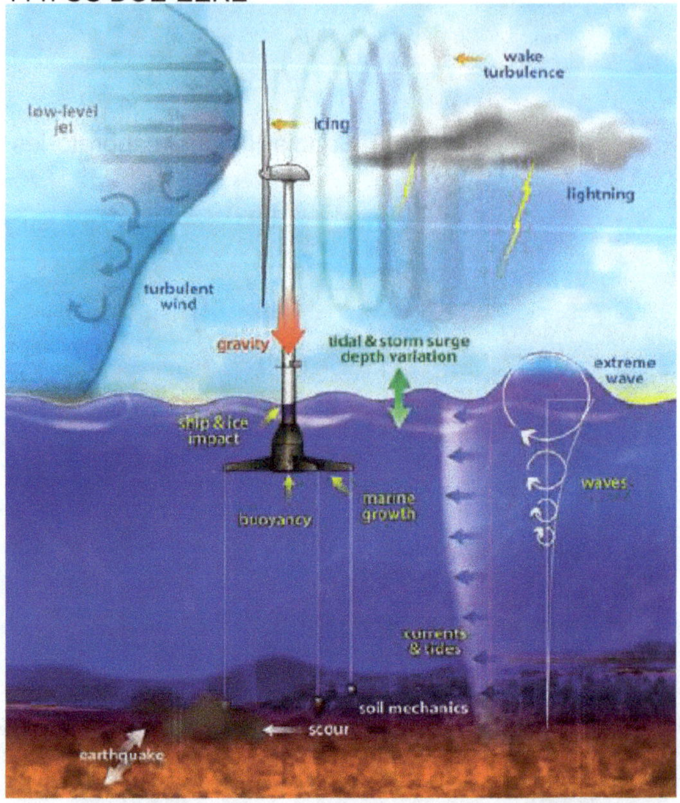

Offshore Installations

Wind can also be harnessed to provide a renewable, environmentally-friendly, economical and low-maintenance small-scale energy source.

- Local applications are diverse:
  - Water pumping: "windmills" have been used for generations to supply water in rural and remote locations for livestock, irrigation, households, and even small communities.
  - Electricity: wind energy systems provide alternatives to fuel generators to power lights, electric fences, and a wide range of appliances.
  - Hybrid applications: wind turbines can be combined with other power sources such as photovoltaic systems, diesel generators and electric power grids to provide cost-effective, non-polluting backup energy.

112. US DOE-EERE
113. Treehugger 114. Enviro-Energy Technologies

Advantages

- Wind is renewable, produced by solar heating of the atmosphere.
- It is an abundant US source of energy to reduce dependence on oil imports.
- Wind is one of the lowest-priced renewable technologies (2.6 - 5.4 cents/kilowatt hour).
- It is a clean source of energy – doesn't pollute like combustion processes.
- Wind turbines don't produce acid rain or greenhouse gases.
- Turbines can be installed on farms or ranches to provide income and local power.
- They can also be combined with other land uses such as livestock.

Disadvantages

- Depending upon site wind conditions, turbines are not always cost-competitive.
- Technology requires higher initial investment than fossil fuel generators.
- Since wind is intermittent, energy is not always available when needed.
- Good wind sites are often remote from areas where electricity demands are high.
- Wind sites can compete with other land uses such as agriculture.
- Noise from blades and visual impacts are sometimes considered to be objectionable.
- Birds are occasionally killed by flying into rotor blades.

Wind Energy Advantages and Disadvantages

115. Cages by Design

Harnessing Water Energy

Hydroelectric power (or hydropower) harnesses mechanical energy from moving or falling water for direct use or conversion to electricity.

- Prior to the widespread availability of commercial electricity, hydropower was extensively used for milling, textile manufacturing and sawmills:
  - Hydroelectric power now supplies about 1,330,000 MW (approximately 16% of world electricity), usually to support peak-load demands since it can be readily stopped and started.
  - Opportunities to expand its use are now constrained in developed nations because prime generating sites are either already exploited or unavailable due to environmental restrictions.
  - Ancient waterwheel principles continue to provide power for local use in some countries and off-grid applications in the US, and led to the development of much more efficient turbines for large and small-scale applications.

Hydropower

Waterwheels were used in ancient China dating back to 31AD to operate bellows used for casting iron, and during the late Middle Ages to pump water from mines, grind ore, run blast furnace forges, and operate metal smith hammers.

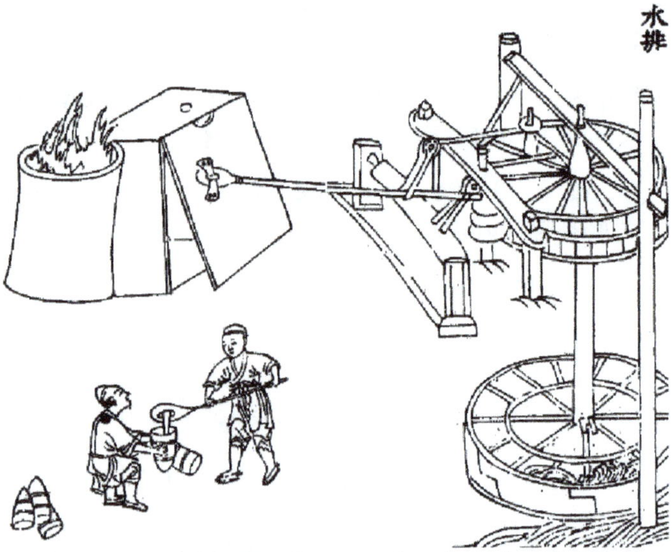

Water Pump in Ancient China

116. Water History

Water Pump in the Middle Ages

Eighteenth century horizontal waterwheels powered gristmills in Spain, and are still operating gristmills in Turkey using water from urban runoff that is diverted into elevated tanks that release water from above.

Eighteenth Century Waterwheel Gristmill

117. Water History

Recent Waterwheel Gristmill in Turkey

118. Computer Science at St. Andrews

119. Lienhard Schulz

Large-Scale Dam

Small-Scale Waterwheel

Hydropower sources and technologies range from very large-scale types used to supply regional power grids, to very small-scale "mini-hydro" and "micro-hydro" devices that serve domestic and stand-alone industrial purposes.

- The relative advantages and limitations of these various types and applications relative to other power options must be considered on a case-by-case basis:
  - Although hydropower, by itself, produces no carbon dioxide or other harmful pollutants, reservoir construction can sometimes cause methane to be released from decaying plant matter that becomes exposed in times of drought.
  - While hydropower can be much cheaper than electricity from fossil fuel or nuclear energy, this availability of abundant low-cost power can attract industries that have negative environmental impacts upon pristine areas.

120. Philip Greenspun

121. Atlantisstrom Germany

122. Marine Current Turbines.

123. Wave Dragon

Dam

Tidal Power

Tidal Stream

Wave Power

Large-scale hydropower applications harness mechanical energy from major dams, ocean waves and tides, and tidal streams, using various types of turbines to produce electricity.

- Many of these technologies are quite new, and some exist only in prototype stages:
  - Large dams that produce massive amounts of hydroelectric power have proven to be important and effective energy sources in many countries.
  - Tidal power, including tidal streams, are relatively new hydropower approaches which are rapidly gaining international interest.
  - Wave power is a promising future source of energy for countries with long coastlines and rough sea conditions.

124. Evgeni Sergeev

Conceptual illustration of a tidal barrage, including embankments, a ship lock, caissons housing, a sluice and turbines.

Tidal Barrage

Tidal Power is classified as a renewable energy source because it is driven by orbital mechanics of the Solar System and provides predictable and inexhaustible energy.

- Tidal barrage plant applications, however, impose certain limitations and consequences:
  - Tides can only generate power about 15 hours each day.
  - Barrages are costly to build, and can present periodic maintenance demands.
  - Estuaries often have high sediment volumes moving through them from rivers to seas that accumulate and impair barrage operations.
  - Sewage concentrations can lead to bacterial growth, impacting local ecologies.
  - Large fish populations are often sucked up through turbines, and fish lifts/ladders are typically not successful.

Barrages are similar to very large dams, built across river estuaries that flow into a sea, and are comprised of enclosed embankments constructed of massive concrete blocks that contain sluice gates, turbines and ship locks.

- Sluices and freewheeling turbines open to allow basins to fill during high tide, and close when the tide goes out so that water pressure drives the turbines until the cycle repeats:
  - Efficiency of tidal generation depends upon the amplitudes of tidal swell (to create heads) which can exceed 30 feet (10m) in some locations.
  - As with wind power, the selection of practical locations is limited to places where high-amplitude tides exist, including river and fjord estuaries in the former USSR, Canada, Australia, Korea, and the UK.

125. Andy Darvill

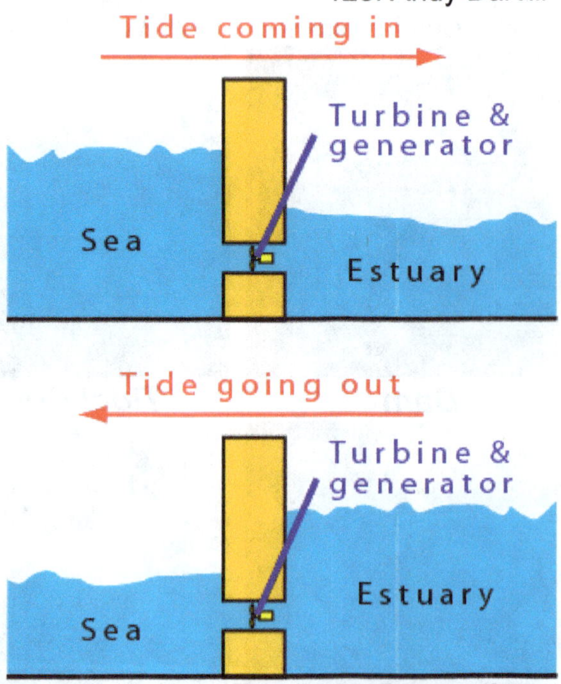

Barrage Operations

126. ePrairie, Inc.
127. Genersys ireland
128. Naional Renewable Energy Laboratory
129. US DOE-EERE

Off-Grid Electrical Power

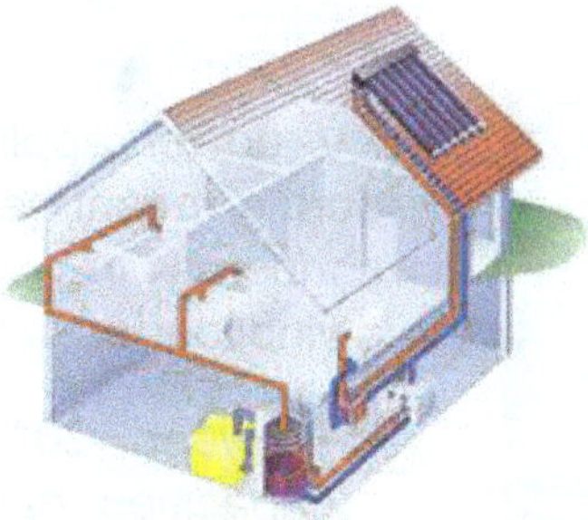

Water and Space Heating

Solar-Thermal Power Plant

Solar Photovoltaic Power Plant

Solar energy has come into widespread use, most particularly for low-power applications where other sources are not available:

- Solar radiation is converted into electricity or applied directly as a heat source for a variety of purposes which include:
  - Electricity generation using photovoltaic systems for roadside emergency telephones, off-grid home power, irrigation water pumps and other services.
  - Space heating, hot water and cooking using direct solar thermal radiation.
  - Solar thermal concentrators that produce steam or heat gases to drive engine-turbines that produce electricity.
  - Solar photovoltaic concentrators that convert radiant energy directly into electricity.

Solar electric or photovoltaic (PV) systems, directly convert sunlight into DC electricity.

- These systems are very reliable (so long as sunlight exposure is available) and have been used extensively on Earth and in space.
  - The devices are lightweight, modular and foldable to facilitate transportation, and are simple to install in remote locations.
  - They require no bulky and expensive fuels, offering clean and inexhaustible sources of power.
  - Although used for supplementary power nearly everywhere, PV systems are most applicable to locations which afford frequent access to sunshine.
  - Some utility jurisdictions enable PV customers to receive credits for transferring excess electricity into the grids their systems are attached to.

130. Pacific Environment

Converting Sunlight to Electricity

131. Michigan Solar Solutions

PV power applications have steadily expanded during recent years in response to environmental concerns in combination with rising costs of dwindling fossil fuel supplies.

- Between 2010-2024, worldwide solar energy capacities have increased at a rate of about 20% annually, although shortages of refined silicon for PV cells has slowed growth somewhat since that period:
  - PV popularity has benefited from reductions in manufacturing costs, where average lowest retail prices for large arrays have dramatically declined from $4.00 - $0.20 per watt between 2005 and 2024.
  - PV markets have also been boosted by many jurisdictions that provide customer tax and rebate incentives, often enabling system installations to recover costs in 5-10 years.

Small-Scale Commercial Applications

PVs are commonly used for low-power off-grid applications using inverters to convert DC electricity to AC (resulting in a 5%-15% energy penalty).

- The effective marriage of solar power and inverters has opened up great opportunities for rural electrification in areas not served by public utilities:
  - Typical installations draw electricity from a co-located PV array, inverter, battery bank and switch gear system located in an outbuilding or other central distribution point.
  - Power from a wind generator or other source is often tied-in with PVs to provide hybrid system capabilities.

132. Sacred Power Corporation

PV-Wind Generator Hybrid System

PV systems are becoming vitally important in threshold and developing nations where more than two billion people lack access to an electricity grid.

- Low population density regions with very limited energy consumption pose special problems due to large costs and efforts required to create public services:
  - Decentralized PV systems offer practical advantages for such areas as an alternative to kerosene lamps, disposable batteries or diesel generators.
  - Photovoltaics, in combination with wind turbines, motor generators or small hydropower plants, can offer sustainable, economical and socially acceptable solutions to energy needs.

133. German Energy Agency

Village System Installation

134. Fraunhofer Institute for Solar Energy Systems

Storage Batteries in Indonesian Village

A major requirement for off-grid PV applications is to provide means to supply or store power for use at night and on cloudy days.

- Battery backup and storage systems are the common method, but present certain limitations and disadvantages:
  - Batteries lower the efficiency of a PV system because only about 80% of the energy that goes into them can be reclaimed.
  - Batteries usually need to be replaced every 5-10 years.
  - They also take up a considerable amount of floor space, have poor efficiency in cold weather, require periodic maintenance, and present potential safety problems.

Geographic and power limitations pose significant constraints upon widespread PV utilization.

- Of the total 1,366 watts per square meter (W/m²) of solar radiation that encounters the Earth's upper atmosphere, the average incoming amount (or "insulation") that reaches the surface in the contiguous US ranges between 125-375 W/m² (or 3.9 kWh/m²/day):
  - While traveling through the atmosphere, about 6% is reflected back to space and about 16% is absorbed, resulting in a peak irradiance at the equator of slightly more than 1,020 W/m².
  - Average atmospheric conditions (clouds, dust and pollution) further reduce insulation by about 20% through reflection, and about 16% through absorption.
  - Atmospheric conditions also affect the quality of insulation reaching the surface by diffusing incoming light and altering the spectrum.

135. William M. Connolley, British Antarctic Survey

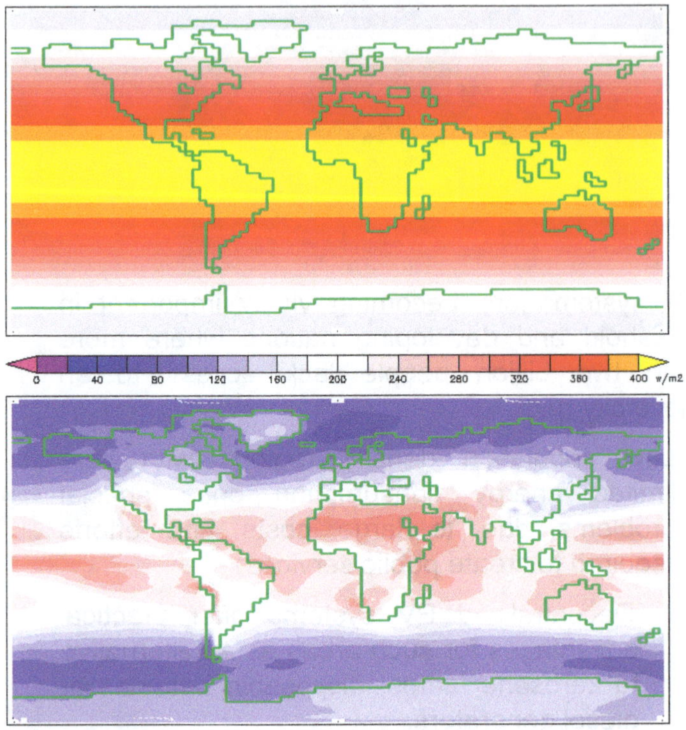

Theoretical annual mean insolation, at the top of Earth's atmosphere (top) and at the surface on a horizontal plane.

Solar Radiation: Atmosphere and Surface

Advantages

- Sunlight energy is free following initial PV installation costs, and is non-polluting.
- Installations can operate with little maintenance after initial setup is accomplished.
- Systems are particularly beneficial in remote locations where public utility connections are not available.
- Grid-connected systems can displace highest-cost electricity during times of peak demand.
- Grid-connected systems can sometimes transfer excess electricity to the grid for energy credits.
- PVs can often be combined with wind power and other energy sources to optimize power production.

Disadvantages

- Installation and replacement costs may require several years to recover.
- Solar panels have limited power density, ranging from only about 17%-20% efficiency.
- High-latitude regions and locations with frequent cloud cover and dust limit effectiveness.
- Solar power is not available at night or during rainy/cloudy periods when electricity may be needed.
- Batter power storage, if needed, imposes large energy penalties, costs and space requirements.
- Inverters to convert DC to AC electricity also impose significant energy efficiency penalties.

PV Installation Advantages and Disadvantages

Concentrating solar power (CSP) systems use lenses or mirrors that track the Sun to focus a small beam of radiation onto a thermal collector or PV cells to produce electricity.

- Within the US more than 1.5 GW of CSP capacity currently exists, and has been operating reliably for more than 15 years.
    - CSP systems range in size from as small as 5 kW (a typical US home can meet most energy needs with about 5-15 kW), to more than 100 MW for utility grid-connected applications.
    - Since the systems make use only of direct sunlight (not diffuse light), use is limited to sunbelt locations.
    - Many provide thermal storage for power during nighttime and cloudy periods.

136. James Fraser

Electrical Power Production

137. US Dept. of Labor - OSHA

Concentrating Solar Power Tower

Solar thermal power CSP systems heat fluids and gases to high temperatures that drive steam or gas engine-turbines.

- These systems use various types of mirrored sunlight collectors and concentrators that focus high-temperature radiation onto heating element targets:
    - Power towers (or "heliostat" power plants) use flat, moveable mirrors to focus energy upon a collector tower that reflects it back to an absorber to store heat for later use.
    - Dish and trough collectors track the Sun to concentrate and focus energy that drives steam or sealed-gas engine-turbines to produce electricity.

138. Henrik Storch

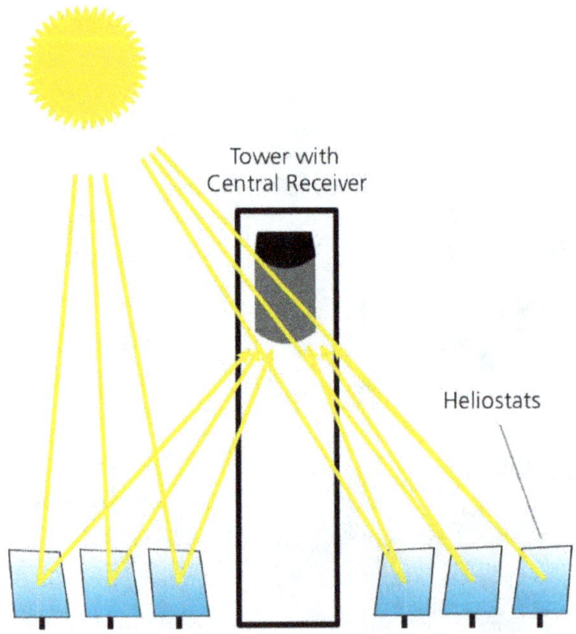

Concentrating Solar Power Tower

Power tower CSP systems are comprised of numerous large sun-tracking heliostat mirrors that focus sunlight onto a hyperbolic receiver at the top of the tower which reflects it back to a second aiming point.

- The concentrated energy heats a transfer fluid or molten nitrate salt in the receiver to generate steam power or hot gas to drive sealed closed-cycle turbine-generators:
  - Power towers can typically produce between 600 KW - 150 MW of electricity, with an average 250 KW - 5 MW in continuous operation.
  - Thermal-electric efficiency may range from 25%-50%, and overall solar-electric conversion turbine-engine efficiency may range from 13%-25%.

Solar dish-engine CSP systems are stand-alone electric generators that use a solar concentrator as the primary component.

- The reflective concentrator surface is made of glass mirrors that reflect about 95% of the sunlight that strikes them:
  - The receiver transfers collected heat either to a bank of tubes containing a gas (usually hydrogen or helium) that serves as a transport medium, or to heat pipes that use an intermediate boiling fluid that condenses to transfer heat to an engine-turbine.
  - Although dish-engine systems are not yet commercially available, ongoing demonstrations indicate good potential to generate up to 25 kW of power.
  - Individual systems can be connected together to increase capacity, and can be combined with other systems for continuous hybrid power-generation.

139 US DOE-EERE

Concentrating Solar Dish-Engine

Trough-type CSP systems use a collector field comprised of many parabolicly-curved trough-shaped reflectors that transmit thermal energy onto inside receiver pipes.

- Parallel rows of the troughs are aligned on a north-south axis to track the Sun from east to west during the day, requiring no mechanical pointing mechanisms:
  - Current systems can generate 80 MW or more of electricity.
  - Thermal storage can enable electricity generation to continue several hours into the evening in combination with other energy sources to supplement low solar radiation periods.
  - Trough systems are the simplest and least expensive of all CSP devices, and may eventually be competitive with fossil fuel energy.

140. Energy

Concentrating Parabolic Troughs

Concentrating solar photovoltaic power (CSPP) is one of the most efficient and cost-effective forms of solar power today.

- CSPP reduces the amount of expensive PV material required over conventional systems, while maximizing the amount of sunlight collected:
  - Dish mirrors concentrate and magnify light (potentially many hundreds of times) with capacities up to many MW of electricity, and possible efficiency rates exceeding 35%.
  - Like other concentrating systems, CSPP is limited to use in the sunniest areas.
  - Tracking is required to maintain efficiency, sometimes requiring sophisticated devices that restrict practical use to electric utility, industrial and large building applications.

141. US DOE-EERE

Concentrating Solar Photovoltaic Tower

142. Geothermal Energy Association

Harnessing the Earth's Heat

Natural geothermal reservoirs of low-to-moderate temperature water (68ºF – 302ºF [20°C - 150°C]) provides sources of heat for direct use or conversion to electrical power in many regions of the world.

- Geothermal electricity is produced by steam or binary-cycle water generators supplied by near-surface wells heated by Earth's magma and hot rock:
  - Because they are most abundant, moderate-temperature sites running binary-cycle power plants are most commonly considered as electricity producers.
  - Technology advancements will be necessary to make these sources of electricity more cost-competitive with traditional energy options before geothermal power becomes a key element of the US energy infrastructure.

Some plants use high-temperature subterranean water to produce electricity.

- Flash steam power plants convert hydrothermal fluid above 360ºF (180ºC) to steam:
  - The fluid is sprayed into a tank held at a much lower pressure than the fluid, causing some of the fluid to vaporize ("flash").
  - The vapor drives a turbine, which, in turn, drives an electrical generator.
  - Any remaining liquid in the tank is flashed again to extract more energy for added efficiency.

143. Renova

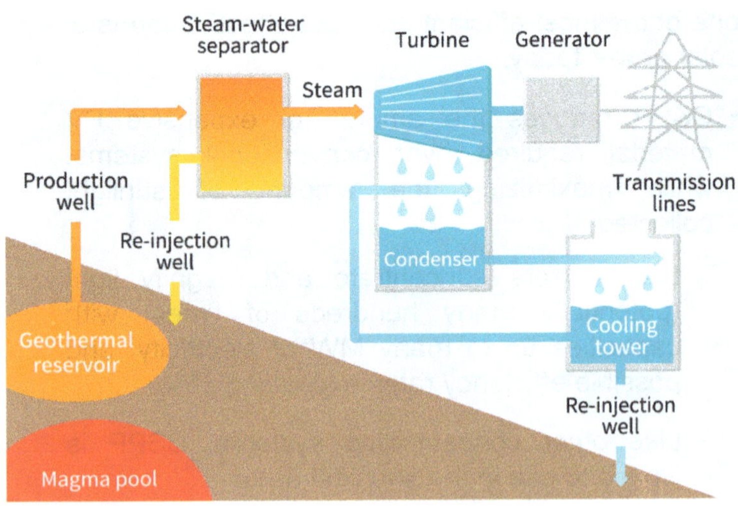

Flash Steam Power Plant

Dry steam power plants use hydrothermal fluids that are primarily steam.

- The steam goes directly into a turbine which drives a generator to produce electricity:
  - This is the oldest type of geothermal power plant, first used at Lardarello, Italy in 1904.
  - The technology remains to be effective, and is used today at "The Geysers" in northern California (the world's largest single source of geothermal power).
  - These plants emit only excess steam and very minor amounts of gases.

144. Geothermal Energy Association

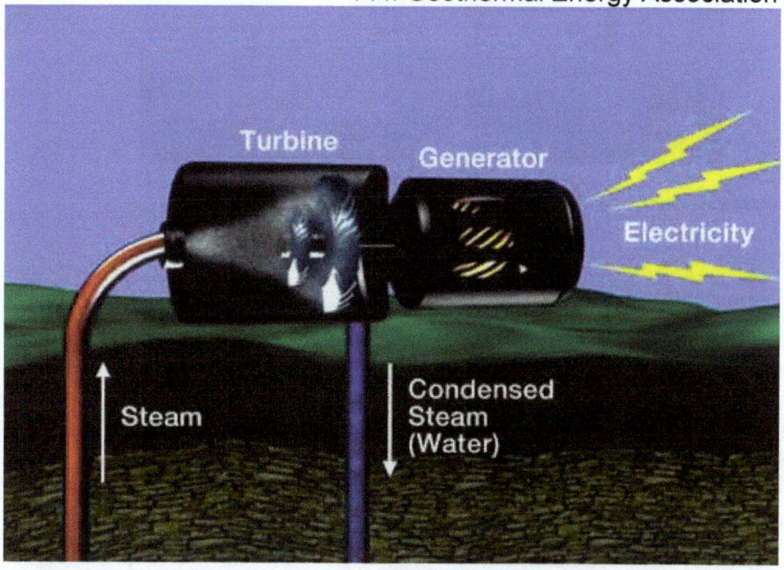

Dry Steam Power Plant

145. Geothermal Energy Association

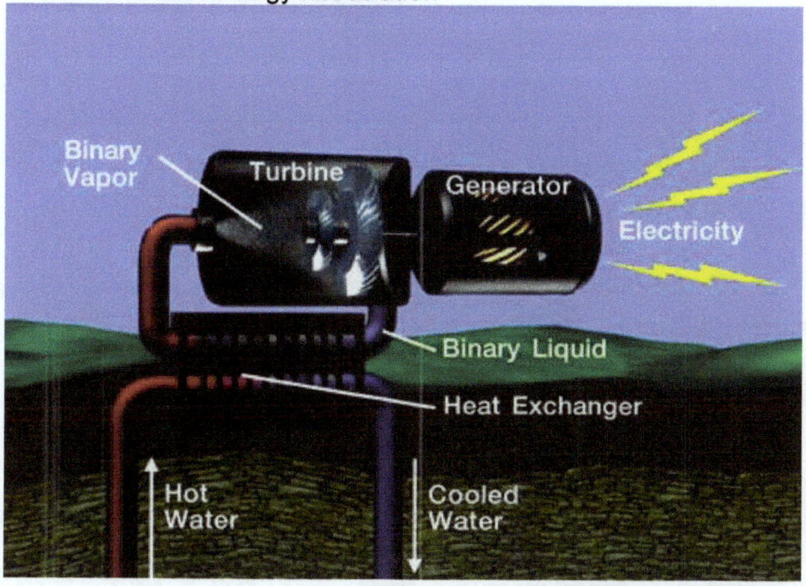

Binary-Cycle Power Plant

Binary-cycle power plants use moderate-temperature water (below 400ºF [200°C]).

- Energy is extracted from the fluid by a second ("binary") fluid with a much lower boiling point than water in a heat exchanger:
  - Heat from the geothermal fluid causes the secondary fluid to flash to vapor that drives a turbine.
  - Since this is a closed-loop system, nothing is emitted into the atmosphere.
  - Moderate-temperature water is by far the most common geothermal resource, and most future geothermal power plants will be binary-cycle types.

146. Picture Newsletter
147. NASA

Nuclear Power Plant

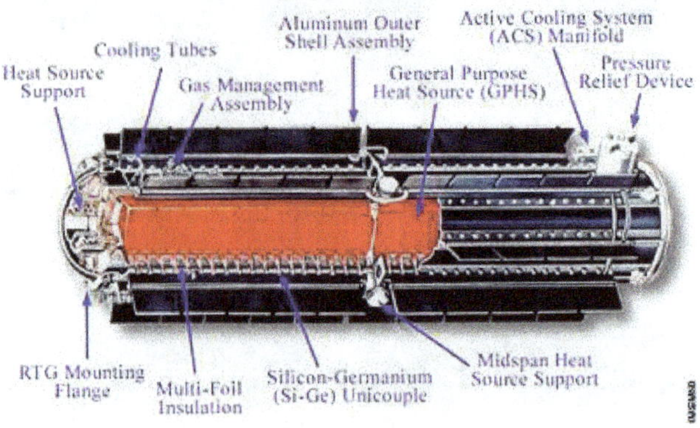

Radioisotopic Thermal Generator

The radioactive decay of certain isotopes (or radio nuclides) releases heat energy through changes in the atomic nuclei that can be used to generate electricity or applied directly.

- Current applications of isotopic energy are used to produce electrical power at large and small levels of scale:
    - Nuclear fission reactors split nuclei of heavy atoms such as uranium to create large amounts of energy to boil water into steam that drives turbine generators that produce commercial quantities of electricity.
    - Radioisotope thermal generators (RTGs) use spontaneous decay of plutonium-238, a heavy radioactive metallic element, to produce much lower heat energy levels used primarily for small-scale space applications.

Nuclear power uses heat from the radioactive decay of a fissile material such as uranium-235 (235U) to boil water into steam to generate electricity.

- The US produces the most nuclear energy, with nuclear power providing about 19% of its electricity, while France produces the highest percentage (68%) from nuclear reactors:
    - International construction of nuclear power plants declined following the 1979 Three Mile Island accident, and again after the 1986 Chernobyl disaster.
    - Renewed international government and public interest in nuclear energy is being motivated by increasing oil prices, improved plant safety designs, and fossil fuel greenhouse gas concerns along with stringent emission standards established by the Kyoto Protocol.

148. Flickr/Bill & Vicki Tracey

Non-radioactive water vapor rises from hyperboloid-shaped cooling towers. Nuclear reactors are located inside cylindrical containment buildings.

Nuclear Power Station

Large and Small-Scale Devices

149. Nuclear Energy Institute

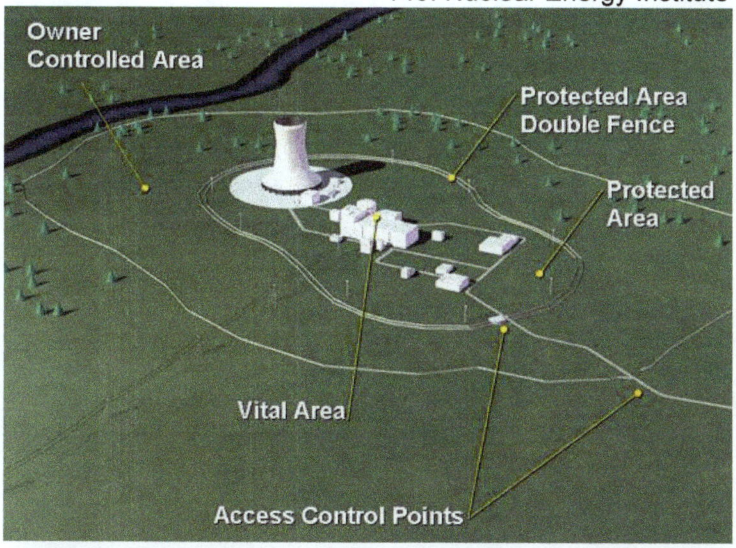

Security at nuclear power plants is regulated and monitored by the US Nuclear Regulatory Commission (NRC). Physical barriers and high-tech surveillance devices control access, and reactors are protected by massive reinforced concrete containment buildings.

Nuclear Plant Security

Until recently, no US nuclear power plant has been ordered without subsequent cancellation for more than 20 years.

- Nuclear power continues to be a controversial subject for several reasons:
  - Opponents cite problems and challenges of safely storing radioactive wastes for indefinite periods.
  - Public fears about possibilities of radioactive contamination by accidents or sabotage are reinforced by global terrorism trends.
  - Proliferation of nuclear weapons by rogue nations under the guise of peaceful energy programs poses a growing threat.
  - Some critics argue that nuclear power is uneconomical in comparison with fossil fuels (although escalating oil prices are changing this condition.)

While nuclear power plants are significantly more costly to build than coal or gas-fueled plants, coal is substantially more expensive than nuclear fuel, and natural gas is substantially more expensive than coal.

- Coal and nuclear plant operating and maintenance costs are comparable, but lower nuclear fuel costs have not yet offset higher investment costs:
  - In many countries, licensing, inspection and certification of nuclear plants add large additional construction costs and delays, resulting in higher interest charges on borrowed construction funds.
  - Streamlining and standardizing of regulatory processes for sighting, licensing and construction without compromising safety can offer strong development incentives.

150. US DOE-EIA

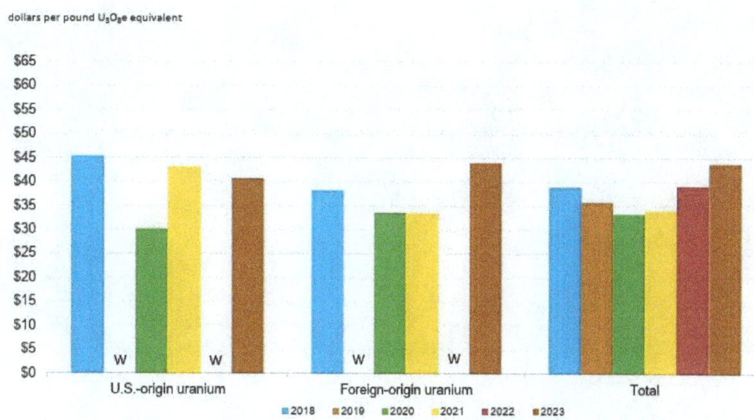

Costs of Uranuim Purchased by Owners and Operators of US Civilian Nuclear Reactors by Origin and Delivery Year, 2018-2023.

As of December 2024, there are 419 operable power reactors in the world, with a combined electrical capacity of 378.1 gigawatts

- Approximately 70% of those are more than 30 years old:
  - In the US there are 94 licensed commercial generating units, producing about 94,000 megawatts of electricity.
  - Supported by incentives provided by the US Energy Policy of 2005, two new sites have been selected, and utility companies are expected to finance more.
  - India is constructing at least 11 new nuclear power reactors with a combined capacity of 8,400 MW.
  - The world's largest nuclear power plant is the Kashiwazaki-Kariwa Nuclear Power Plant in Japan, with a net capacity of 7,965 MW.

151. IAEA.

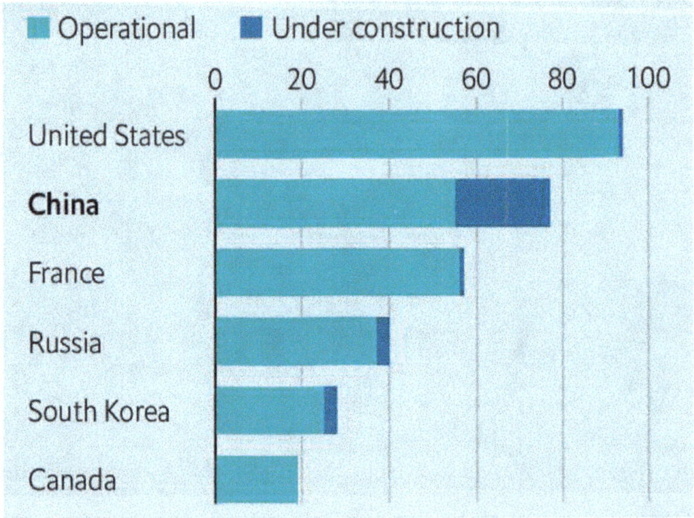

Operating and Under Construction Plants till 2023

152. IAEA

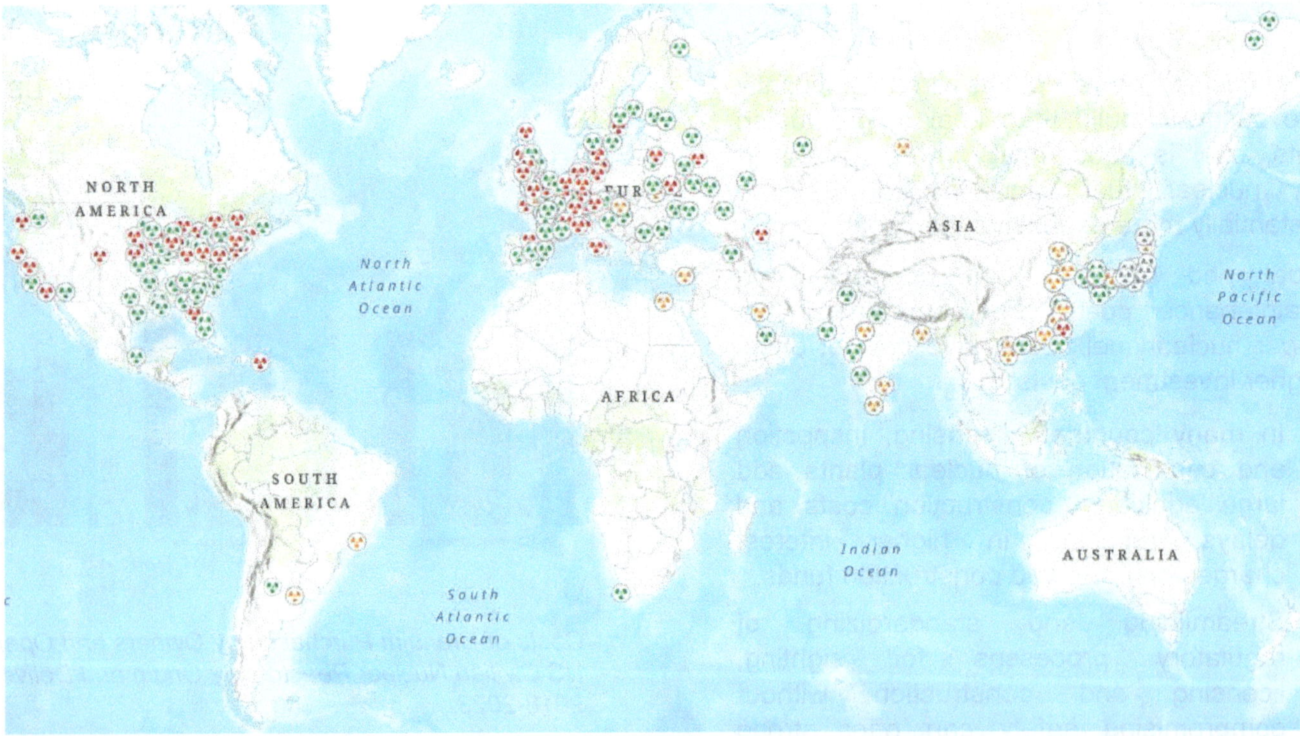

Countries with Nuclear Reactors- 2024

Nuclear power, coal and wind are broadly considered to be the only realistic large-scale energy sources that can be expanded sufficiently to significantly reduce oil and natural gas dependence after production has peaked.

- Current nuclear reactors return about 40-60 times the investor energy needed to produce it, better than coal or natural gas, but less than hydropower:
  - Nuclear power is the only mature technology used in the production of electricity that does not create greenhouse gases that contribute to global warming.
  - In addition, and contrary to popular belief, coal-burning actually releases more radioactive waste into the environment than nuclear power.

A possible obstacle for expanding nuclear power use may be a limited supply of uranium ore.

- Proponents of nuclear power argue that the amount of available recoverable uranium that can be reprocessed from standard reactors using breeder reactors will last hundreds of years:
  - Breeder reactors can reprocess as much as 95% of the original material, but have been banned in the US since the Carter Administration due to a risk of weapons-grade weapon proliferation.
  - A US initiative, the "Global Nuclear Partnership" announced in 2006, will spearhead an international effort to reprocess fuel in a manner that will prevent proliferation, and make nuclear power available to developing countries.

153. IEA.

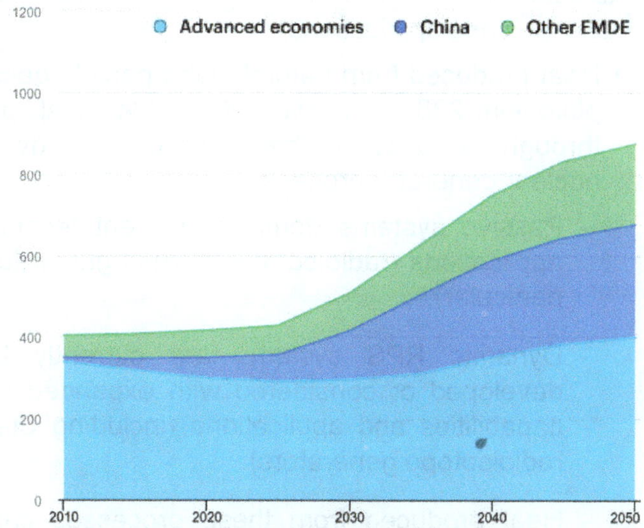

Existing and Projected Nuclear Power Output Gigawatts of Electricity

154. Tungsten

Nuclear fuel can be used many times through reprocessing:

1. Uranium is mined, enriched and delivered to the plant; 2. spent fuel is reprocessed (or 3. stored in a final reposition for geological disposition) and then 4. reprocessed fuel is recycled and reused.

Fuel Cycle

155. NASA / STAIF

Radioisotope power systems (RPSs) are extensively used in space applications.

- Heat produced from natural alpha particle decay of plutonium-238 is converted to electrical power through a passive thermoelectric or dynamic nuclear generator process:
  - Passive systems dominate current technology applications (radioisotope thermal generators in particular).
  - Dynamic RPS systems are currently being developed or considered with expanded future capabilities and applications (including Sterling radioisotope generators).
  - Heat produced from these processes can be useful for thermal control of subsystems in extremely cold space environments.

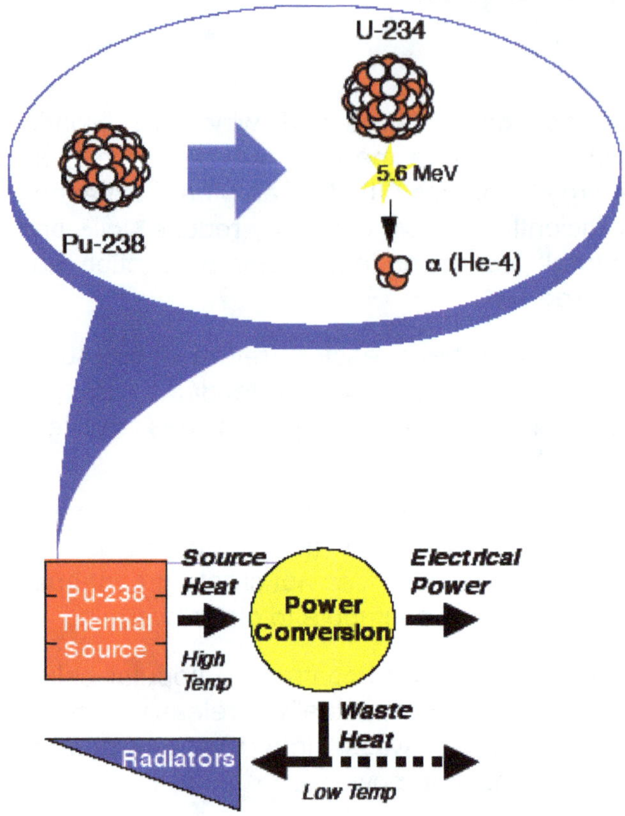

Products of Radioactive Decay

156. Nuclear Power

Radioisotope thermal generators (RTGs) afford a number of important advantages over photovoltaics, fuel cells and other available technologies for deep space and planetary surface applications.

- Unique features include:
  - Steady power delivery independent of distance and orientation with respect to the sun.
  - Operation in shadowed and heavily clouded regions and locations (craters and thick atmospheres).
  - Power generation in extreme environments (Venus and Titan).
  - Operation in high-radiation environments (Jovian space).
  - Long-duration operations (tens of years or more).

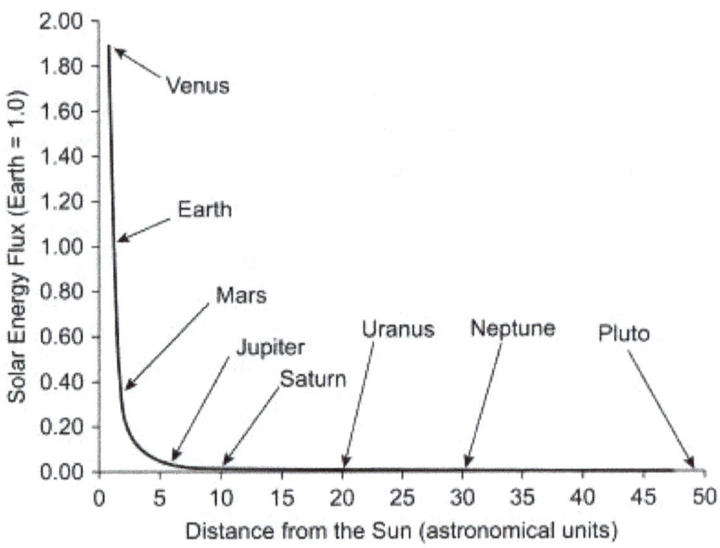

Solar Energy Flux as a Function of Distance from the Sun (AU)

Radioisotope Thermal Generators

A common application of RTGs is to provide power for unmanned spacecraft.

- Since 1961, "System Nuclear Auxiliary Power Program" (SNAP) units have been used for probes traveling far from the Sun where solar panels are ineffective:
  - Viking landers and scientific experiments were left on the Moon by the crews of Apollo 12 – 17.
  - Voyager 1 and 2 traveled beyond Uranus and Neptune, 70 times farther from the Sun than Earth, where it is very cold and there is little solar energy.
  - Pioneer 10 and 11 traveled past Jupiter, Saturn and ultimately, beyond Pluto.
  - The NASA/ESA Ulysses spacecraft flew by Jupiter in 1992, and went into orbit around the Sun.
  - Cassini explored Saturn's rings and moons.
  - New Horizons launched in 2006 will visit Pluto and its moon Charon.

157. NASA

Cassini mission equipment included three RTGs providing a total of 850 watts of power.

Cassini RTG Inspection Before Launch

158. Astronomy

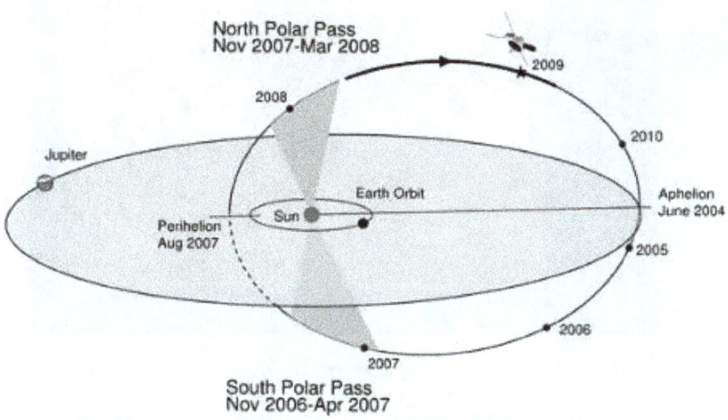

Ulysses Jupiter Spacecraft Trajectory

159. NASA

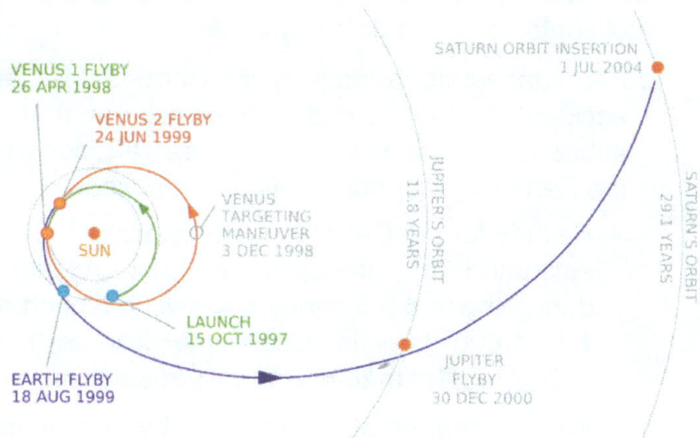

Cassini Saturn Spacecraft Trajectory

Since the initial SNAP-3 mission in 1963, RTGs continue to be an indispensable part of the US and international space programs.

- RTGs have been used in more than 25 missions that have orbited Earth, traveled to other planets and their moons, and explored the surfaces of Earth's moon using roving vehicles:
  - Astronauts on five Apollo missions left numerous RTGs behind to power a variety of "Apollo Lunar Surface Experiment Packages".
  - One SNAP-2 RTG produced 75 watts of electricity to power an experiment package for nearly 8 years.

160. NASA

Apollo 12: Alan Bean Loading a Plutonium Heat Source into an RTG

RTGs using plutonium-238 produce relatively low-penetration alpha radiation, requiring only 18mm equivalent of lead shielding, yet pose a hazard if a fuel container leaks or is ruptured.

- For spacecraft, a primary concern is that an accident might occur during launch, or that a subsequent passage of the spacecraft close to the Earth might contaminate the atmosphere:
  - A study for the Cassini-Huygens probe (1997) estimated the probability of such releases during the first 3.5 minutes following launch at 1 in 1,400; later in ascent into orbit as 1 in 476; and after that, less than 1 in a million.
  - While alpha radiation produced by plutonium-238 will not penetrate the skin, it can be harmful to internal organs if inhaled or ingested (particularly, the liver).

161. US Department of Energy

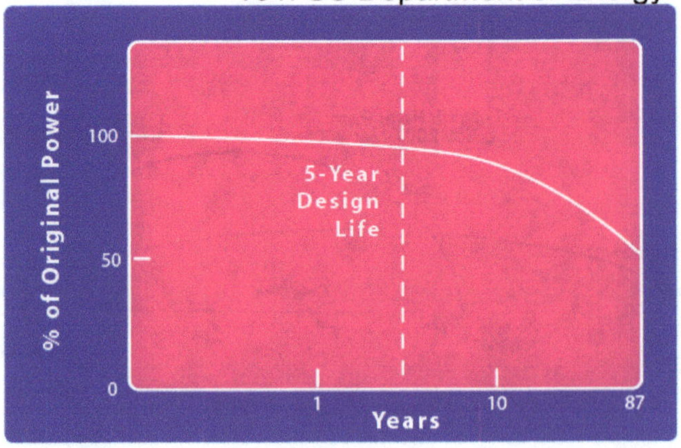

The 87.74 year half-life of plutonium-238 is very short in contrast with the 24,110 half-life of plutonium-239 used in nuclear weapons and reactors. As a consequence of the shorter half-life, Pu-238 is about 275 times more toxic by weight.

87 Year Half-Life of Pu-238

Radiation events are rare, but they do occur.

- Six known accidents have involved RTG-powered spacecraft:
  - April 12, 1964: a failed US Transit-5BN-3 navigation satellite launch burned up upon re-entry near Madagascar with traces of Pu-238 detected.
  - May 21, 1968: a Nimbus B-1 weather satellite in an erratic trajectory was deliberately destroyed (its fuel container was recovered intact).
  - 1969: two Soviet Cosmos missions containing RTG-powered rovers failed, releasing radioactivity. (Five other Soviet/Russian spacecraft failures between 1973 – 1993 carried nuclear reactors.)
  - April 1970: an Apollo 13 failure plunged a SNAP-27 RTG into the Pacific Ocean with its container intact.
  - November 1996: a Russian Mars 96 launch failed (2 RTGs are assumed to have survived re-entry).

Nuclear fuel modules within RTGs must be designed to withstand a variety of accidents:

- **Fire**: Direct exposure to solid propellant fires such as might be encountered during a launch accident must not produce damage or nuclear fuel release.
- **Blast**: Fuel modules must be able to survive high temperatures associated with collisions with debris and atmospheric re-entry.
- **Earth Impact**: Modules must survive impacts at 120 mph (200 km/hr) (approximate top speed for an aeroshell falling to Earth) on sand, water, soil, or rock surfaces without release of fuel.
- **Water Immersion**: Fuel capsules must be able to resist long-term corrosive effects of seawater in the event that ocean recovery is not possible.

RTGs have also been used for a variety of terrestrial applications.

- Uses range dramatically in size and purpose:
  - One of the first terrestrial RTG uses was by the US Navy (1966 –1995) at the uninhabited Fairway Rock Island in Alaska.
  - Some very small plutonium-238 RTGs (plutonium "cells") have been used in heart pacemakers to ensure long "battery life".
  - The former Soviet Union constructed many unmanned lighthouses and navigation beacons power by Beta-M RTGs.
  - It is estimated that as many as 1,000 discarded RTGs that have exhausted their engineered life spans remain in Russia, imposing radioactive contamination risks.

162. Finland Regional Government

Many former Soviet RTGs developed for various purposes are no longer functional and exist in hazardous dilapidated and vandalized conditions.

Discarded Soviet RTGs

Regrettably, no renewable fuel or new technology option offers a panacea to fulfill expanding energy demands or to solve environmental problems associated with prevailing sources.

- All "alternatives" combine distinct disadvantages and limitation with benefits they afford:
  - Biofuels such as ethanol and biodiesel release CO_2 and other greenhouse gas emission just as fossil fuels do, they compete with food crops for agricultural land, and require energy to produce.
  - Hydrogen requires much more energy to produce than it yields through fuel cells, and is difficult to store and transport.
  - Geothermal, hydro, solar and wind power are vitally useful, but site limitations constrain nation-wide energy contributions.

183. Elena's Puerto Rico Travel Guide, USA

Nature uses a huge diversity of different mechanisms and organisms to produce energy for life using solar energy, biomass, recycled heat and other sources. It is clear that we must do the same.

A Need for Diversity

Independence from fossil fuels is not likely to occur any time soon.

- Replacing the vast majority of current US energy that comes from fossil fuel burning using agricultural plant biomass fuels, for example, will be difficult or impossible:
  - Croplands that provide essential food for livestock and humans would need to be multiplied many times and redirected to corn, sugarcane, soybeans and other biofuel feedstocks.
  - Most of that land and its yields would be needed to produce fuel required for the tractors and processing to create the fuel products.

164. FCE

Agricultural biofuels require substantial land and energy to grow and process the plants

Fuel Balances

165. *Larry Bell*

Oil and Natural Gas	**Coal**	**Biofuels**
Advantages: • High energy densities • Many derivative products • Easy to store and transport **Disadvantages:** • Dwindling global supplies • CO_2 and other emissions	**Advantages:** • Good energy densities • Many derivative products • Presently relatively abundant • Cleaner technologies available **Disadvantages:** • Mining impacts upon land areas	**Advantages:** • Resourceful use of biowastes • Can reduce fossil fuel use • Apply current technologies **Disadvantages:** • Compete for land with food crops • Produce CO_2 when burned
Nuclear Power **Advantages:** • Potentially a substantial source • Environmentally-clean energy • Relatively abundant / inexpensive **Disadvantages:** • Possible safety/security risks • Concerns about safe storage	**Hydrogen** **Advantages:** • Derived from multiple sources • Technologies being improved **Disadvantages:** • Low energy density / volume • High costs of production • Difficult to store/transport	**Solar Power** **Advantages:** • Excellent energy supplement • Free energy following installation • Concentrator techs promising **Disadvantages:** • Geographic/weather limitations • High implementation costs
Geothermal Power **Advantages:** • A free energy source • Environmentally-friendly energy **Disadvantages:** • Limited geographic sites • Few high temperature sources • Not currently cost-competitive	**Hydropower** **Advantages:** • Free energy after installation • New technologies developing **Disadvantages:** • Limited geographic sites • Environmental marine impacts • Large systems are costly	**Wind Power** **Advantages:** • Free following installation • Environmentally-friendly energy • Rapidly decreasing costs **Disadvantages:** • Weather-dependent operations • Substantial implementation costs

Key Advantages and Disadvantages of Source Options

Ethanol, mostly produced from corn in the US (and sugarcane in Brazil), is a key biofuel contender to extend fossil supplies.

- Although assessments of net corn ethanol life cycle yields vary considerably, prominent organizations and researchers project a net positive:
  - A study presented in the US Department of Agriculture *Agricultural Economic Report No. 813, 2024*, estimates that for every BTU needed to produce the fuel, there is a 40% energy gain.
  - The USDA report assumes that only about 17% of the production energy comes from fossil fuels (gasoline and diesel), so that for every 1 BTU of fossil fuel consumed, there is a 6.34 non-fossil BTU gain.

166. Bob Swihart

It is much more economical to reduce fossil use than to replace excess consumption with bio fuels.

Corn Ethanol

Considering the Options

167. Tyler Hamilton Foundation

Cellulose ethanol production is accomplished using a variety of biorefinery hydrolysis and fermentation processes to break down fibres into glucose fuel. Extensive use of the plant material increases energy yield over corn ethanol but also adds process costs.

Cellulosic Ethanol

Ethanol produced from plant cellulose is the same product as corn ethanol but is created using different feedstocks and processes.

- Instead of using only starch from the plant kernel, the fuel yield is increased by using other parts more fully:
  - An important advantage is the ability to utilize a large variety of biowastes, including corn stover, cereal straws, sugarcane bagasse, saw dust and paper pulp, as well as fast-growth grasses.
  - The left-over lignin dry mass can be burned to create steam, or gasified to produce electricity.
  - Although process enzymes add substantial production costs, technology advancements and a variety of process byproducts are expected to improve economic factors.

Biodiesel is often produced using oil from soybeans blended with petroleum diesel (typically a 15/85 mix) and is rapidly gaining popularity.

- This fuel can be used in standard diesel engines with comparable energy efficiencies to petrodiesel (about 80% vs. 83%):
  - Although efficiency yields are slightly lower due to energy required to convert soybean oil to fuel, biodiesel can dramatically increase the energy produced per unit of fossil fuel consumed (about 3.2 units vs. 0.83 for petrodiesel).
  - Biodiesel reduces fossil fuel life cycle consumption roughly proportionally to its concentration; an approximate 19%.reduction for a 20% biodiesel / petrodiesel blend (B20), and about 95% for 100% biodiesel (B100).

168. US DOE-EERE

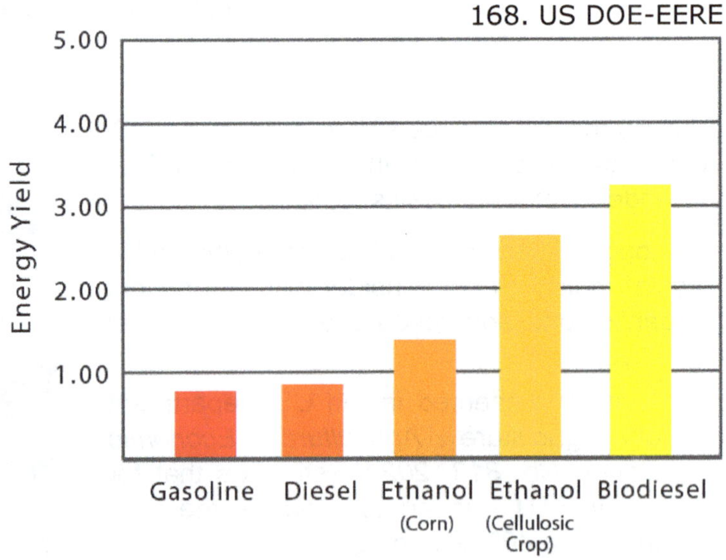

Plant oil blended with petroleum diesel can reduce fossil fuel consumption along with fossil (old-source) CO_2 emissions per unit of engine work.

Energy Yield / Unit of Fossil Fuel Consumed on a Life Cycle Basis

Biodiesel

Nuclear power which currently provides about 19% of US (2024) electricity represents a large potential fossil fuel replacement source.

- Expansion of this option can be expected to gain widespread interest provided that important obstacles can be overcome:
  - Public confidence regarding safety from nuclear accidents such as the Chernobyl and Three Mile Island events must be restored.
  - Environmentally-safe long-term storage of hazardous wastes must be assured.
  - Safeguards from growing terror threats and theft of nuclear materials will be needed.
  - Enactment of security measures that enable breeder reactors to be reinstated to reprocess spent materials will greatly extend supplies if global weapon proliferation risks can be prevented.

169. US DOE-Office of Civilian Radioactive Waste Management

Nuclear power must be considered as an important option to reduce fossil fuel dependence, but public safety and security concerns must be addressed.

Nuclear Power

170. Geocities

Hydrogen is regarded to be an energy carrier rather than a source because of net deficit power it produces relative to power needed to create and convert it to electricity.

Hydrogen

Hydrogen requires more energy to create than it yields, and imposes additional cost and efficiency penalties for compression, liquefaction, transport, bulk storage, transfer to vehicles and fuel cell inefficiencies.

- When hydrogen is produced using fossil fuels, there are also pollution penalties:
  - Electricity produced by burning natural gas in a large unit produces CO_2 emissions just over 0.96 pounds/kWh (435g/kWh).
  - When hydrogen is produced by reforming natural gas, CO_2 emissions resulting from creating it are about 0.74 pounds/kWh (336g/kWh).
  - If the product hydrogen is then passed through a fuel cell to create electricity, CO_2 emissions are about doubled to 1.94 pounds/kWh (880g/kWh) since fuel cell efficiency is at best about 50%.

171. National Renewable Energy Laboratory

Utility-scale solar power systems are large, expensive, and are limited to regions with ideal Sun and weather conditions that tend to be remote from population centers.

Solar Power

While solar energy must be regarded as a very important resource to reduce fossil fuel energy consumption through home and business space and water heating economies, it is not likely to offer a major contribution in supplying national electrical power needs.

- The most effective and promising technologies, collecting solar power systems (CPSs) are limited due to a scarcity of suitable locations with adequate and dependable direct sunlight:
  - Optimal commercial PV systems require large, expensive arrays that require several years for investment cost recovery.
  - Many of the best sites are in remote areas distant from population centers where power is needed, requiring construction of distribution systems that add costs and cause transmission line losses.

Geothermal sources currently produce 0.4% of total electricity in the US .

- Future growth is severely limited by an extreme scarcity of sites with suitably high temperatures (225º - 600ºF [107° - 315°C]):
  - Of the approximately 3,700 MW of electricity produced in the US, about 1,600 MW is generated at The Geysers, a plant located about 90 miles north of San Francisco.
  - Other smaller plants are located or under development in California, Nevada, southeastern Oregon, southwestern Idaho, Arizona, and western Utah.
  - Geothermal electricity is presently not very competitive with fossil and nuclear-fueled power due to much lower temperatures and energy conversion efficiencies.

172. US DOE-EERE

Geothermal power is limited to certain western US locations, and is presently not cost-competitive with fossil or nuclear due to lower energy and conversion efficiencies.

Geothermal Power

Hydroelectric power constitutes nearly 27% of renewable energy in the US, about 5.86% of total electrical power.

- Expansion of capacities from conventional approaches that harness falling water is not a major option in the US and developed countries because most primary sites are already being exploited or are unavailable for environmental reasons:
  - While construction of surface reservoirs has slowed considerably since about the 1980s, hydropower is the dominant electricity source in Idaho and Washington.
  - Since water can be stored as potential energy, an important role is to provide hour-to-hour peak load power adjustments in combination with other electrical sources.

173. US Bureau of Reclamation

Although most US inland sites are already being used or are prohibited for environmental reasons, new technologies to harness wave and tidal power may open up future resource opportunities.

Hydropower

Wind provides about 10.3% of US renewable energy, this source is growing at a rapid rate averaging about 7.7% per year.

- Wind power is recently realizing cost reductions, and in some locations new wind farms are costing less than coal or nuclear-fueled plants with equivalent capacities:
  - Although current production costs per kWh appear to be leveling, rising fossil fuel costs are making wind more and more cost-competitive.
  - Growth is being supported by new state energy policies and the success of "green marketing" across the country.
  - Interest in small stand-alone systems is also growing, with significant advancements in technology and increased adoption

174. Berkeley Lab.

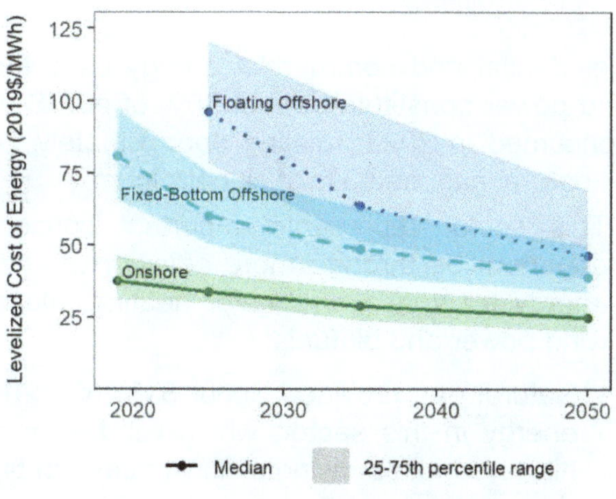

Energy Production Costs for Large Commercial Wind Projects

Wind power use is growing rapidly throughout many parts of the US due to lowering costs and environmental benefits. Small units can operate with wind speeds as low as 6 mph, and large ones at 13 mph.

Wind Energy

California proposed to obtain 33% of its electricity from renewable sources by 2020 and surpassed this goal, achieving 34.5% from RPS.

- It will be challenging to close an approximate 80% gap if this goal is applied nationally:
  - Biofuels can help to extend fossil fuel resources for heating and transportation, but are not promising electricity producers.
  - Nuclear power is not generally characterized as a renewable source, and new plant permitting and construction will require many years.
  - Hydrogen is a net electricity drain, not a producer.
  - Solar electricity is not regarded to be a major power option, and is weather-dependent.
  - Geothermal and hydropower sites are limited and constrained by environmental concerns.
  - Wind energy is growing, but constitutes only a small fraction of a percent of power capacity.

175. US DOE - EIA

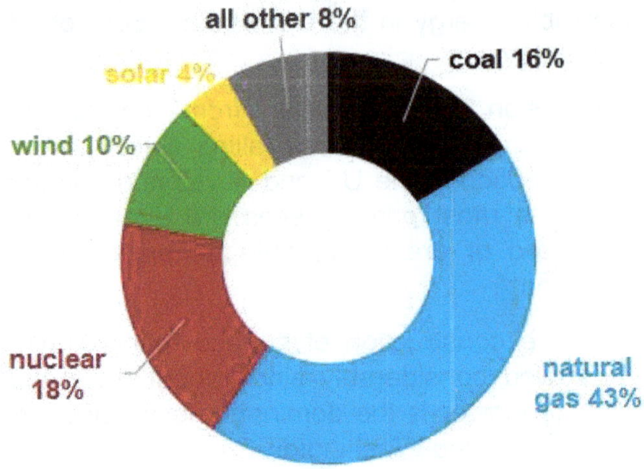

During 2023, 60% of all US electricity was produced from fossil fuels and 18% from nuclear plants, with 20% from renewable sources, including Hydroelectric power.

US Electrical Power Industry Net Generation, 2023

Residential and commercial energy uses for heat and power constituted about 20% of all US energy consumed in 2023, totaling approximately 24% of all natural gas, and 3% of all petroleum.

- This sector represents important conservation opportunities through more efficient use of solar energy for space and water heating, along with wind power and biofuels:
  - Natural gas provided about 51% of all heating energy in this sector, which will become hard hit as supplies dwindle and prices continue to rise.
  - Most of the remaining residential and commercial energy (about 10%) was supplied by petroleum, competing with supplies needed for transportation and petrochemical production.

176. Derek Jensen
177. Enoch Lau
178. US Coast Guard
179. Dennis Mojido

Residential and Commercial Sector

180. Texas A&M University
181. State of New York
182. US DOE-EERE
183. US DOE OCRWM

Industry and Electrical Sectors

Coal and natural gas currently supply about 45% of all energy used by industry, and about 59% of all electrical power.

- Transitioning from these sources to nuclear which provides 19% of all US electricity can stretch these resources and reduce greenhouse gas emissions considerably.
  - Industry presently consumes about 35% of all natural gas, and constitutes about 24% of the energy used in this sector.
  - Coal, which is much more abundant, provides about 16% of all electrical power, presently consuming approximately 90% of all US electricity used.

Public, commercial and personal transport accounted for 30% of 2024 US primary energy consumption.

- Most (89%) of this energy was provided by petroleum, constituting about 47% of the source use:
  - Increases in vehicle efficiencies and their utilization can substantially reduce petroleum consumption, and biofuels can extend supplies even more.
  - Fuel cell-powered vehicles can only reduce petroleum and natural gas consumption if non-fossil fuel sources are used to produce the hydrogen, and without major depletions of electricity for other purposes.

184. Carplus UK
185. Ntnl Railway Historical Society
186. Phoenix International
187. Mobile Review

The Transportation Sector

188. US DOE-EIA

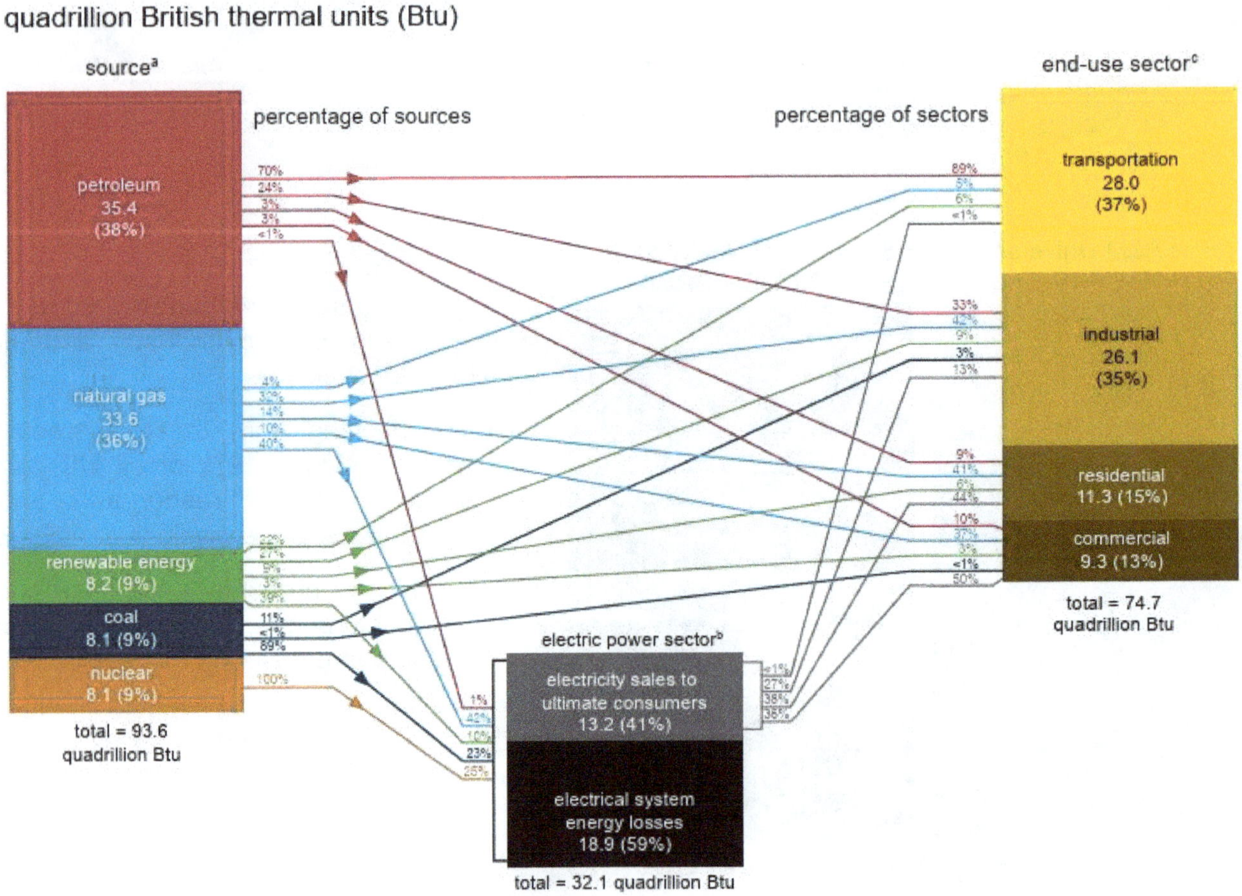

US Primary Energy Consumption by Source and Sector, 2023 (Quadrillion Btu)

While all are important, no energy option or combination can be realistically expected to satisfy needs of uncontrolled consumption.

- Continuing growth trends must be contained and hopefully reversed, applying resourceful conservation principles to do more with less:
  - **Reduce** waste and pollution through more energy-efficient building design, lifestyle habits and technologies.
  - **Recycle** natural energy and material resources using renewable, sustainable approaches that preserve and extend fossil fuels.
  - **Reuse** production process heat and chemicals to maximize yields, economies and resource transfer benefits for other purposes.

189. Larry Bell

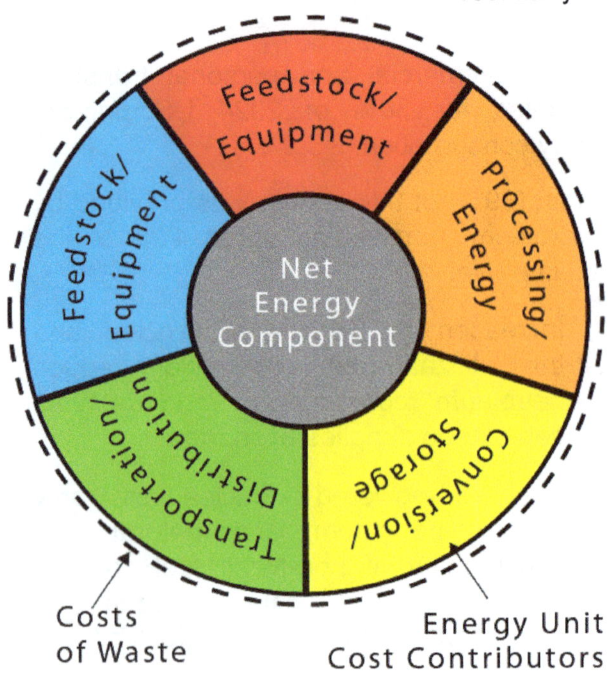

Each unit of energy conserved saves costs in the form of energy, money and pollution to create it.

Value of Conservation

190. Image Source 191. Amemco
192. UCMP Berkley 193. NASA 194. Penn. State U.
195. ABC Body Building 196. RADeCORE

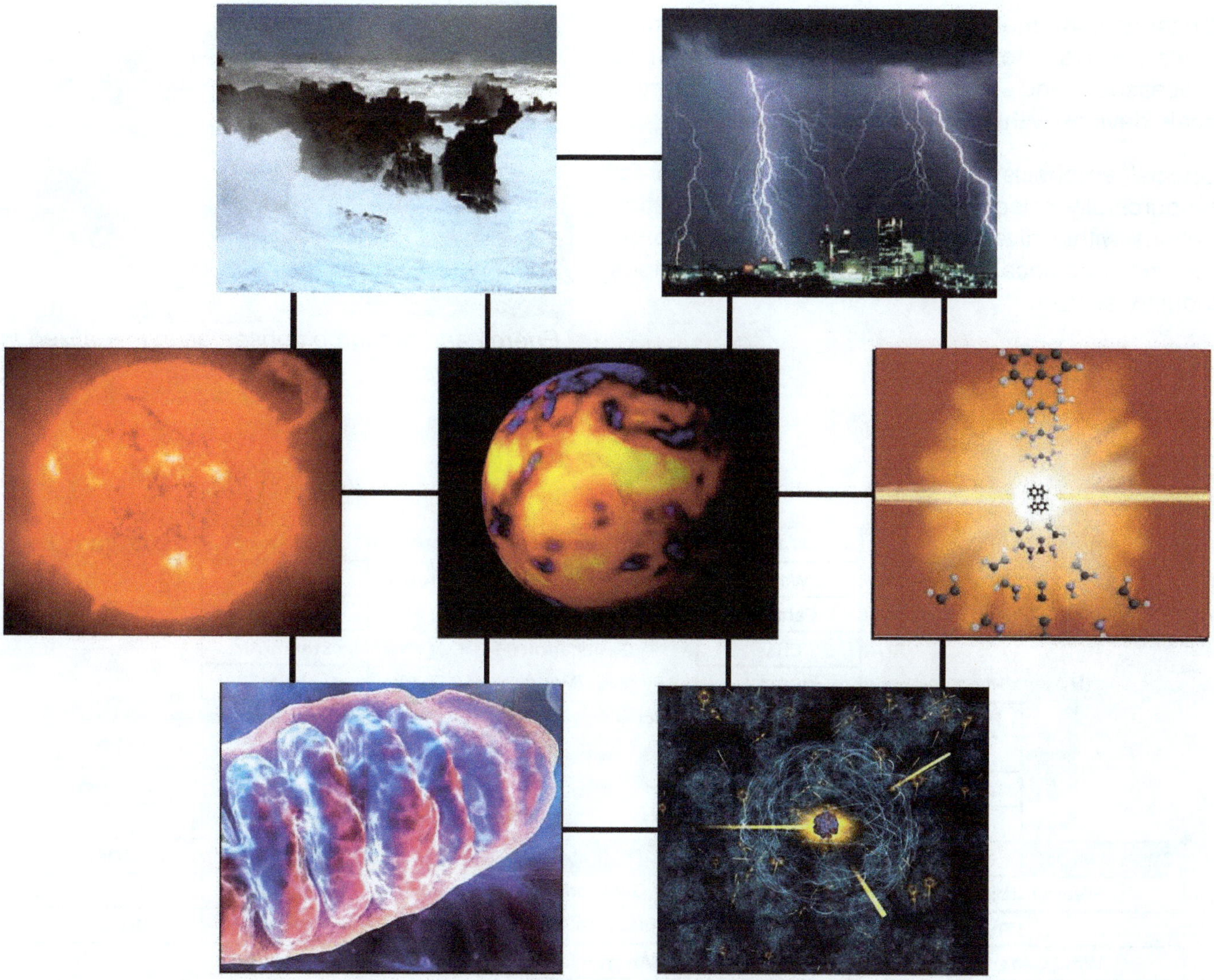

PART II: MECOPTICS FRAMEWORKS

MECOPTICS proposes a broad, top level framework to consider ways that a variety of energy and material assets can be processed and utilized as efficiently as possible.

- MECOPTICS highlights basic natural and economic principles that underlie conservation possibilities, factors necessary to make them effective, and ways to expand benefits:
  - Conservation methods include low-tech, small-scale ways to enhance naturally-occurring processes, and technologically-intensive, large-scale devices with prodigious capacities.
  - Special emphasis is directed to opportunities for resourcefully selecting and combining conservation options within diversified, comprehensive national, regional and local networks that connect various industry sectors.

197. PBS
198. U of Strathclyde
199. Aero Ecology
200. FCIPT
201. Sower Project
202. Ymparisto

Energy and resource conservation is realized in many forms.

203. Larry Bell

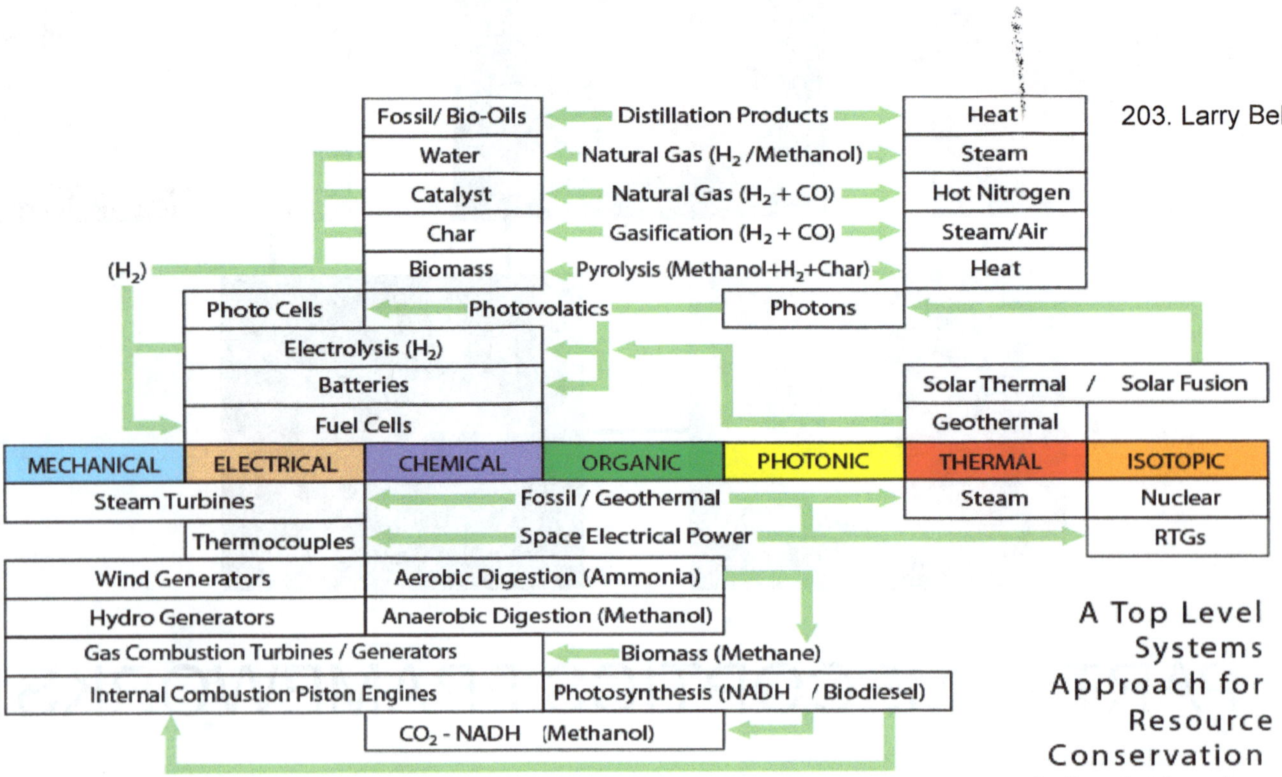

A Top Level Systems Approach for Resource Conservation & Optimization

MECOPTICS

Mechanical • Electrical • Chemical • Organic • Photonic • Thermal • Isotopic • Connectivity and Synergy

204. Larry Bell

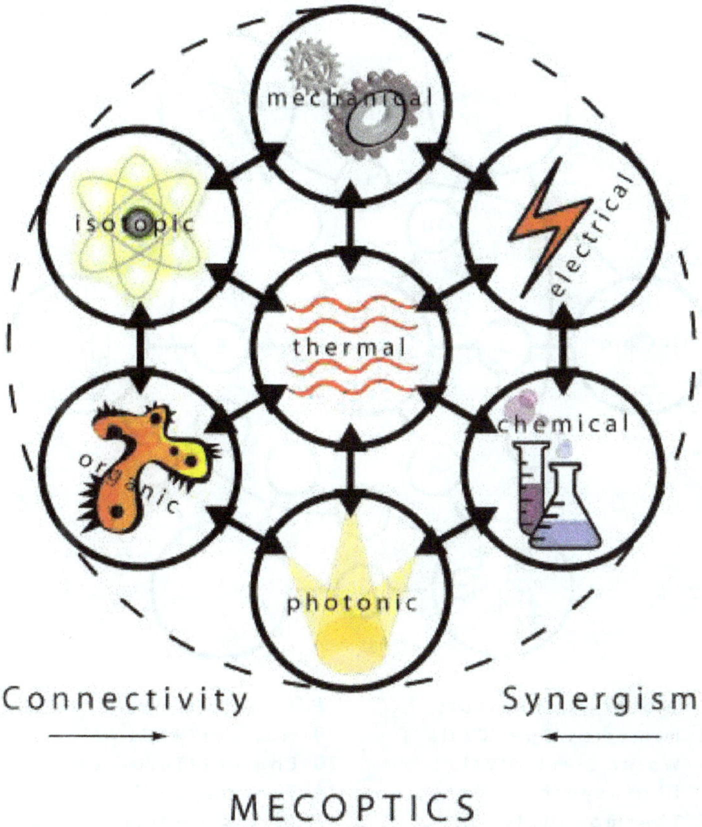

The term "MECOPTICS," incorporates initials of seven general categories of processes that create and convert natural resources:

- **M** echanical processes provide kinetic energy to power vehicles and electric generators.
- **E** lectrical processes produce electromagnetic energy for motors, lighting, heating, and a wide range of other uses.
- **C** hemical processes convert organic and inorganic feedstocks to gaseous and liquid fuels, commercial materials and other products.
- **O** rganic processes use microorganisms to transform plant and animal biomass wastes into methane and fertilizers.
- **P** hotonic processes convert sunlight into leafy plant materials, electricity and energy for water and space heating.
- **T** hermal processes harness and utilize heat energy for power generation and biomaterial processing.
- **I** sotopic processes create heat energy through atomic decay to provide power for terrestrial and space uses.

205. Larry Bell

Mechanical

These systems collect or produce kinetic energy that can be converted to electrical power.

Electrical

These systems generate, store and/or transfer electricity for power and heating.

Chemical

These systems process organic and inorganic materials into other forms through molecular and atom reactions.

Organic

These systems use microorganisms to combine or breakdown molecules and compounds into derivative products.

Photonic

These systems convert solar radiation into electricity, heat or means to create photosynthetic products.

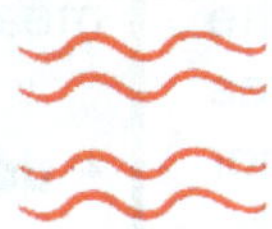

Thermal

These systems use heat energy to produce chemical reactions and to drive turbines and engines.

Isotopic

These systems produce heat energy through the natural decay of radionuclide, atoms that can be converted to electricity.

For purposes of convenience and comprehension, MECOPTICS processes and devices are grouped into general categories that characterize important features and uses. Each type typically works in combination with others to accomplish desired outcomes.

Processes and Devices

The MECOPTICS Model

By recognizing fundamental attributes of important processes and devices, we can begin to visualize on a general systems level how the energy and material inputs and outputs of each can link together with others to share beneficial relationships.

- As illustrated in the diagram, thermal processes and influences often have central roles essential to the workings of others:
  - Heat is energy in its purest form that originates from solar fusion radiation, the natural decay of radioisotopes, chemical reactions associated with molecular transformations, and electro-mechanical phenomena.
  - This valuable resource can be used directly, stored in other forms, and converted to provide electrical and mechanical power.

206. Larry Bell

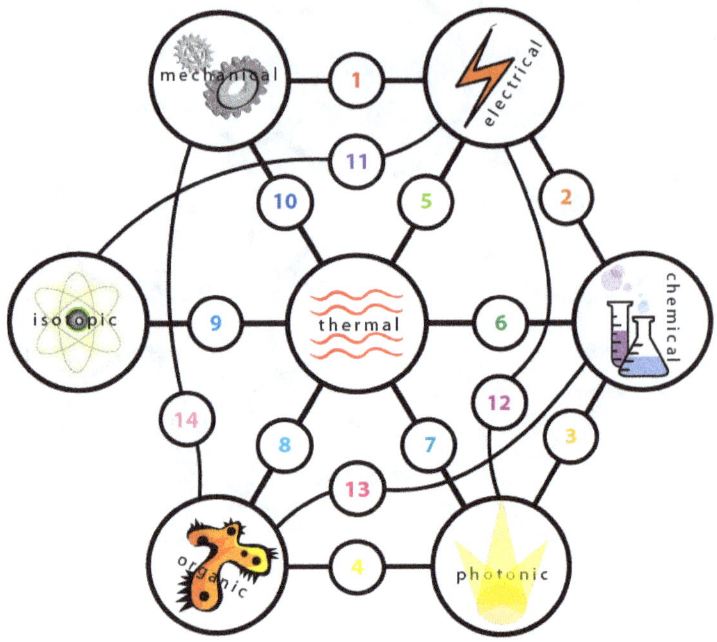

1 Motors/Generators
2 Batteries/Fuel Cells
3 Water Electrolysis
4 Photosynthesis
5 Thermocouples
6 Distillation
7 Water/Space Heaters
8 Pyrolysis/Gasification
9 Nuclear/RTGs
10 Engines/Turbines
11 Thermocouples
12 Photovoltaics
13 Biodigesters
14 Cultivation/Processing

Representative Processes and Devices

Heat is the universal medium of exchange that forms the basis of energy economics.

- This "thermodynamic" energy transfers the capacity to do work or raise temperatures back and forth between all of the other MECOPTICS processes and devices by exciting molecules and atoms to move more rapidly and become hotter:
  - Each of these transfers (or conversions) causes some heat energy to be dissipated into the surroundings, often reducing availability to do what we desire.
  - Some types of conversions are much more "expensive" then others, and all add to costs.

Heat energy is a fundamental resource that enables mechanical and electrical processes to do work, and is measured in a variety of ways:

- **Joule** (**J**): the amount of force required to move a certain weight a certain distance (1 J= 0.737 foot pounds)
- **Watt** (**W**): the power or rate of work accomplished (1 W= 1 J/second)
- **Horsepower** (**hp**): a measure of power needed to lift 550 pounds off the ground for one second (1 hp= 746 W)
- **British thermal unit** (**Btu**): the heat energy needed to raise the temperature of one pound of water one degree Fahrenheit (1 Btu = 1,055 J)

Measuring Heat Energy Products

All matter is composed of molecules made of atoms which are connected together by attractions between electrons around the atoms.

- When molecules break apart to create other molecules, a change in the motion of their bonding electrons cause them to either absorb stored heat from the chemical reaction, or to release some of their stored energy to the product molecule:
  - When a molecule contains more energy than its product molecule can absorb, the reaction is "exothermic", and excess heat is released out into its surroundings.
  - When the product molecule requires more heat than the original molecule can provide, the process is "endothermic" and outside heat must be added for the process to occur.

207. UT Environmental Sciences Institute

Fire is an exothermic chemical chain reaction that produces heat and light when a combustible fuel is rapidly combined with a reactant (oxygen).

An Exothermic Chemical Process

Bodies lose or gain heat energy through conduction, radiation and convection:

- Conduction transfers heat from a hot to a cold body through free electron diffusion as adjacent atoms vibrate against each other or are exchanged:
  - This mostly occurs in solids, such as metals, where atoms are densely packed and are in constant contact to increase collisions.
- Radiation occurs through electromagnetic atom vibrations in the heat spectrum:
  - All objects radiate heat unless they are at absolute zero (virtually impossible), and insulators and reflectors are used to keep heat in or out.
- Convection transfers thermal energy through circulation of a heated medium (typically from or to fluids or gases):
  - A heated, less dense fluid receives heat, causing a cooler, denser fluid to move in and replace it.

208. Larry Bell

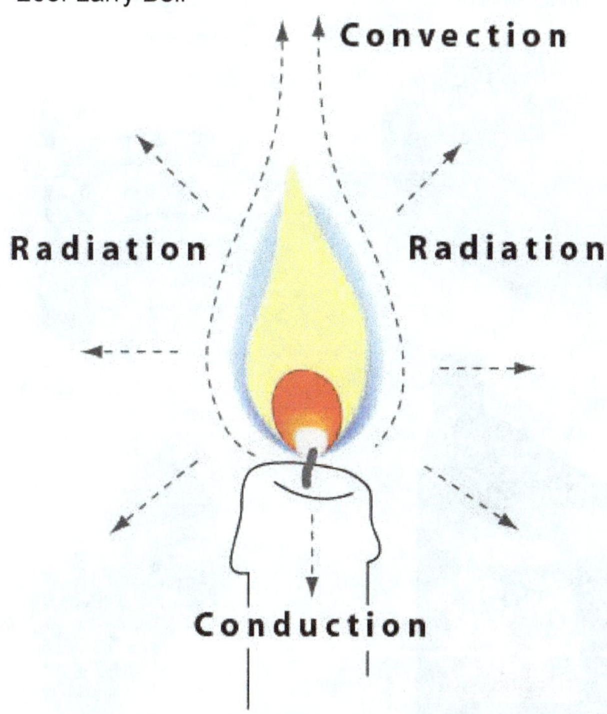

Conduction, Radiation and Convection

Mechanical systems always convert some amount of their kinetic energy through friction to heat, and electrical devices release kinetic energy of electron collisions with atoms for heat.

- Conversion of kinetic energy to heat, and transport of electricity from a source to a use-destination both present efficiency penalties that can be reduced, but never eliminated in the form of "waste heat".
  - Waste heat is the energy created by a machine or technical process for which no useful application is utilized after "free energy" has been extracted.
  - Free energy is the amount of mechanical or other energy that is extracted from a process to do useful work.

209. Sierra Pacific Innovations

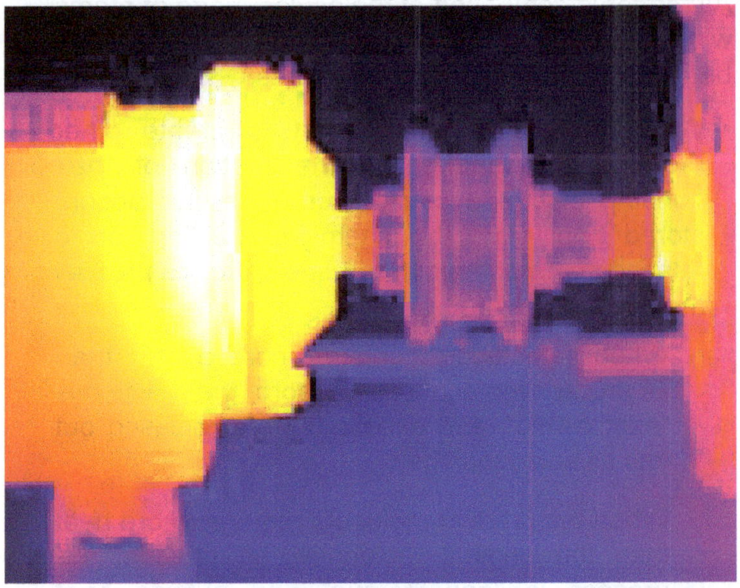

Thermal Image of an Electric Motor

Energy conservation reduces waste heat by making conversions and transfers as efficient as possible, and reinvesting product heat energy to do other work.

Mechanical Heat Losses

210. Silberwolf
211. Silberwolf
212. ConocoPhillips
213. Keyosk

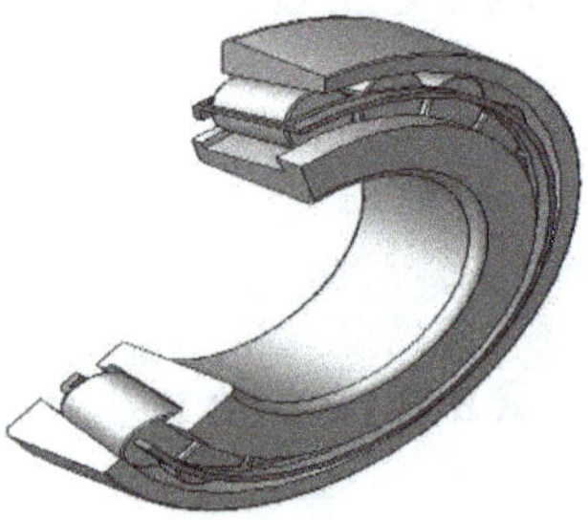

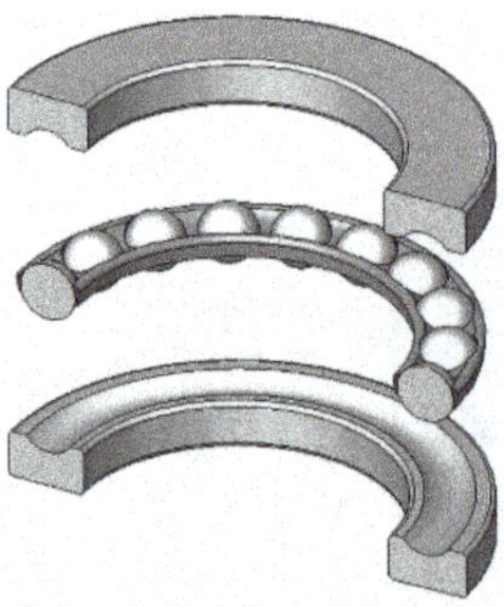

Reducing Frictional Heat Losses

Friction caused by relative motion between two contacting surfaces excites atomic vibrations (electromagnetic forces) that convert kinetic energy into heat.

- Frictional heat energy losses can be reduced by minimizing contact surfaces and making them slippery, but can not be entirely eliminated:
  - Roller bearings and ball bearings are used to convert sliding friction into smaller rolling friction.
  - Lubricants such as oil, graphite or Teflon convert "solid friction" into "liquid friction".
  - Replacement energy must always be put back in to perpetuate the mechanical process.

214. JLC Electromet

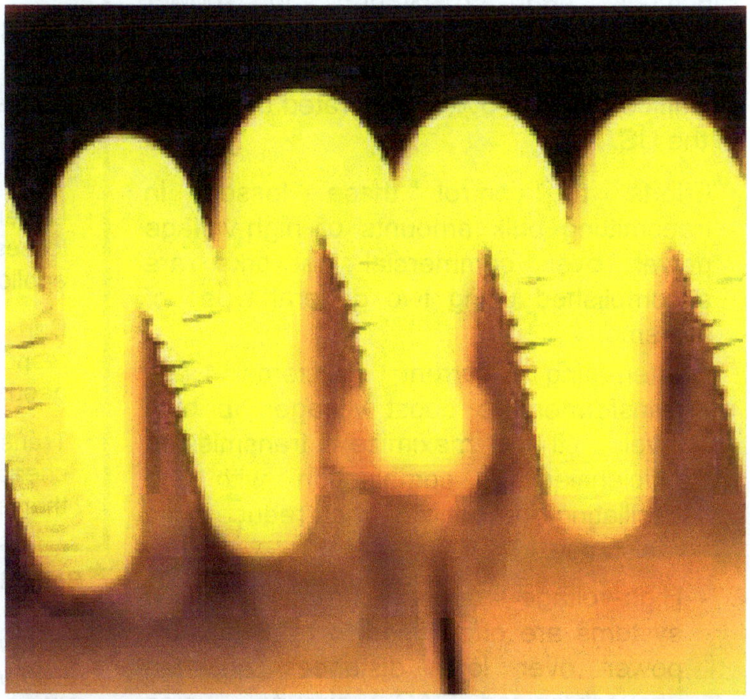

Electric current passing through a conductor transfer kinetic energy that agitates atoms and convert the kinetic energy to ohmic (resistance) heat.

Electrical systems produce large amounts of by-product heat through electrical resistance of the conductor (ohms) that oppose passage of an electric current (amperes) in a circuit:

- Electron collisions in the conductor (ohm resistance) cause the atoms to vibrate more violently about their respective equilibrium positions. The resulting agitation releases energy (heat) proportional to the square of current strength times the conductor resistance as stated in Joule's law:

  Q = $I^2 \cdot R \cdot t$ where:
  Q = heat by constant current (I)
  R = resistance
  t = time

Heat from Electrical Resistance

Heat losses in electrical devices caused by resistance drain energy away from intended purposes and can also impair or burden some equipment operating functions.

- The amount of resistance ("Joule heat" or "ohmic heat") can be reduced by converting a direct current (DC) to an alternating current (AC) using an "inverter":
  - Conversion to AC redistributes current density near the surface of the conductor rather than throughout its cross section so that fewer electrons collide as they pass through. (This is called the "skin effect.")
  - When heat is desired as the output product, conductors are selected with high electric resistance and are often coiled to maximize heat transfer (release).

215. Larry Bell

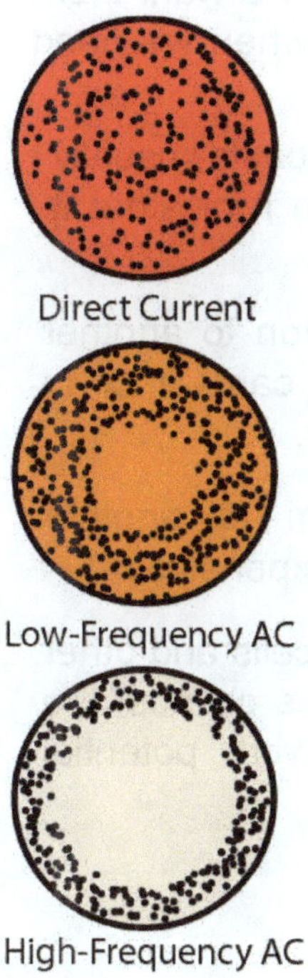

DC Conductor

DC currents distribute electron flow throughout a conductor, causing more collisions with atoms (more heat).

AC Conductors

AC currents distribute electron flow towards the outer surfaces of conductors, producing less resistance (heat). Higher frequencies increase this benefit.

Heat Reduction Through AC

Electrical heat resistance in power transmission lines is a major source of energy losses, potentially accounting for efficiency penalties of about 5% of generated capacities in the US.

- Efforts to control these losses in transmitting bulk amounts of high-voltage power over commercial networks are accomplished using two general types of systems:
  - Alternating current systems use transformers to boost voltages to high levels that maximize transmission efficiencies in combination with the oscillating frequencies that reduce ohm resistance.
  - High-voltage direct current (HVDC) systems are often used to transmit bulk power over long distances between unsynchronized AC circuits using advanced solid-state inverters that convert to AC at connecting station interfaces.

AC Systems	HVDC Systems
• Almost universally used by US commercial power distributors.	• Useful for long-haul bulk power transmission in remote areas.
• Lower costs / transmission losses for shorter applications than HVDC.	• Enable power transmission between unsynchronized AC distribution systems.
• Can change voltage to respond to particular power needs.	• Reduce line costs due to use of 2 rather than 3 conductors.
• Transformers are currently cheaper and more reliable than solid-state devices.	• Minimize power losses associated with corona discharge of ions/electrons.*

** Both AC and DC transmission lines can produce coronas that cause significant power losses as their electric fields tear electrons from surrounding un-ionized air and positive ions, or electrons are attracted to the conductor while charged particles drift away. HVDC systems can experience about half the loss of AC lines of the same power.*

Applications and Advantages of AC and HVDC Power Transmission Systems

Energy sources often aren't located where we want them to be, in the form we desire, or available when we need them most.

- Substantial economic and efficiency costs usually occur when we move energy, transform it or store it for future use:
  - Transporting energy from one location to another invariably requires more energy to carry it or to compensate for losses along the way.
  - Energy conversion from one form to another imposes penalties that can be very expensive.
  - Energy storage using batteries, fuel cells and other devices imposes conversion penalties on both the input and output sides along with potential transport losses.

216. Jean-Pierre Daniel
217. Ben Ostrowsky
218. US Bureau of Reclamation
219. Adrian Pingstone
220. Popular Mechanics
221. NASA

Transportation, Conversion and Storage

Energy costs per units of power delivered to consumers by each source are fundamentally influenced by the respective transportation, conversion and storage systems used.

- Electrical power economies associated with renewable sources are particularly sensitive to conversion and storage methods:
  - Wind and hydro efficiencies are determined by the way kinetic energy is captured and transformed, and whether the electricity is used directly, transferred to a power grid, or stored in batteries or as hydrogen.
  - Geothermal efficiency depends upon whether the heat will be used directly or converted to electricity.
  - Hydrogen efficiency depends upon the source and production method, the conversion form (electrical or chemical), and the conversion/use technologies (at source and end-use).

222. Larry Bell

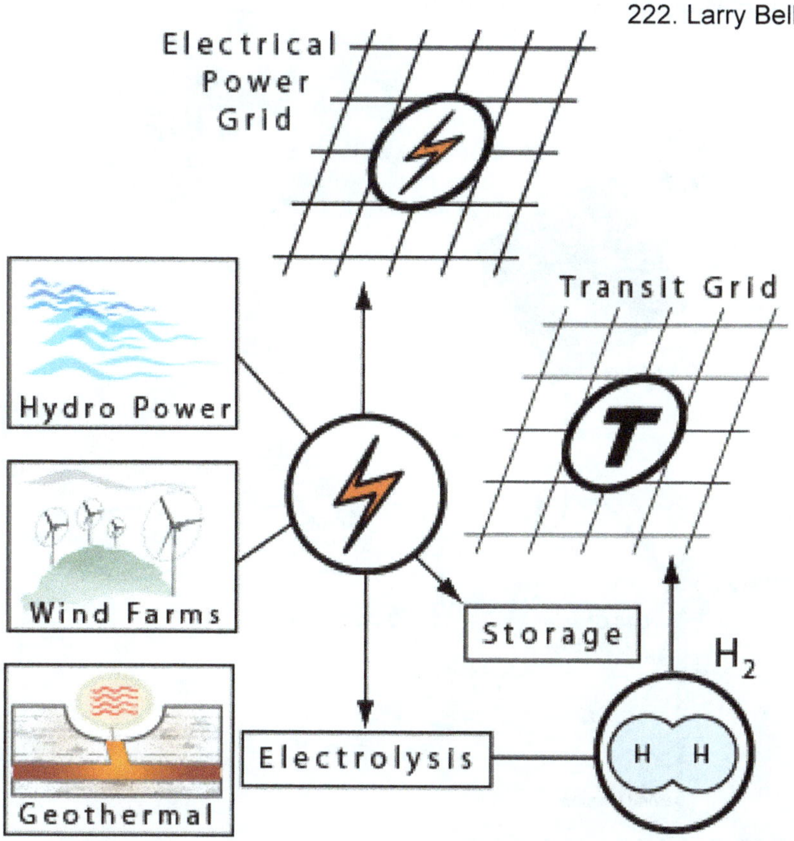

Energy Transfer Grids

223. Larry Bell

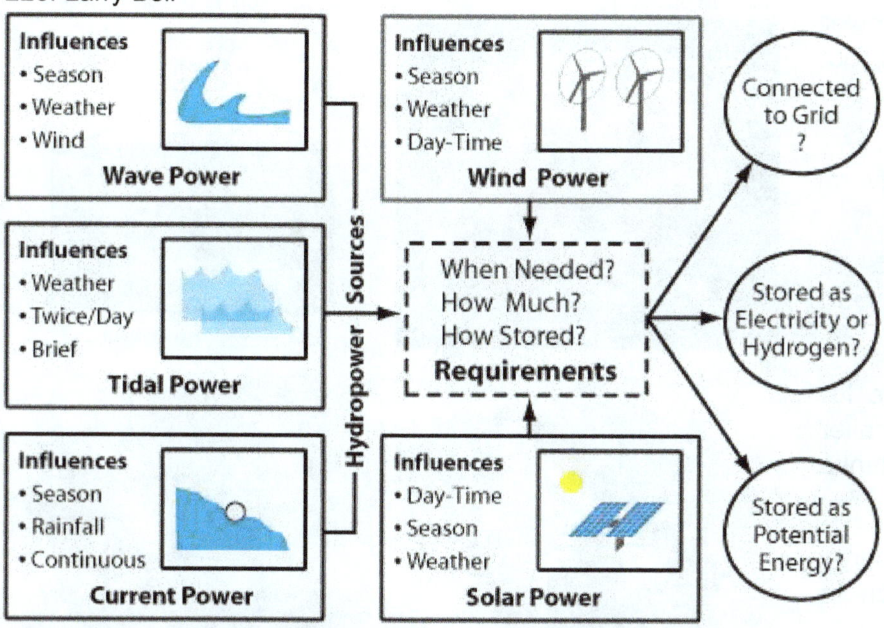

Energy Use/ Storage from Intermittent Sources

Wind, solar and most types of hydropower are intermittent electricity sources.

- Different strategies can be applied to provide more continuous, useable power:
  - Two or more types can be combined together to increase power levels and continuity.
  - They can be connected to a two-way electron power grid, potentially being credited for supply contributions.
  - They can store energy for times needed in the form of electricity (batteries) or hydrogen (fuel cells).
  - They can also power pumps to raise water, storing potential energy that can be released to drive turbines.

224. Enviro-Energy Technologies

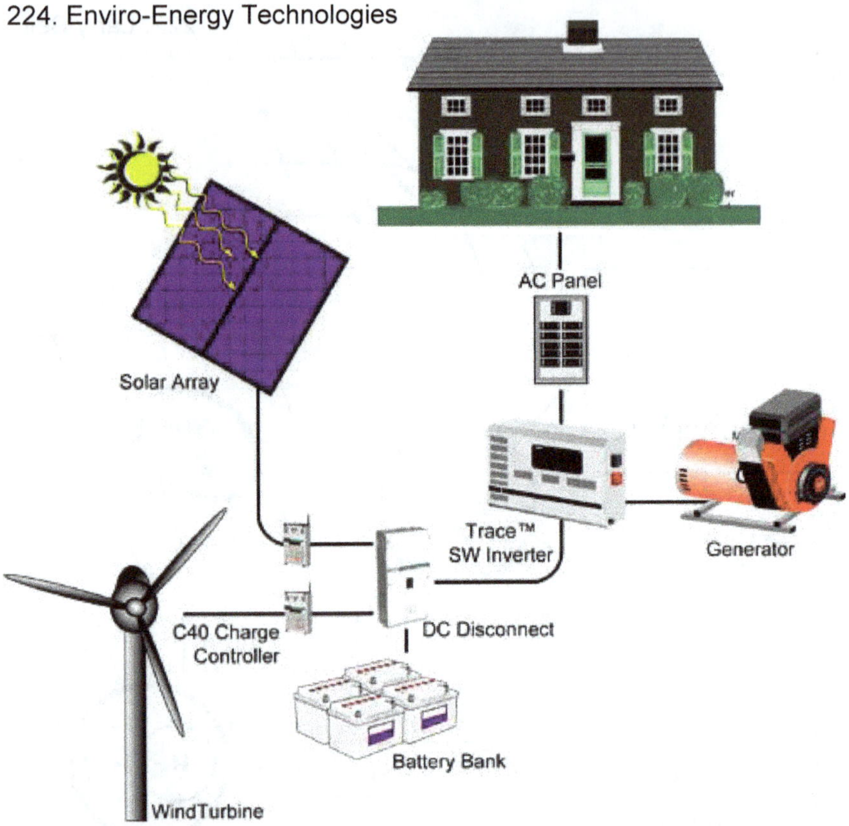

Off-grid electrical systems provide continuous electrical power to thousands of consumers who live and work far beyond the reach of power lines.

- Since wind and sunlight are not always present when needed, battery banks to store electricity are vital to these systems:
  - DC power that is generated and stored is usually converted to AC for general use.
  - PV systems and wind turbines are often combined in off-grid applications to realize benefits of both technologies.
  - Batteries can be viewed as "buckets of electricity" that are emptied and re-filled by inverter/charger devices, with capacities measured in Amp Hours.

Off-Grid Electrical Systems

Hydrogen fuel cells can provide electrical power plus heat for a variety of off-grid applications, but at high energy processing, containment and transportation costs.

- This fuel presents special challenges for several reasons:
  - H_2, the smallest molecule of all, tends to escape from gas pipelines, making transport by this means very problematic.
  - Increasing gas pressures to compensate for low energy/volume density using smaller containers requires stronger/heavier tanks and energy for compressors.
  - Liquid hydrogen (LH_2) imposes added energy costs for liquefaction and tank insulation to prevent boil-off.
  - Hydrogen is highly combustible and explosive, necessitating special safety precautions.

225. Purdue University
227. Micro-vett

226. E-Marine, Inc.
228. University of Birmingham

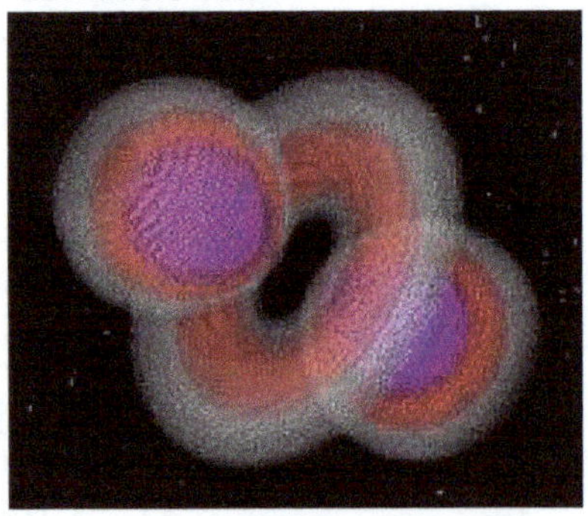

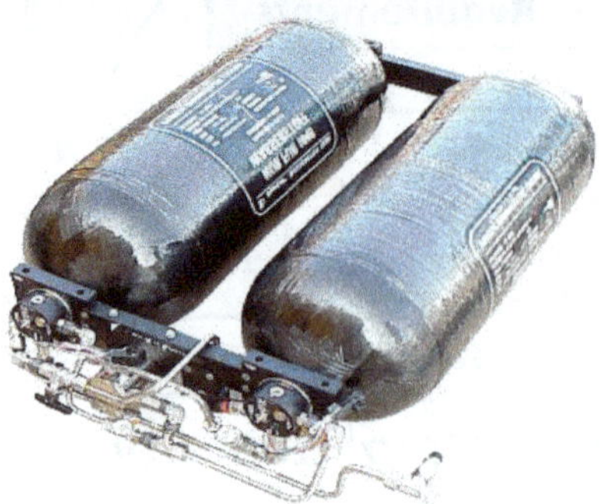

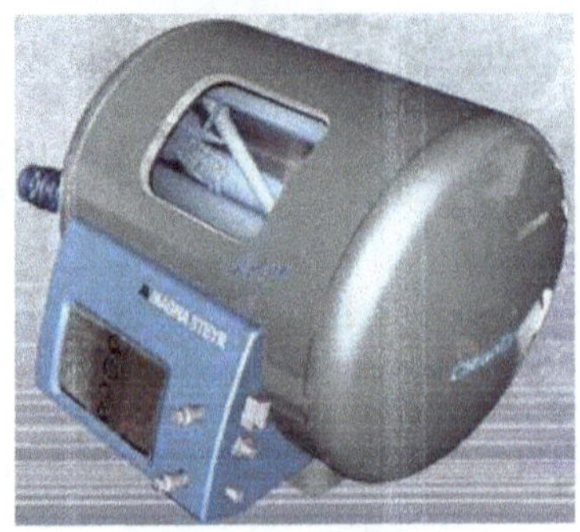

Hydrogen Transit and Containment

Advantages:

- Provides very high storage densities as a liquid with mild pressurization and cryogenic constraints.
- Can be stored at room temperature and pressure when mixed with water.
- Can be reformed to produce H_2 with no harmful waste, and can be mixed with existing fuels to burn efficiently.
- Can be used as a suitable fuel in conventional engines with slight modifications.

Disadvantages:

- Is very energy-expensive to produce.
- Is a toxic gas with a potent odor at ambient temperatures.

Different ways to transport hydrogen from production through distribution are being pursued within the energy industry.

- One approach is to store hydrogen within ammonia (NH_3) for later release using a catalytic converter at the destination:
  - Since ammonia is the second most commonly produced chemical in the world, a large infrastructure for production, transportation and distribution already exists.
  - Although ammonia burns poorly at atmospheric pressures, under compression in automobile engines it is a reasonable fuel that burns efficiently.

H_2 Storage in Ammonia

Metal hydrides can be used to store hydrogen as liquids or solids.

- Sodium borohydride, lithium aluminum hydride and ammonia burane are among leading candidates:
  - Sodium borohydride and ammonia burane can be stored as liquids when mixed with water, but must have very high concentrations for desirable H_2 densities, requiring complicated water recycling within fuel cells.
  - Sodium borohydride will enable less expensive fuel cells (without platinum contacts) to be used, but is energy-expensive and will require recycling plants.

Advantages:

- Some are easy-to-fuel liquids at ambient temperatures and pressures.
- Can also be constituted as pellets.
- Have good energy density/volume.

Disadvantages:

- Have much worse energy density/weight for standard cars than gasoline tanks with the same energy.
- Hydrides present safety hazards for H_2 storage densities (above 10% of total weight).
- A hydride will need to be recharged with H_2 either onboard a vehicle, or at a recycling plant.
- Fuel cell design will need to be appropriately correlated with the particular hydride used.

H_2 Storage in Metal Hydrides

While solid metal hydride hydrogen storage is promising for automotive applications, it will require lithium storage vessels that are much larger and more expensive than gasoline tanks for standard cars.

- Assuming lithium storage with the same hydrogen energy content is contained in a conventional automobile gasoline tank:
  - The lithium/solid metal hydride tank will require three times the volume of the gasoline tank (45 gallons vs. 15 gallons [170 liters vs 57 liters]).
  - The solid metal hydride fuel will weigh approximately 600 lbs (272 kg) (vs. 150 lbs [68 kg] for gasoline).
  - The lithium tank material will cost more than \$40/lb (vs. less than \$1/lb for the steel gasoline tank).

229. Larry Bell

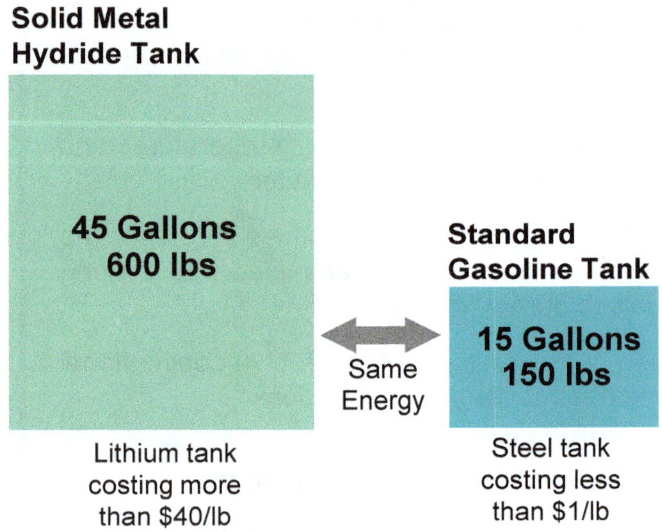

Solid Metal Hydride H_2 Storage

An alternative to metal hydrides is to use regular hydrocarbon fuels as hydrogen carriers with a small hydrogen reformer to extract H_2 as needed by a fuel cell.

- On-board reformers transform a liquid fuel (gasoline, diesel or ethanol) into a hydrogen-rich gas (reformate) that can be used to supply a fuel cell:
  - First, the reformer breaks down long hydrocarbon molecule chains into simpler molecules (i.e. hydrogen, water and carbon).
  - The gas is then purified until it is suitable for use in a fuel cell (typically a 45kW cell).
  - The reformate emerges with about 15% H2, 20% CO, 1% methane, and the rest nitrogen and other gases.
  - Maximum power output is limited by a hydrogen production rate of about 1.4 grams/second.

Advantages

- Can run on multiple hydrocarbon fuel types allowing motorists to choose the cheapest/most available options.
- Avoids storage problems/complexities associated with high-pressure H_2 and cryogenic LH_2 approaches.
- Hydrocarbon fuels are currently established within the energy production and distribution networks.
- Direct methane fuel cells do not require reformers to operate (but have lower energy densities than conventional fuel cells).

Disadvantages

- Reformers can add costs and complexities along with associated system maintenance/longevity problems.
- Current reformers tend to be slow, limiting power generation.
- Present fuel cells tend to be less energy-efficient than internal combustion engines (offering advantages to methanol/ethanol).

On-Board H_2 Fuel Cell Reformers

230. Larry Bell

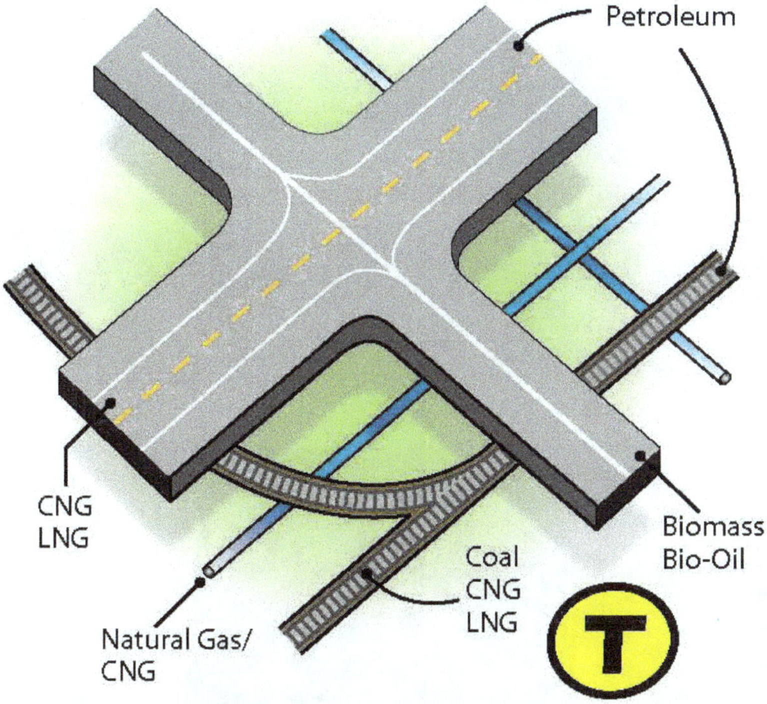

Fuel Distribution and Storage

Rising petroleum costs add expenses for transporting fuels and other materials to end-users, thereby compounding cost escalations.

- Transport economies to reduce energy consumption and costs can be realized in three basic ways:
  - Reduce public and commercial fuel and material consumption using more energy-efficient vehicles and conservation-minded living and travel practices.
  - Select the most energy-efficient transportation mode for bulk fuels and materials (pipeline, rail or truck).
  - Convert fuels to electricity at or near the supply sites to enable power transmission by wire.

Storage and transportation add substantial cost and complexity to natural gas supplies

- No single approach can address all needs and conditions:
  - Pipelines are economical for moving gas on land, but are not practical across oceans.
  - Many existing pipelines in North America are operating near capacity, causing shortages in some regions (particularly colder locales).
  - Special tanker ships transport LNG across oceans, while tank trucks carry LNG and CNG over shorter distances to end-users or distribution points such as pipelines for further transport.
  - Additional facilities for liquefaction or compression at production points, and end-user facilities for gasification or decompression of the gas, expand the requirements and costs.

231. Pacific Summit Energy
232. Peker Construction, Inc.
233. Enter Stage Right
234. Kleenheat Gas Australia

Natural Gas Storage and Transportation

Natural gas and biomass methane are highly flammable and explosive, and must be handled carefully.

- Containment and transportation must take safety precautions into account:
  - Local production at the end-use site must provide proper ventilation to ensure that gas does not collect in hazardous concentrations that can present explosive risks or cause suffocation by displacing oxygen.
  - Since the gas is odorless and colorless, special leak detection means are essential.
  - Although the gas can be transported by pipeline over both short and long distances, for shorter distances or to areas where pipeline infrastructure is limited the distribution is usually accomplished in a compressed form (CNG).
  - Compressed gas is usually contained in robust, high-pressure tanks made of steel or composites, and is dehydrated to avoid water corrosion of metal parts.

235. Rawhideleasing

Steel Tanks

Composite Tanks

Natural Gas / Methane Containment

236. Western Natural Gas

237. SGC Engineering

Natural Gas / Methane Transport

238. US DOE-EIA

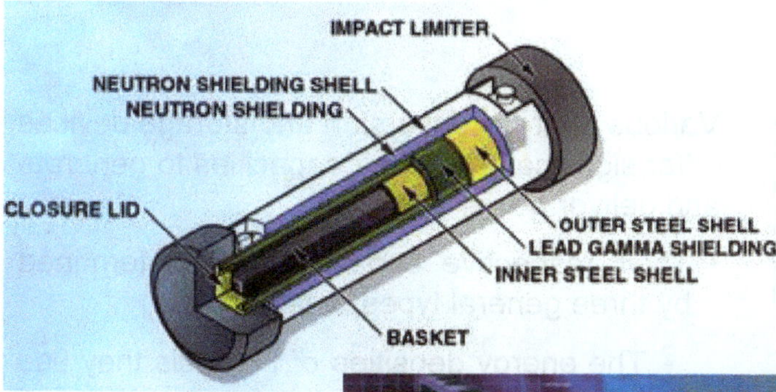

Truck Container Cutaway

Spent Nuclear Fuels

The US Department of Energy has proposed to create a centralized storage facility at Yucca Mountain in Eureka County, Nevada to provide long-term containment of spent nuclear fuels.

- Although the site was approved by the 105th Congress in October 1997, the plan has not yet been approved to proceed:
  - DOE's Office of Civilian Radioactive Waste Management has estimated that more than $96 billion (20088 dollars) will need to be spent over the next 100 years to develop, operate and close the Yucca Mountain project.
  - The waste would be contained in steel canisters buried 984 feet (300 meters) below the surface and designed to last at least 1,000 years, using the mountain to provide a natural shielding barrier.

Safe transportation and containment of spent nuclear fuels presents particularly difficult issues and controversies.

- Current radioactive waste is stored at temporary locations, including water basins in nuclear power plants and dry surface locations:
  - Although most nuclear waste is solid and relatively non-corrosive, a significant amount of old liquid waste still exists which can corrode metal storage tanks internally.
  - Interim storage using many older types of tanks has presented a troublesome leak history.
  - Continuing accumulations of spent materials are creating major disposal problems where former plant storage areas are no longer accepting additional materials.

239. Yucca Mountain Information Office

The proposed facility would ultimately contain up to 40,000 metric tons of spent nuclear material, and be licensed under renewable terms of 100 years.

Proposed Yucca Mountain Nuclear Waste Storage Facility

240. Colorado State U.
241. Alan D.
242. Milica Sekulic
243. US DOE-EERIE

For internal combustion engines, fuel efficiency is essentially the same as thermal efficiency.

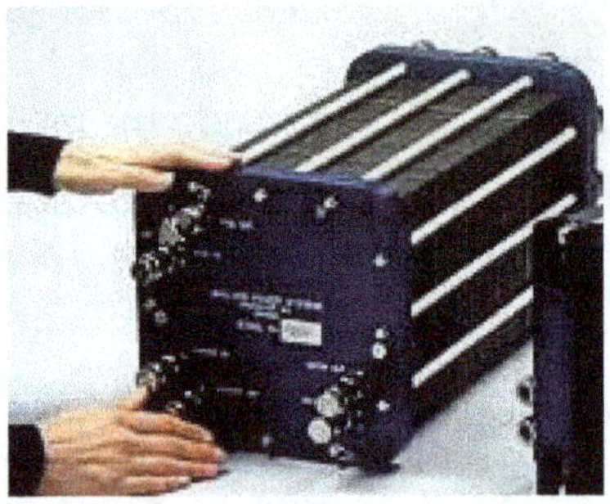

Unlike batteries which store chemical energy, fuel cells require continuous replenishment of fuel and reactants.

Conversion and Storage Devices

Various energy conversion and storage devices differ significantly in their capacities to generate and deliver power efficiently.

- Their respective capacities are determined by three general types of factors:
  - The energy densities of the fuels they use influence systems that apply internal combustion processes.
  - The effectiveness of electro-chemical transfer devices influences operations of batteries and fuel cells.
  - The amount of "free energy" that can be extracted to do work, vs. heat energy that is wasted, influences all systems.

> ***Power*** *is the rate at which work is done (e.g., kW hours or Joules/second).*
> ***Energy density*** *is the amount of energy stored in a unit weight or volume (e.g., W hours/kg or W hours/liter).*

Power and Energy Density

Batteries and fuel cells are the most efficient ways to convert chemical energy into electricity because they bypass intermediate thermal and mechanical stages associated with internal combustion engines and thermionic (e.g. steam turbine) converters.

- Maximum efficiencies of all heat-driven engines are constrained by "Carnot" limits which typically reduce energy outputs vs. inputs by 30%-40%:
  - Since fuel cells convert chemical energy directly to electricity, their efficiencies can exceed Carnot maximums, even when operating at relatively low temperatures where thermal systems are most efficient.
  - Combined-cycle heat and power systems which recycle heat in electricity production operate at higher efficiencies than other engines (55%-60%), but are limited to high-power applications.

244. Larry Bell

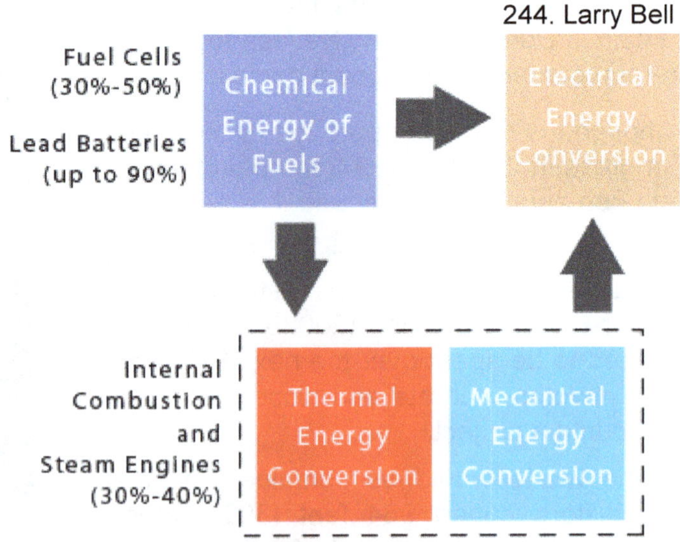

Direct chemical-to-electrical conversions are more efficient because they avoid the need to convert from chemical-to-thermal- to mechanical-to-electrical.

Electricity Conversion Efficiencies

The energy content (or "heat of combustion") contained in a fuel is the amount of heat energy produced when it is burned.

- Relative efficiencies of energy content are measured in terms of either fuel or energy efficiency:
  - Fuel efficiency refers to either thermal or fuel energy available in a carrier fuel to that which can be converted to kinetic energy or other work on the basis of inputs vs. outputs per unit (such as miles per gallon).
  - Energy efficiency is similar to fuel efficiency, but input units are measured in heat terms (such as Mega Joules or BTUs), and input-output relationships (energy intensity) is measured in terms of work product (such as kWh for electricity).

Fuel Types	Heat Values			
	MJ/L	MJ/kg	BTU/Gal.[2]	Octane[3]
Gasoline	32.90	45.00	125,000	97-98
Diesel	35.80	45.50	137,380	N/A[4]
Gasohol[1]	28.06	43.54	120,900	93-94
Ethanol	21.1	27.00	84,400	129
Methanol	15.60	22.61	62,800	123
LPG	25.00	34.39	95,475	115

1. 10% ethanol / 90% gasoline
2. US gallon (0.833 x imperial gallon)
3. Research Octane Number (RON)
4. Applies to spark ignition engines only

Energy Contents of Various Fuels

245. Chevron: Fuels/ bulletin, fuel economy Passenger car in EPA urban cycle (F7P75)

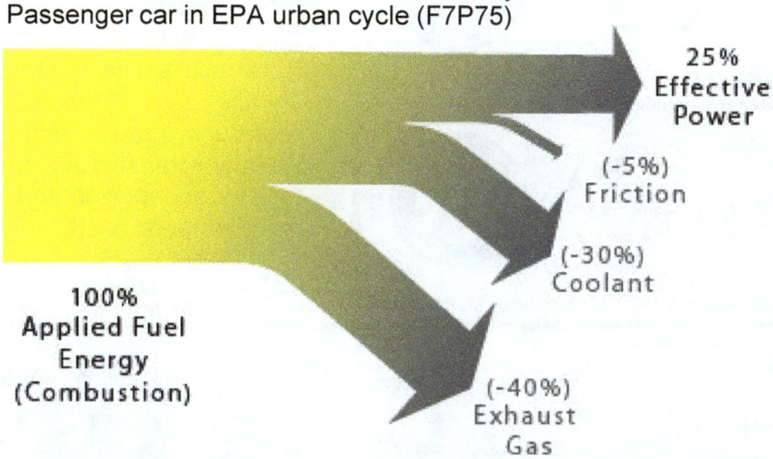

The gasoline combustion engine is relatively inefficient in the actual driving cycle. It converts about 65% of the gasoline to heat, and only about 30% to mechanical energy. Of this, only about two-thirds of the mechanical energy reaches the wheels.

Typical Energy Split in Gasoline Internal Combustion Engines

Generally, diesel engines are about 20-30% more efficient than gasoline engines in terms of miles per gallon.

- Although diesel fuels weigh somewhat more than gasoline per volume, the fuel also contains about 10%-20% more energy per volume, particularly at low idle speeds that trucks usually have:
  - Higher diesel compression ratios offer an advantage.
  - Heat of combustion contained in diesel fuel also occurs at higher values than in gasoline engines, causing the liquid to give up more latent energy for more complete and efficient conversion.
  - All internal combustion engines are relatively inefficient, and reducing vehicle weight is a key improvement strategy.

Efficiencies of conversions from chemical to mechanical energy depend upon the types and numbers of processes used.

- Determinations of real power costs must take the energy and financial expenses of fuel sources into account:
  - Overall fuel cell efficiencies are impacted by costs of producing and transporting the hydrogen, the form of its use, and characteristics of the system that produces the electricity.
  - Internal combustion and steam engine efficiencies are determined by fossil fuel and/or biofuel costs along with engine type, design and application.

Conversion Devices	Conversion Processes	Wheel Efficiencies
Diesel Fuel Engines	Chemical-Mechanical	Approx. 33%
Gasoline Fuel Engines	Chemical-Mechanical	Approx. 25%
Generators/ Motors	Mechanical-Electrical	30%- 60%
Fuel Cells Compressed (CH_2)	Chemical-Electrical	Approx. 22%
Fuel Cells Liquid (LH_2)	Chemical-Electrical	Approx. 17%

Although direct chemical conversions tend to be most efficient for producing electrical power, internal combustion engines are most efficient to produce mechanical power for automotive purposes because they avoid an intermediate electric motor stage. Fuel/hydrogen costs influence relative overall economies.

Mechanical Wheel Power Efficiencies

246. Larry Bell

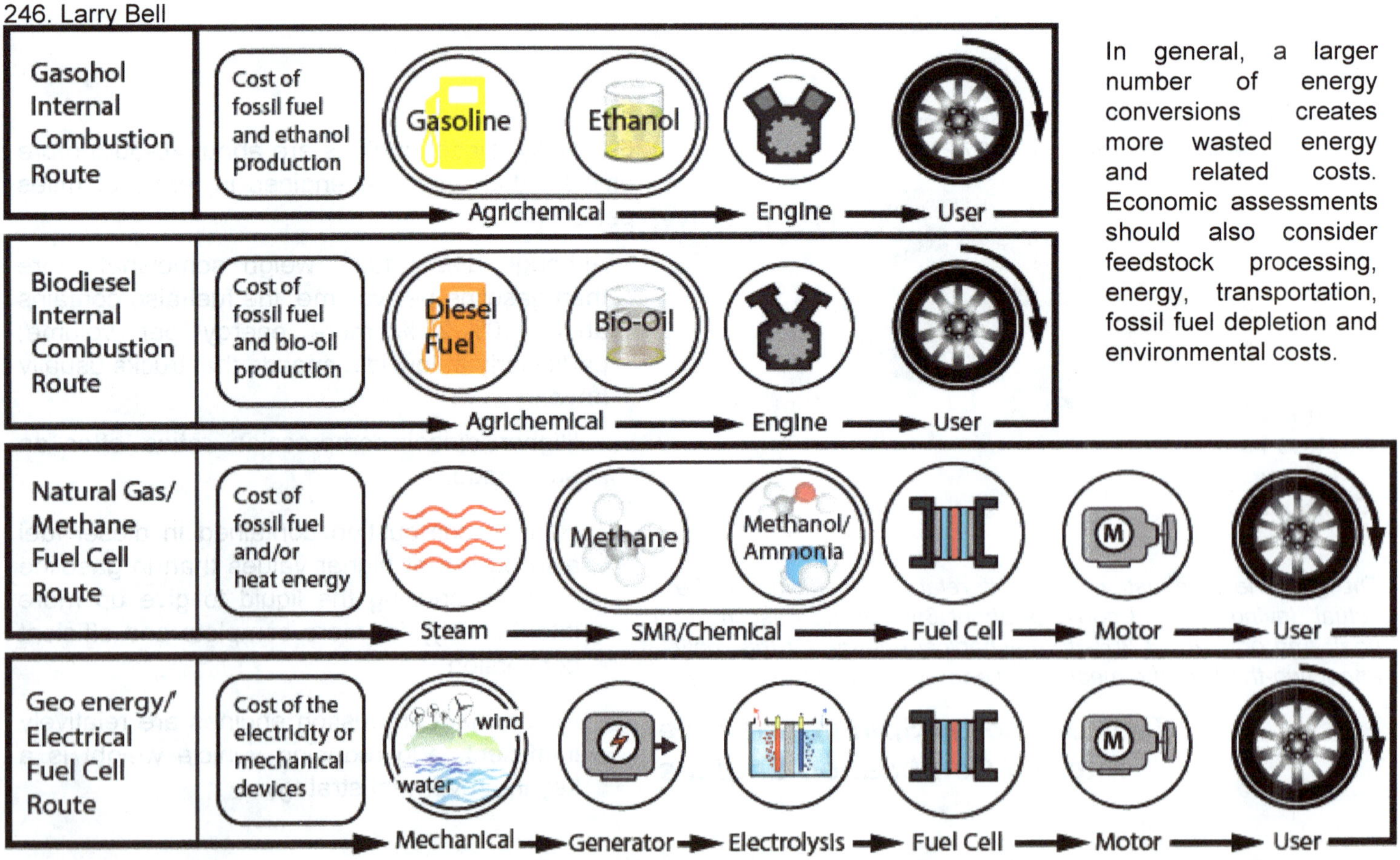

In general, a larger number of energy conversions creates more wasted energy and related costs. Economic assessments should also consider feedstock processing, energy, transportation, fossil fuel depletion and environmental costs.

Mechanical Power Conversion Efficiencies and Cost Penalties

The sources and processes for obtaining hydrogen have primary impacts upon fuel cell economics.

- As an energy carrier rather than resource, the net loss of output vs. input can only be offset by economies on the energy supply side:
  - Production of hydrogen using steam methane reforming imposes economic and energy costs for natural gas or methane recovery processes to provide the feedstock, along with heat energy to create steam.
  - Production using electrolysis of water requires large amounts of electricity, competing with power for other uses.
  - Additional energy is required to compress the H_2, liquefy it, or incorporate it into a chemical transport such as methanol or ammonia.

247. Larry Bell

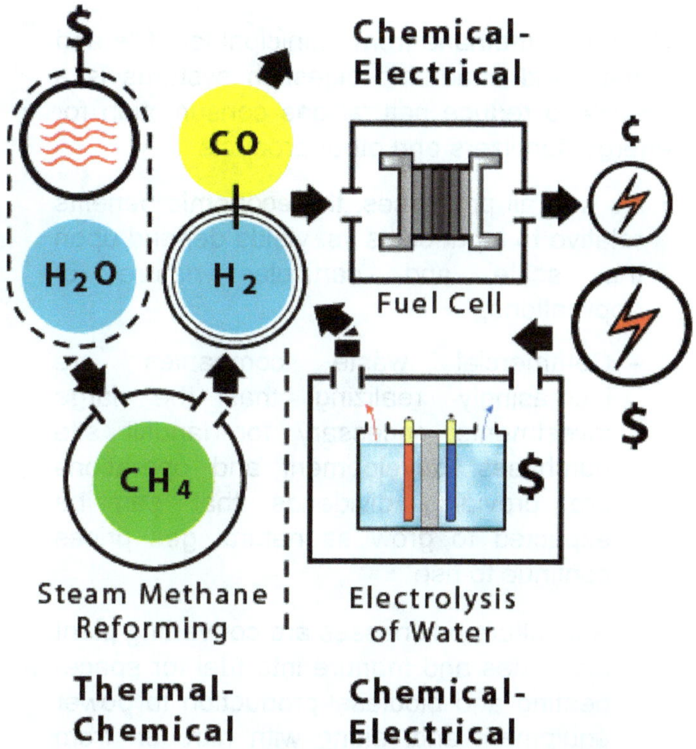

H_2 Production Economics

248. Larry Bell

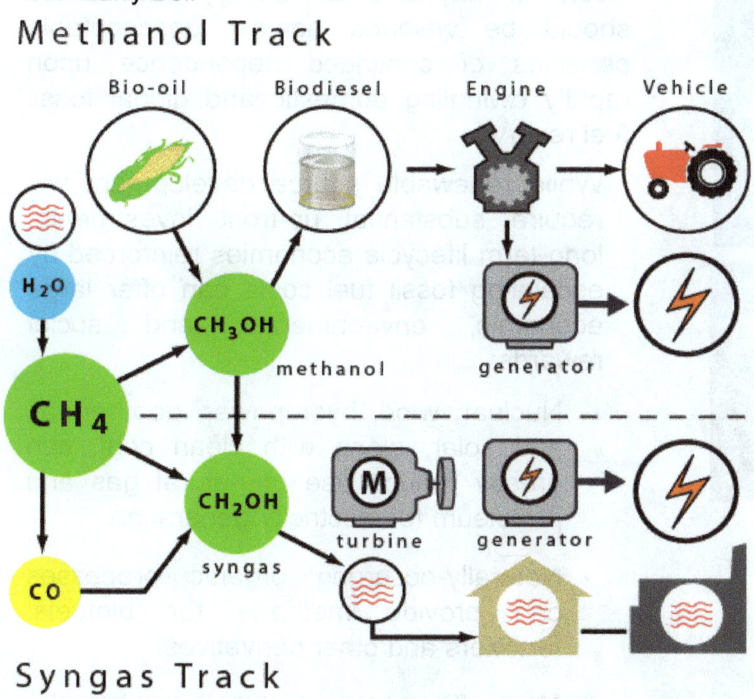

Methane Reduction for Methanol and Syngas

The economics of producing hydrogen from natural gas or methane from biodigestion processes must be compared on a case-by-case basis with other energy options that apply the same steam methane reforming process.

- In addition to H_2, SMR can also produce syngas for direct use, or as feedstock for methanol.
  - Methanol plus plant bio-oil produces biodiesel which can fuel engines to provide mechanical and electrical power, or the methanol can be converted into fertilizer to grow food and oil crops.
  - Syngas can be burned to provide heat for space and industrial processes, or can drive turbines and generators to provide mechanical and electrical power.
  - Syngas is also created as a product of coal gasification (coal gas).

249. Larry Bell

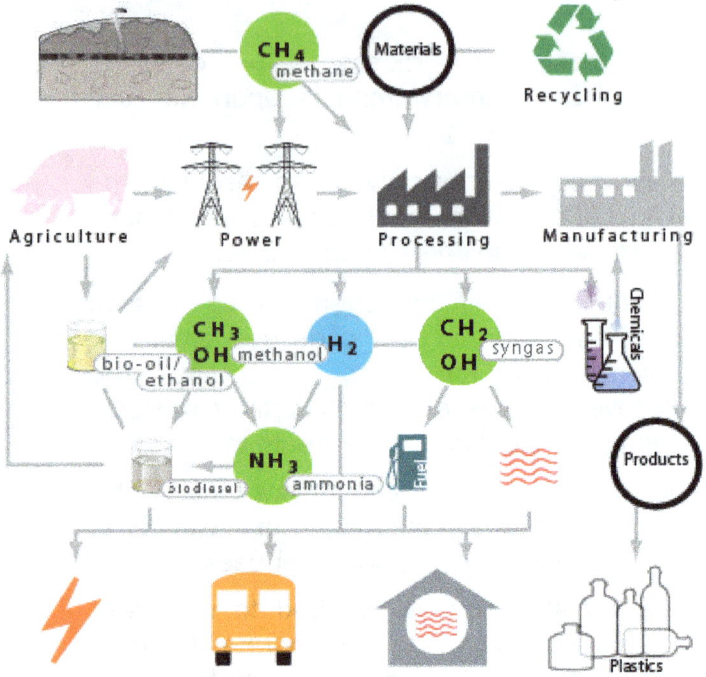

Methane from Landfills and Agriculture

Capturing methane from municipal landfills and constructed anaerobic digestion systems is a means to reduce natural gas consumption for energy, fertilizers and other products.

- As with all processes, the economic benefits relative to input costs vs. yields depend upon the scale and particular nature of applications:
  - Commercial waste companies are increasingly realizing that the large investments necessary for landfill site purchases, development and operations are providing dividends that can be expected to grow as natural gas prices continue to rise.
  - Agriculture businesses are converting plant biowastes and manure into fuel for space-heating and biodiesel-production to power equipment and, along with nitrogen from composting, into fertilizers.

250. The Pasty
252. U of Warwick

251. Subiaco
253. MSNBC

Costs of any and all energy alternatives should be weighed against unacceptable penalties of continued dependence upon rapidly dwindling domestic and global fossil fuel reserves.

- While renewable source development will require substantial up-front investments, long-term lifecycle economies reinforced by escalating fossil fuel costs can offer large economic, environmental, and social rewards:
  - Nuclear, wind, hydropower, geothermal, and solar, along with clean coal, can greatly reduce use of natural gas and petroleum for electricity generation.
  - Naturally-occurring organic processes can provide methane for biofuels, fertilizers and other derivatives.
  - More efficient transportation and industry technologies can conserve energy, and reduce environmental impacts.

254. Larry Bell

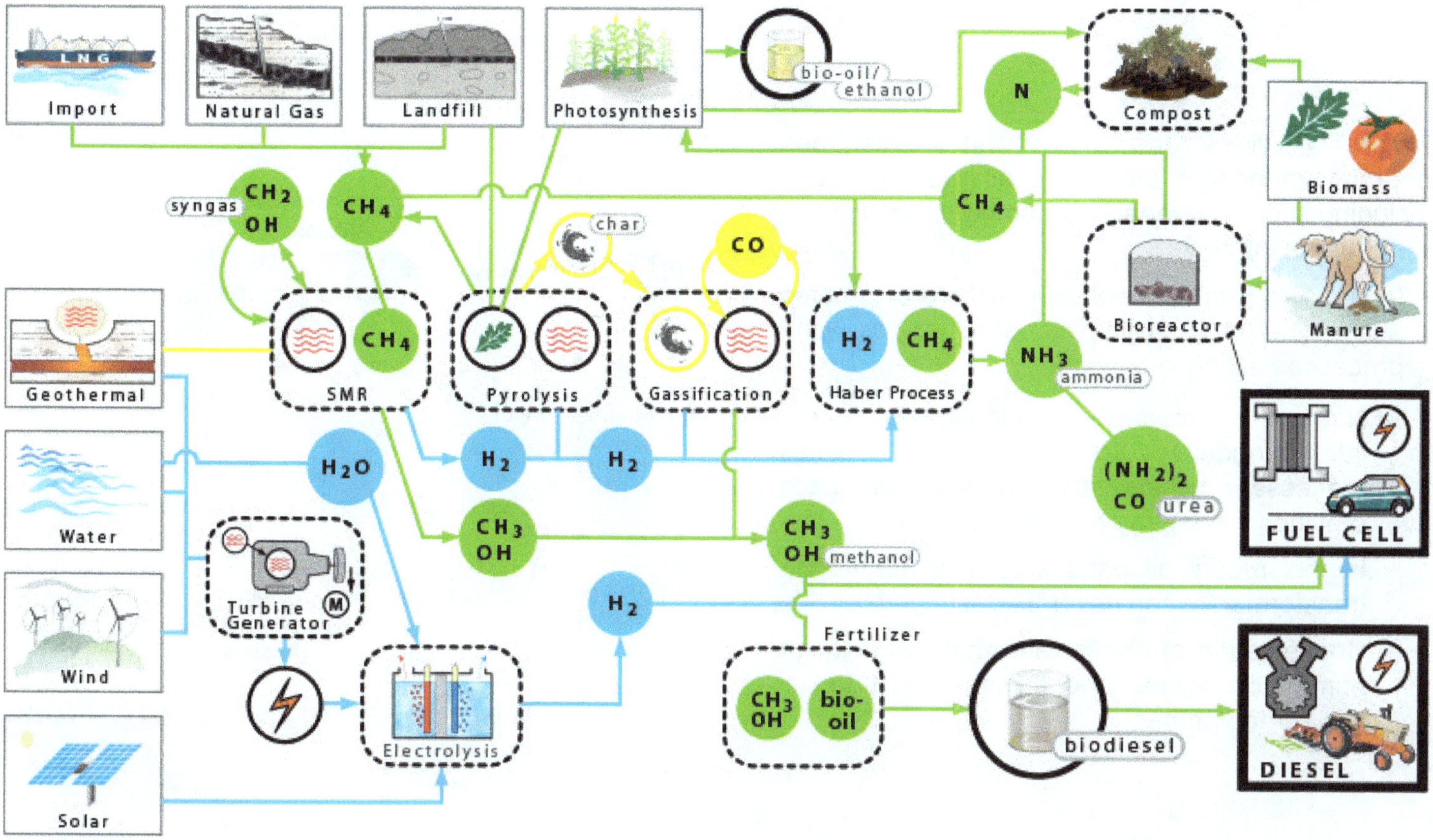

Unsustainable rates of fossil fuel depletion cannot be reversed by any single alternate energy source.

- Real progress towards energy conservation and greenhouse gas reduction can only be achieved through a combination of means:
    - Nuclear and renewable energy use must be greatly expanded.
    - Petroleum and natural gas must be shifted away from electrical power, and reserved for transportation and petrochemicals.
    - Wind, water, solar and geothermal development and use must be accelerated.
    - Biofuels and methane recovery must increasingly replace oil and natural gas for fuels and chemical feedstock.
    - All strategies require that we consume less and use resources more efficiently.

255. Larry Bell

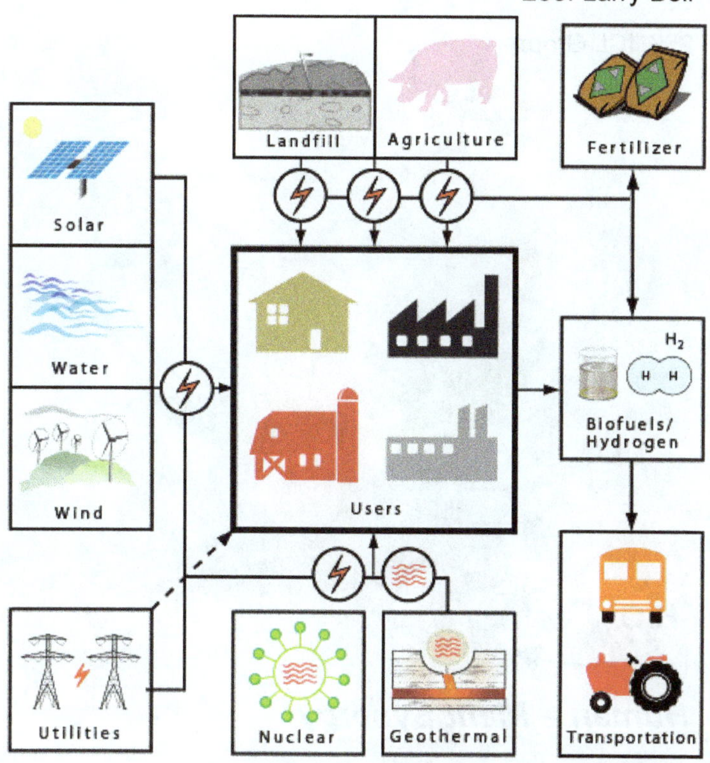

Combined Resource Utilization

Nature provides the basic ingredients and operating principles that make all systems work; including those we adapt to comply more closely with human interests.

- In Nature's grand model, all parts are connected and synergetic, with complex forces and processes acting in concert:
  - Nature's model demonstrates the interdependency of large and small processes, where the product of each provides the energy and feedstock for others.
  - In this model all parts are used, and while transformable, can never be destroyed.
  - Conservation of energy and material through human processes and devices must follow this same approach.

256. Robert G Woodmansee, Colorado State University

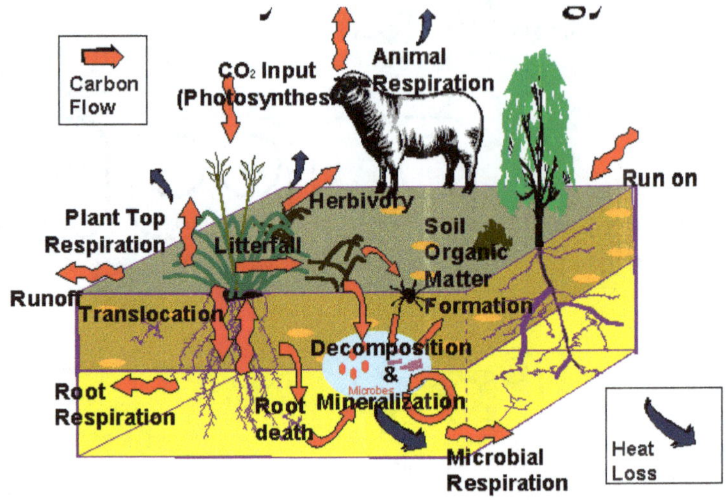

Carbon Cycle and Energy Flow

257. ICL Group

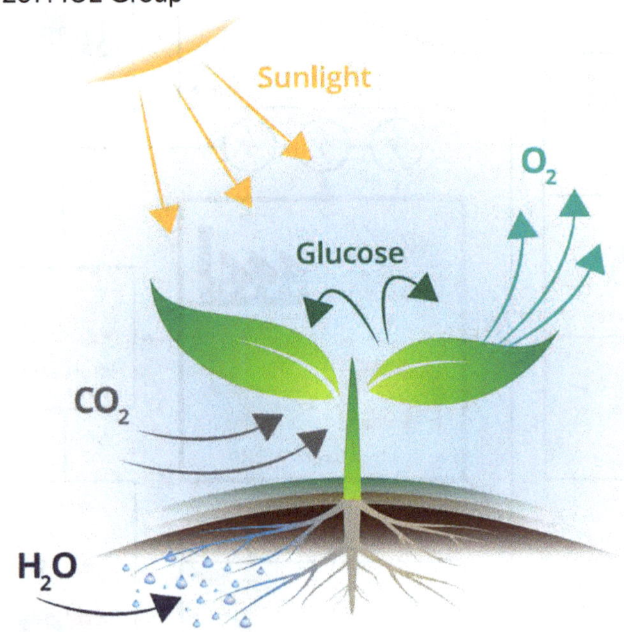

Human – Plant Synergy

Synergetic exchanges between plant and animal respiration processes are an example of the way Nature recycles energy and chemical materials.

- Green plants use photon energy to convert water and carbon dioxide into glucose carbohydrates for growth energy, and periodically release excess oxygen that we and other creatures require to live:
  - During the growth seasons, plants take up more CO_2 than they expel from sources such as the decomposition of biomass, various ocean and atmospheric sources, and fossil fuel combustion.
  - Then, during the rest of the year, plants release more CO_2 than they use, causing regional atmospheric levels to rise.

Viewed as synergetic systems from a human benefit perspective, we might regard plants as devices that process photonic and chemical energy into food and gaseous respiration products that we depend upon to exist.

- The processes of photosynthesis and plant respiration accomplish this in two stages that apply light-dependent and light-independent reactions:
  - During the day, chlorophyll mechanisms use stored ATP and NADP molecules to capture electrons and split water into oxygen and hydrogen components, producing NADPH in the reaction.
  - At night, light-independent ("dark") reactions use ATP and NADPH to convert CO_2 into sugar glucose carbohydrates, the chemical energy that fuels plant metabolism.

258. Larry Bell

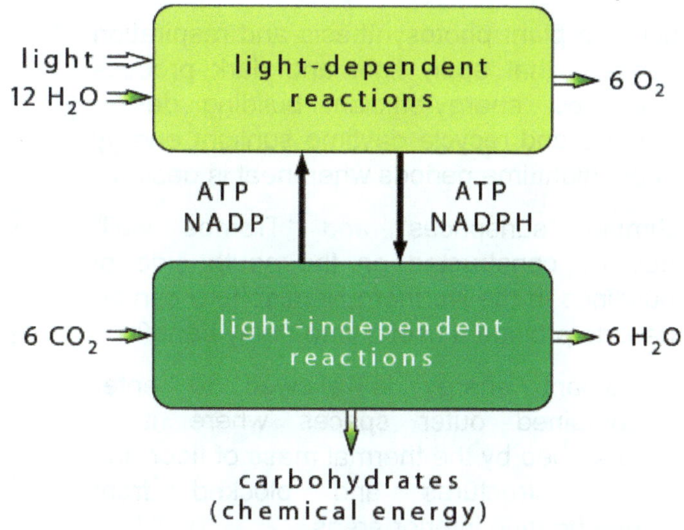

Plants use photosynthesis to convert light and water into oxygen during the day, and respiration (an opposite process) to convert CO_2 and NADPH into NADP, water and carbohydrate energy at night.

Plant Photosynthesis and Respiration

From a plant's point of view, the CO_2 that we and other natural sources produce can be perceived as a feedstock, along with water-derived oxygen, to enable it to grow.

- It is fortunate for us and other aerobically-supported life that the excess ("waste"), discarded oxygen, is recycled for other useful purposes:
  - Oxygen released into the atmosphere combines with hydrogen to be recycled back to leaves and soil in forms of humidity and rain.
  - Other plants grow, die and decompose, providing CO_2 that is recycled back for photorespiration by new plant life to perpetuate the cycles.
  - This time-tested model has proven success over many millions of years, and is one that we can apply in many ways.

259. Agri. Tech

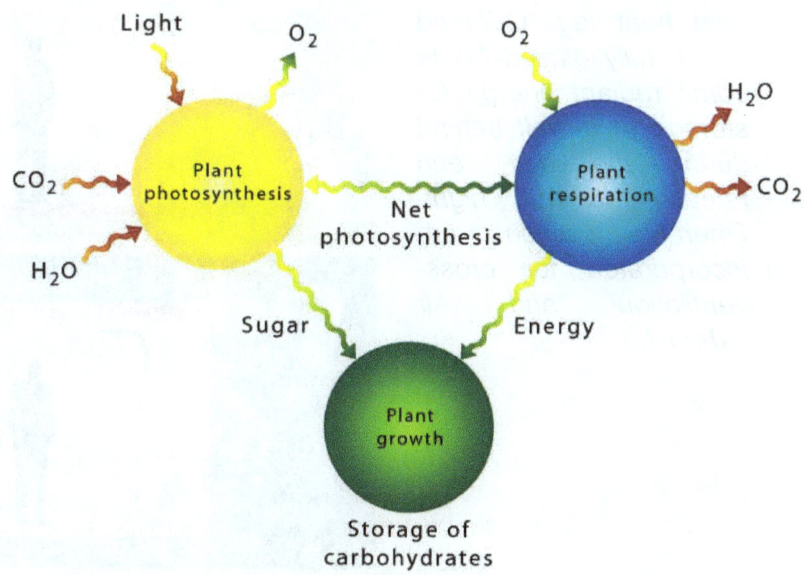

Plants use photonic energy to convert water and gases into chemical carbohydrate energy that they require for growth, and we other creatures require to survive, releasing excess oxygen in the process.

A Natural Conservation Model

Much like plant photosynthesis and respiration processes that apply light and dark process advantages, energy-efficient building design can store and recycle daytime sunlight energy for cool nighttime periods when heat is desired.

- Simple "sunspaces" and "Trombe wall" devices constructed on the south side of buildings in the Northern Hemisphere can be used to provide a useful time-delay benefit:
  - Radiant energy is allowed to enter contained outer spaces where it is absorbed by the thermal mass of floor and wall structures and blocked from overheating interior areas.
  - At night the warm mass radiates heat inside where it circulates by convection to conserve energy.

260. Energy Encyclopedia

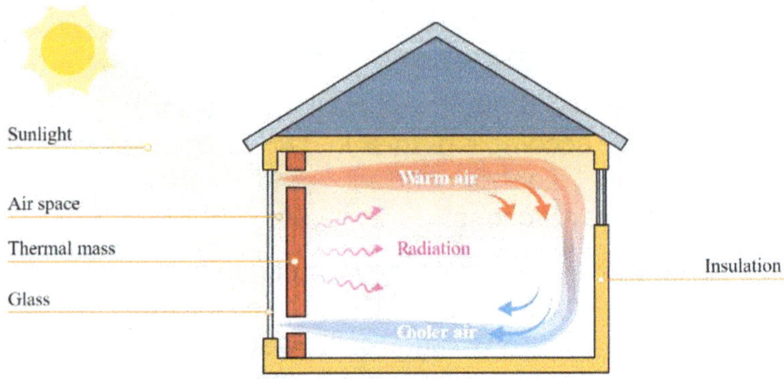

Heat Conservation in Building Design

Sunspaces and Trombe walls make use of solar gain heat lag. Oriented south, fully glazed fronts admit radiant energy for storage in a wall behind during the day, and release it at night. Operable sections are incorporated for cross-ventilation and air exhaust.

261. Heating, Cooling, Lighting, N. Lechner

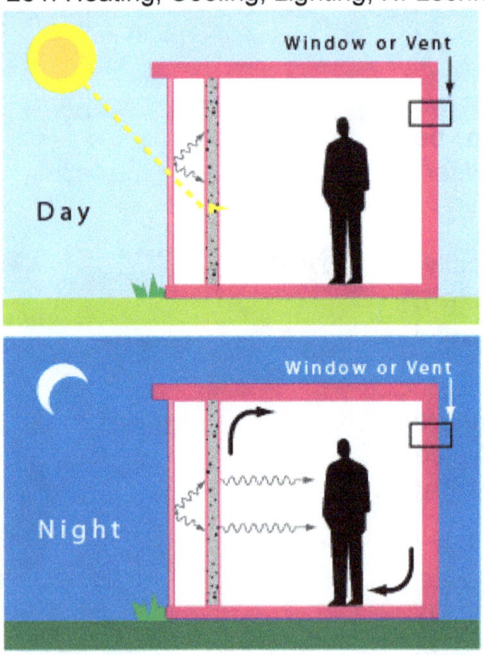

Trombe Wall Cycles

262. Heating, Cooling, Lighting, N. Lechner

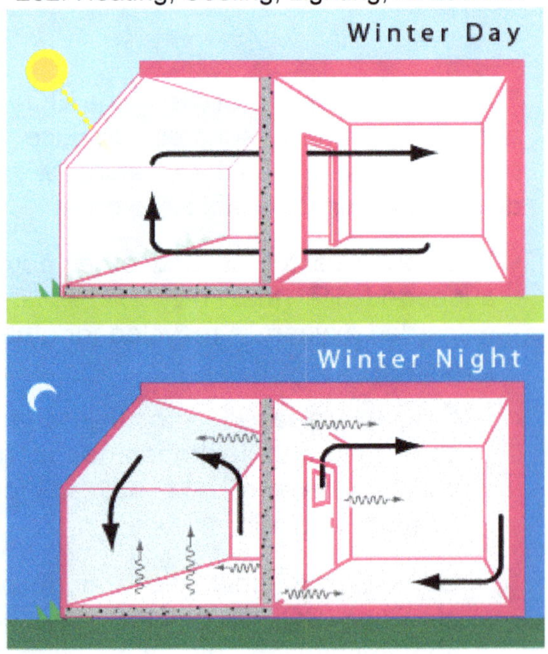

Daily Sunspace Cycles

263. Heating, Cooling, Lighting, N. Lechner

Sunspaces can be augmented by "solar furnaces" that eliminate or reduce energy required for water and space heating. South-facing collectors heat water or air that is circulated through storage tanks or rockbeds.

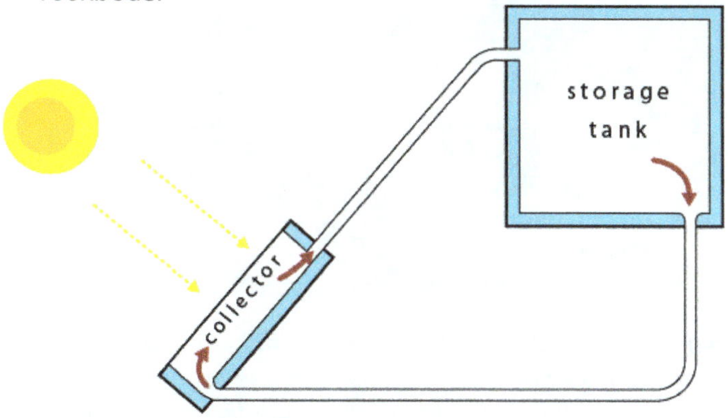

Solar Furnace Hot Water System

264. Heating, Cooling, Lighting, N. Lechner

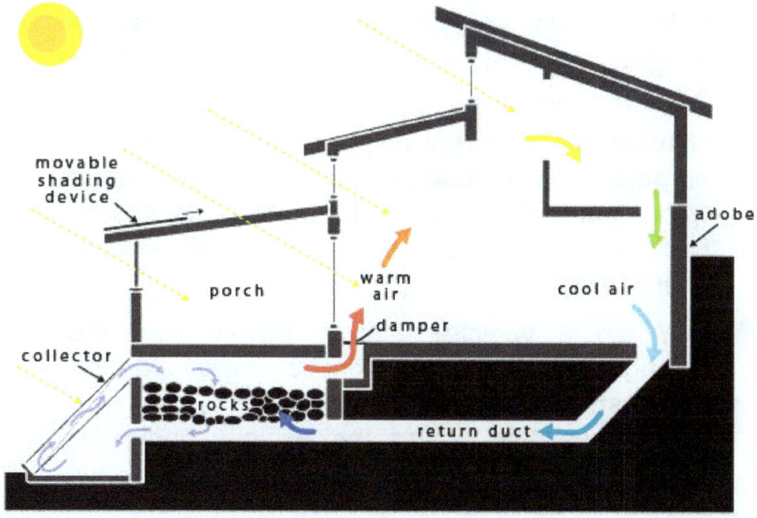

Hot air rises up from a storage bed to provide warm air at night, and cooled return air flows back to the collector in a loop. An operable vent on a north wall exhausts hot air and allows cross-ventilation.

Solar Furnace Space Heating System

265. Heating, Cooling, Lighting, N. Lechner

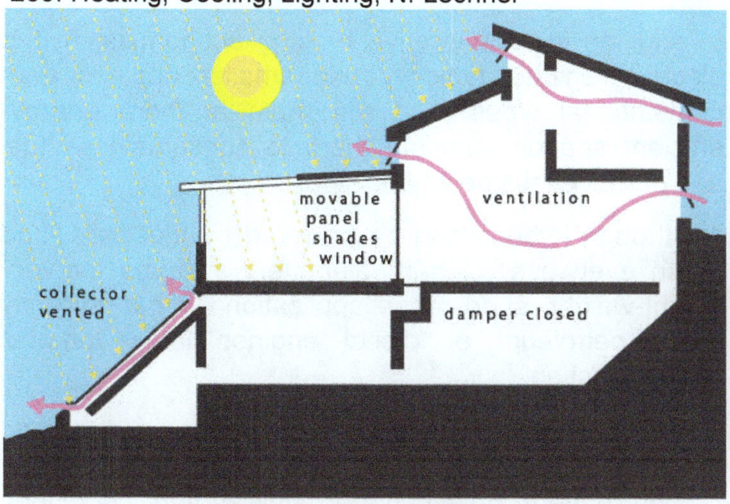

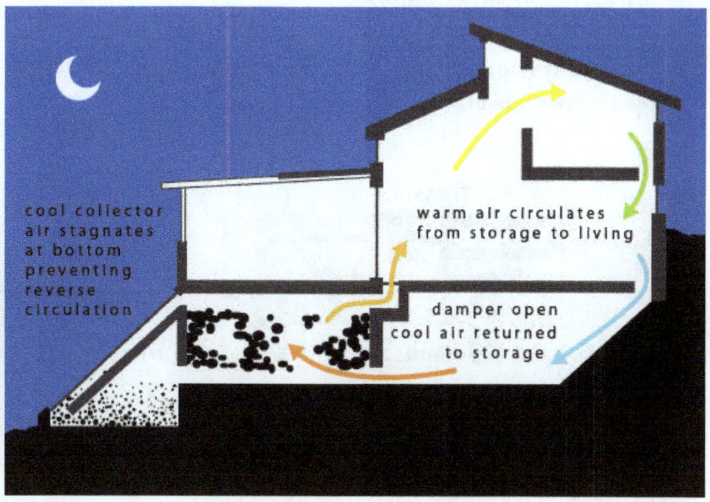

Summer Day

During the summer day, the sunspace is opened to the interior spaces.

- The rock storage vent is closed, causing the furnace to act as a thermal flue.
  - Hot air is exhausted form the furnace.
  - Cool air from the north side circulates through interior spaces by natural cross-ventilation.

Winter Night

During the winter night, the sunspace is closed from interior spaces.

- Rocks provide the heat collector:
  - The rock storage vent is opened, allowing warm air to circulate from this area throughout interior spaces by natural convection.

Hybrid electric vehicles that use regenerative braking to recycle product energy back into power are another example of natural economy.

- Hybrids use two separate power systems, usually an internal combustion engine in combination with a generator/motor unit that converts energy back and forth between kinetic and electro-chemical forms:
  - When a vehicle receives power from the engine, some of that kinetic energy can be transferred to generate electricity to recharge batteries that power the vehicle when the engine is less efficient.
  - The generator/motor unit also helps slow the vehicle when desired, again transferring the kinetic braking energy to a storable electrical form.

266. DIME C, University of Salerno

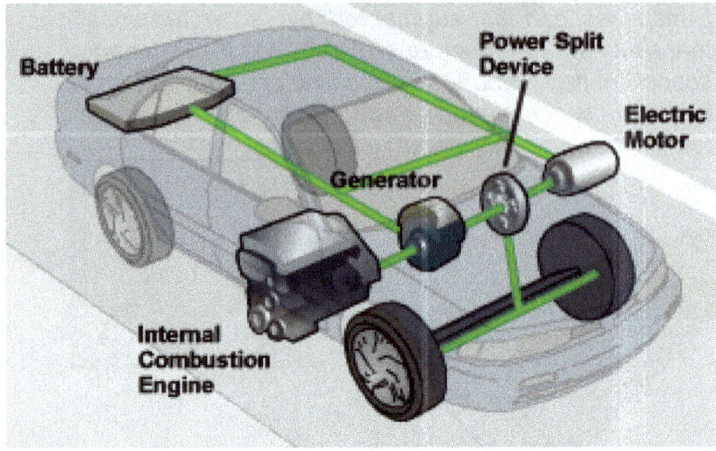

Hybrid electric vehicles configured with engines and electric motors connected in parallel can use regenerative braking to slow the vehicle and convert kinetic energy that would otherwise be wasted in the form of heat into storable electricity.

Recycling Kinetic Energy

267. University of Salerno

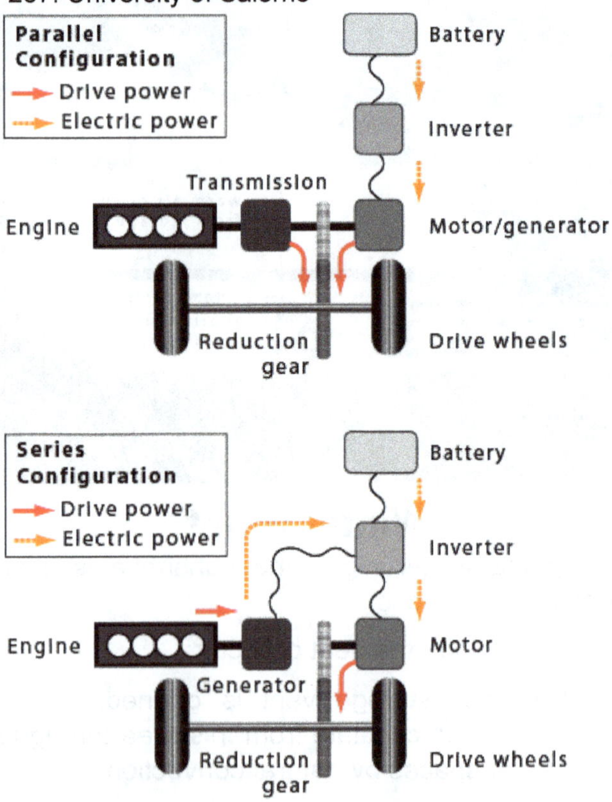

Hybrid Electric Vehicles

In addition to regenerative braking economies, hybrid electric vehicles typically use smaller engines than conventional types, and run them at more energy-efficient speeds, causing them to consume less fuel and produce less pollution.

- Although hybrids can be powered by fuel cells, and can even use "plug-in" all-battery systems for very light-weight, short-range applications, most currently use petroleum or diesel engines in a "parallel configuration":
  - Parallel arrangements separate the engine and electrical systems so that a vehicle can run entirely on battery power under slow urban conditions when an engine is not very efficient.
  - An alternate "series configuration" uses an engine only to recharge an electrical drive system, eliminating a regenerative braking option.

268. theade

Combined power and heat (CHP or "cogeneration") is still another simple energy recycling example that applies natural conservation principles.

- CHP is becoming commonly used to generate electricity and heat from a single fuel source, and to capture waste heat for a variety of other uses:
    - In some cases the "waste" output heat is recycled to reduce energy required to sustain the electricity-generating process.
    - In other instances the excess heat is captured and used to provide hot water and steam, to drive air conditioning systems, or is exported for other production processes.

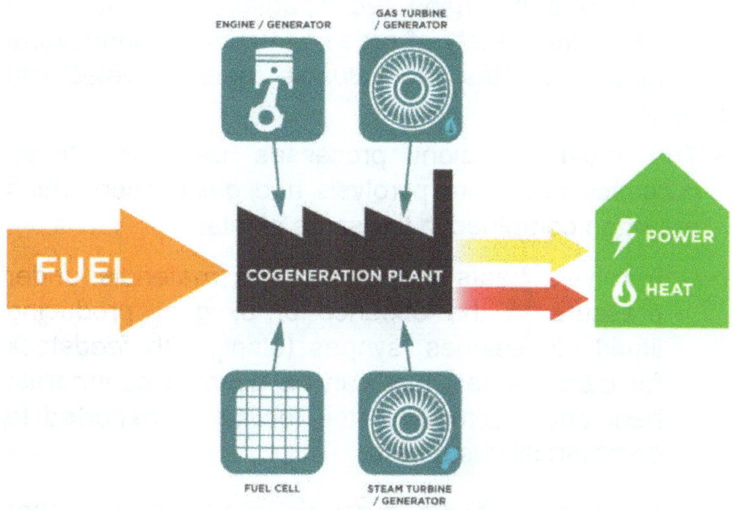

CHP Energy Recycling

269. Kelcroft E&M Consulting Engineers

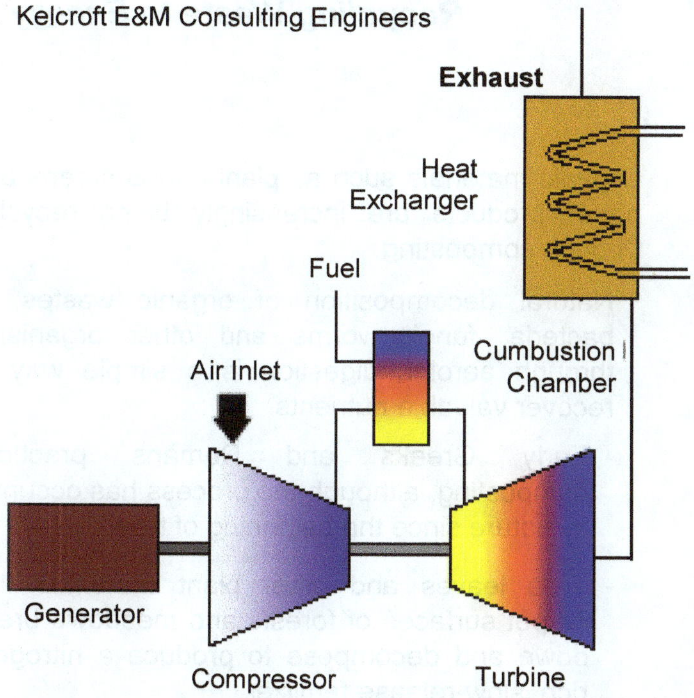

Cogeneration plants can yield efficiencies of 85% or more, reducing energy consumption, CO_2 pollutants and power transmission losses.

Cogeneration Benefits

CHP cogeneration plants can be powered by a variety of different fuels, with high output efficiencies.

- By recycling large amounts of heat energy that conventional fossil fuel plants produce, CO_2 combustion products are minimized.
    - Cleaner operation enables these systems to meet stringent environmental standards.
    - Cleaner operations also enable the plants to be located closer to population centers where power is needed, resulting in shorter transmission distances with reduced line resistance losses.

270. Larry Bell

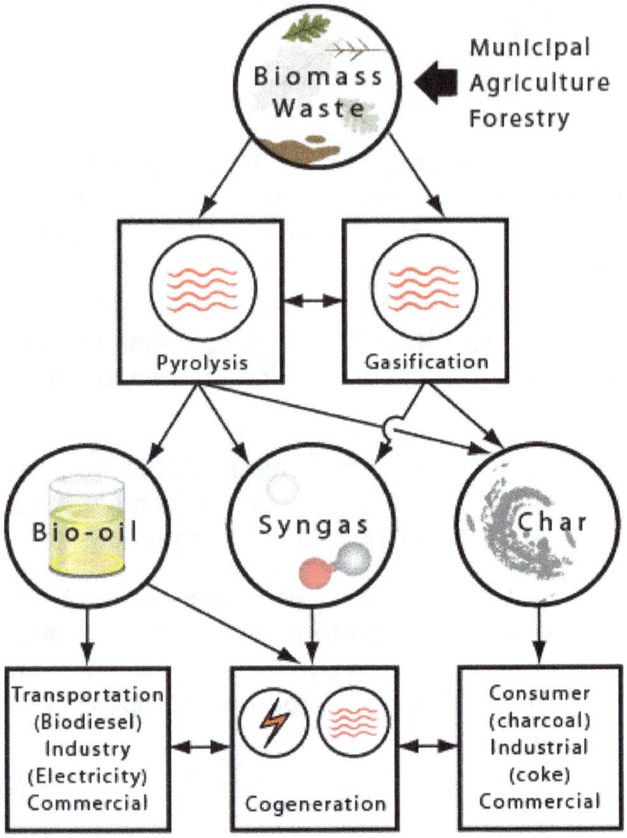

Recycling Waste to Energy

Cogeneration can be used in combination with other available and emerging technologies to convert and recycle a wide variety of organic municipal, agricultural and forestry wastes into valuable fuels and electrical power.

- The most prevalent processes used for these purposes today are pyrolysis and gasification which are often combined in the same plants:
  - Pyrolysis heats the carbon-rich materials under pressure in the absence of oxygen, producing liquid oil residues, syngas (along with feedstock for carbon char) that can be used to cogenerate heat and electricity in the facility, or exported to commercial customers.
  - Gasification heats pyrolysis products at higher temperatures in an oxygenated environment to also produce syngas that can be used for cogeneration, along with charcoal and coke which can be exported.

271. Larry Bell

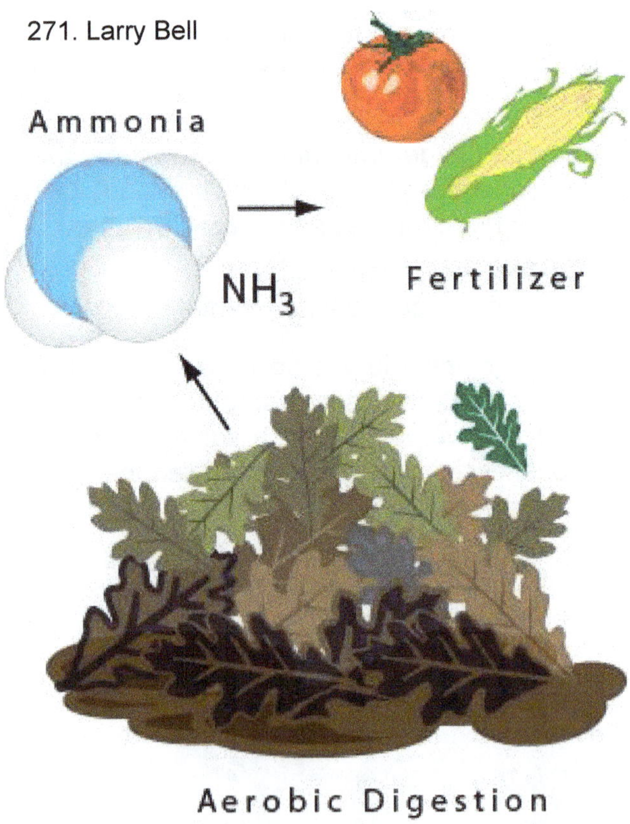

Recycling Waste into Nutrients

Organic materials such as plants, food scraps and paper products are increasingly being recycled through composting.

- Natural decomposition of organic wastes by bacteria, fungi, worms and other organisms through aerobic digestion is a simple way to recover valuable nutrients:
  - Early Greeks and Romans practiced composting, although the process has occurred in Nature since the beginning of time.
  - Tree leaves and other plant materials that carpet surfaces of forests and meadows break down and decompose to produce a nitrogen-rich, slow-release fertilizer.
  - The same processes are employed in environmentally-sound gardening practices.

Conservation principles we can apply using Nature's connected and synergetic lessons are not limited to certain technical categories, industry sectors or geographic scale boundaries, but extend across broad aspects of life and enterprise.

- By observing the workings of Nature, we can become more aware of larger potentials, and more fully foresee natural consequences of our choices and actions:
  - We can consider ways that processes and products of different industry and business sectors can be beneficially coordinated.
  - We can recognize possibilities for managing resources and production infrastructures more effectively at national, regional and local levels.
  - And we can examine opportunities for applying sensible and rewarding conservation practices in our everyday lives.

272. Larry Bell

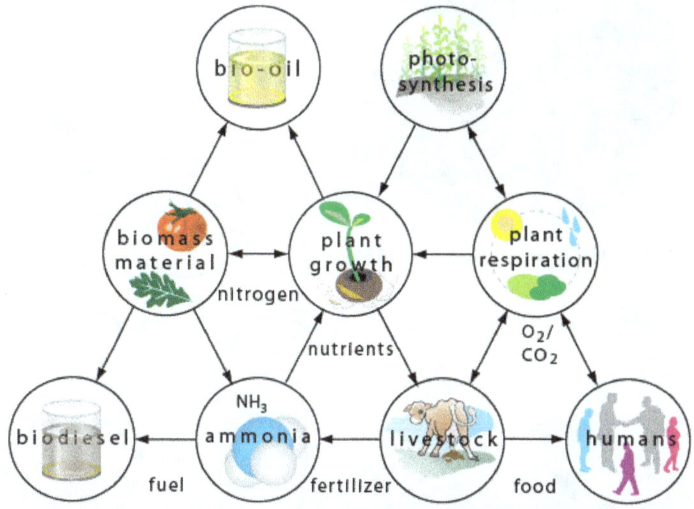

In Nature, all parts and processes connect and interact in diverse and complex ways. Broader awareness of the interrelationships can provide better understanding of conservation opportunities, and the consequences of choices and actions.

Outgrowths of Photosynthesis

273. Larry Bell

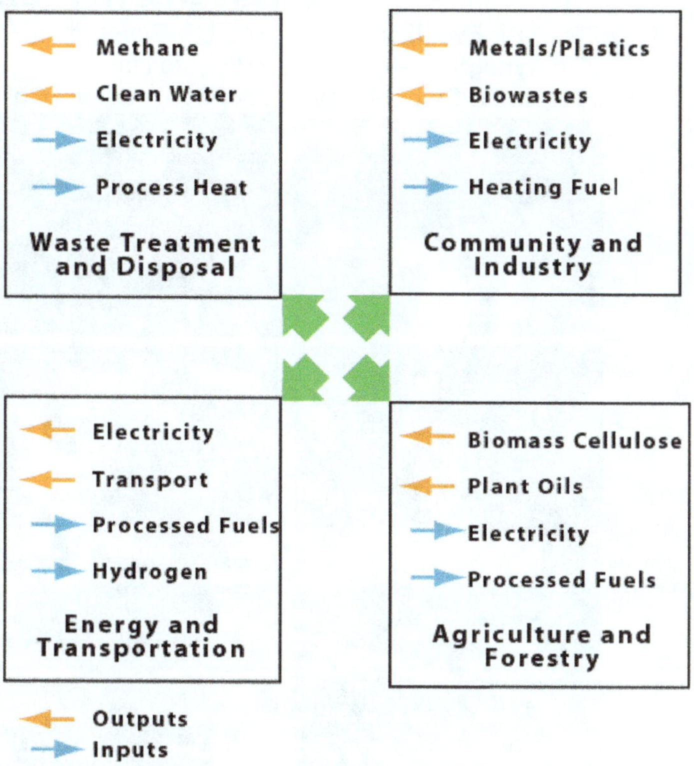

Conservation Stakeholders

All public and business sectors have important stakeholder interests in clean resource development, use and conservation.

- Coordinated networks connecting these sectors afford opportunities to combine capabilities and exchange services for mutual and far-reaching benefits:
  - Various organizations and activities generate waste outputs that can be used as input feedstock for others.
  - Private and corporate energy consumers can create markets for vehicles and heating systems that use cleaner, more efficient and sustainable fuel sources.
  - Transportation and utility industries can respond to consumer demands with energy-saving products and distribution infrastructures.

274. Larry Bell

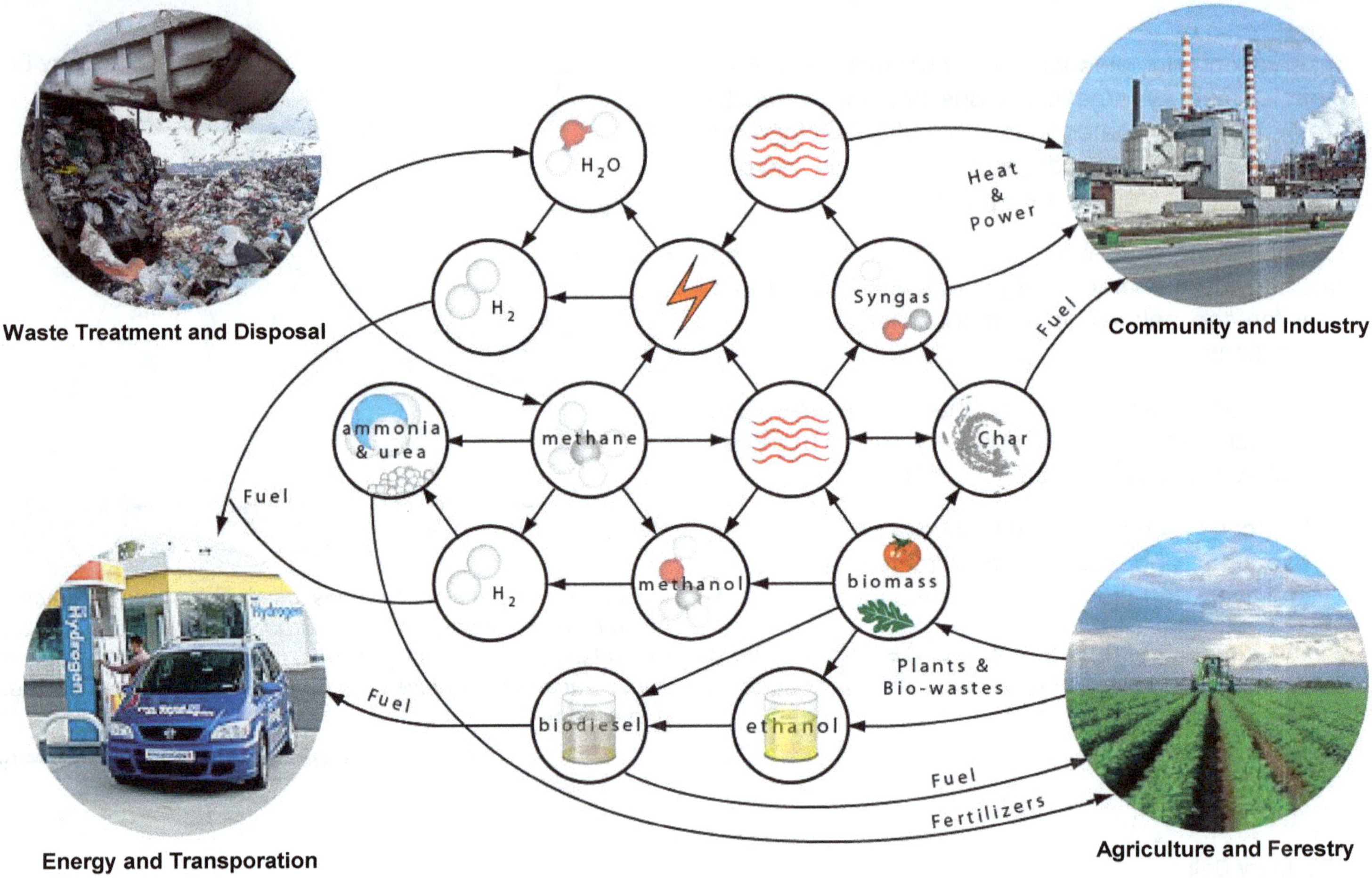

A Resource Network

275. SKF Evolution 276. US Government
277. Tynagh Energy 278. Auto Junction
279. Ex.Br. Student Gov. 280. Ogata Lakota Coll.

The MECOPTICS framework emphasizes conservation principles that influence appropriate selection and use of available resource options.

- To facilitate clearer understanding of alternatives, it can be helpful to consider basic ways that key processes and devices convert resources into useful forms and outcomes, their capabilities, and their limitations:
  - "Processes" refer to methods through which energy and materials interact in a controlled manner to produce electrical power, gases and chemical compounds that we desire through natural and human-assisted means.
  - "Devices" refer to natural and human systems that enable processes to occur.
  - Most generally, these processes and devices work in concert with others to accomplish connected and synergetic functions.

281. Larry Bell

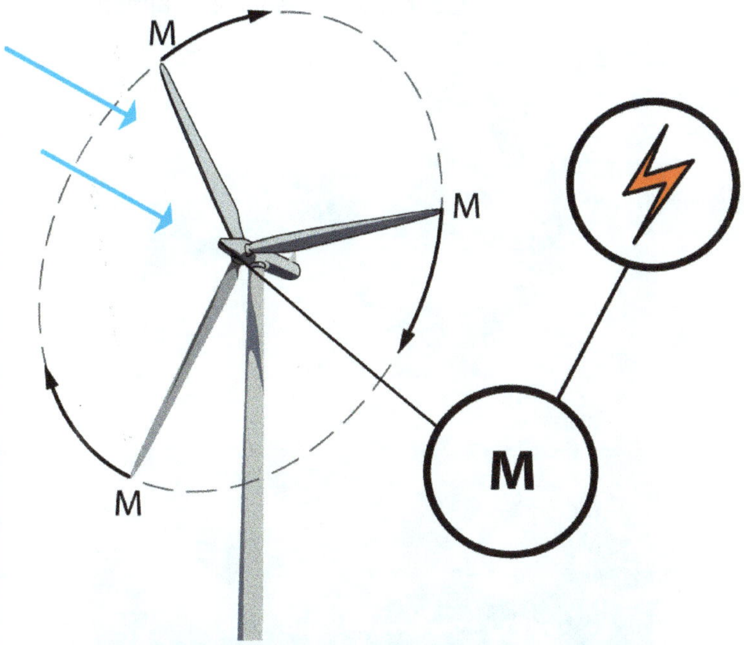

Since the theoretical energy generated by a wind turbine equals the cube of wind speed, small increases in wind speed produce large increases in power.

Wind Turbines

Wind turbines are devices that capture kinetic energy with propeller-like blades mounted on a rotor and convert it via generator devices into electrical power.

- The blades act much like airplane wings:
  - When the wind blows, low-pressure air forms on the downward side of the blades creating "lift".
  - The lift force is actually much stronger than the wind's force against the front of the blades (called "drag").
  - The combination of lift and drag causes the rotor assembly to spin like a propeller, turning a generator shaft to create power.
  - The theoretical amount of energy generated is a function of the cube (third power) of wind speed, so that if wind speed is doubled, power increases by a factor of eight (although in actual practice it is less).

Approximately 2% of the Sun's energy reaching Earth is converted to wind energy (about 50 – 100 times more than the energy converted to biomass by all plants).

- The Earth's surface heats and cools unevenly, creating atmospheric pressure zones that cause air flow from the high- to low-pressure areas:
  - Daily and seasonal changes have important influences on wind conditions.
  - In Iowa (for example) seasonal winds are strongest in early spring and weakest in summer; daily winds are usually strongest in the afternoon and lightest in early morning.
  - Strongest winds generally prevail out of the southwest to northwest.

282. Natural Resources Canada

Upward Air Currents

Putting wind power to use is not a recent idea.

- Civilizations have been harnessing wind energy from the beginning of recorded history:
  - Wind propelled boats on the Nile in 5,000 BC.
  - By 200 BC, simple windmills in China were pumping water, and vertical-axis windmills ground grain in the Middle East.
  - By the 11th century, windmills were used extensively in the Middle East for food production, and returning merchants and crusaders brought the technology back to Europe.
  - The Dutch adapted windmills for draining lakes and marshes in the Rhine River Delta.
  - Early settlers introduced the technology to America in the late 19th century, first to pump water, and later, to provide electricity.

283. Natural Resources Canada

Traditional Rural Windmill

284. Derek Jensen

Small Independent Unit

Contemporary wind turbines range in size from small ones used in rural areas to supplement electricity from public grids, to large units that are linked together for multi-megawatt, utility-scale applications.

285. SatCon Technology

Wind Farm

Wind turbines come in different blade orientations and configurations.

- Modern systems generally fall into two basic types of groups:
  - Horizontal-axis types include three-bladed versions located upwind with blades facing into the wind, and two-bladed versions located downwind.
  - Vertical-axis turbines include an eggbeater-style Darrieus model (named after its French inventor).
  - Horizontal-axis types are commonly used today, and are the primary focus of current US Department of Energy (DOE) research and development.

286. American Wind Energy Association

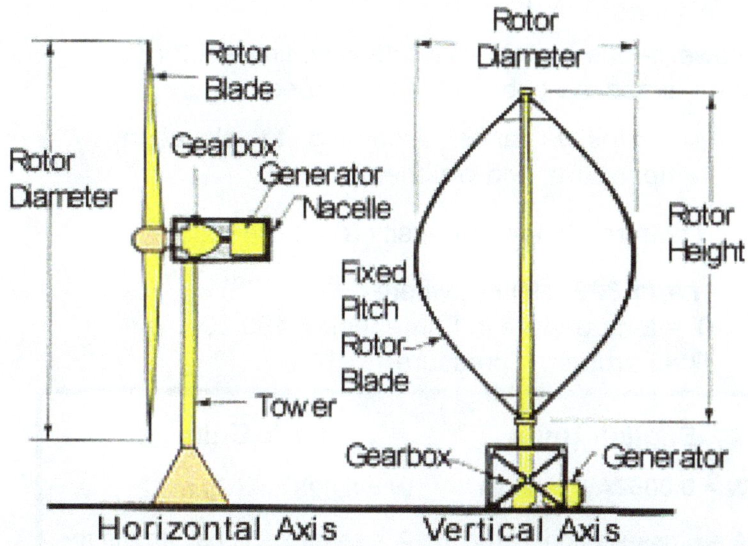

Wind Turbine Configurations

287. MIT

Familiar Horizontal-Axis Types

288. DOE-EERE

Vertical Axis System

The theoretical amount of available wind power is determined by the equation **W = ½rAv³**, where:

W = power; **r** = air density; **A** = rotor area; **v** = wind speed

- This means that:

Power = one-half times air density, times rotor area, times the cube of the wind speed:

- Air density varies according to elevation, temperature, and weather fronts
- The formula for air density (r) is:

r = (1.325 × P) / T, where:
T = temperature in Fahrenheit + 459.69
P = barometric pressure

English Units	Metric Units
W = 0.0052Av³ where:	**W = 0.625Av³** where:
A = cross-sectional area in ft^2 swept by blades	A = cross-sectional area in m^2 swept by turbine blades
V = wind speed in miles/hr	V = wind speed in meters/sec

Power Measured in Watts

289. Dragonfly Energy

Air density (the number of molecules in a unit volume of air) influences the mass that moves turbine blades. Density decreases slightly with humidity and altitude. Air is more dense in winter than in summer, producing more power with the same wind speeds. Density is also less at high mountain altitudes (lower air pressure), but wind speeds are higher

Influences on Rotor Power

290. Iowa Energy Center

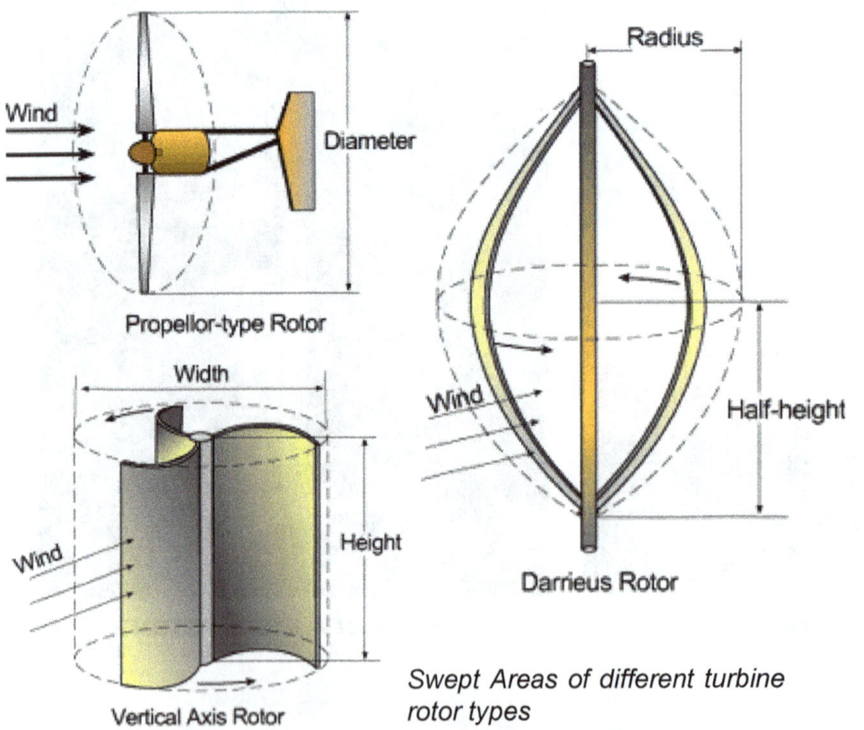

Swept Areas of different turbine rotor types

Wind Turbine Rotor Influences

Although the power equation shows an exponential increase with wind speed, the actual power increase is more linear than predicted.

- Since turbines are not perfectly efficient, the power curve associated with each turbine is most significant:
  - Since the area swept by turbine blades is circular, **Swept rotor area A = πr².**
  - Assuming a wind speed of 15 mph, standard air density, and a rotor radius of 10 ft, the wind power (English units) is:

W = $0.0052Av^3$
W = $0.0052(314ft^2)(15\ mph)^3$
W = 0.0052(314)(3375)
W = 5,511 watts (5.511 kilowatts).

Most commercial-size wind turbines used today are 3-blade horizontal-axis types with 164 - 328 ft (50 - 100 m)-long blades weighing up to 12,000 pounds (5,443 kg) each or more, and producing 1.5 - 3 MW of electricity.

- The turbines are often grouped at distances of 5-15 times blade-diameter apart:
  - Mechanical wind energy converted into rotational power is transferred to generators that produce electricity which can be used directly, fed into a transmission grid, or stored in batteries.
  - Mechanical transmissions keep the generators operating efficiently throughout a range of wind speeds.

291. International Network for Sustainable Energy

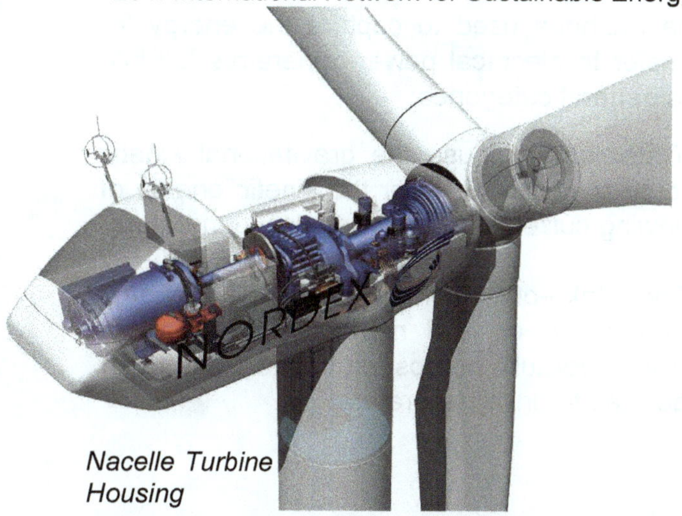

Commercial Horizontal Axis Turbine

Large wind turbines have a variety of mechanical and electrical components that convert kinetic energy to electricity.

- Most of these mechanisms are contained in a housing called a "nacelle":
  - **Anemometer**: measures wind speed and sends data to a controller.
  - **Controller**: starts up/shuts off the machine at maximum speeds.
  - **Generator**: often an AC 60-cycle induction type.
  - **Rotor assembly**: includes the blades that capture wind energy and a spinning hub.
  - **Gears**: connect low-speed and high-speed shafts that drive the generator.
  - **Wind vane**: orients the turbine to the wind using a yaw drive.

292. USDOE-EERE

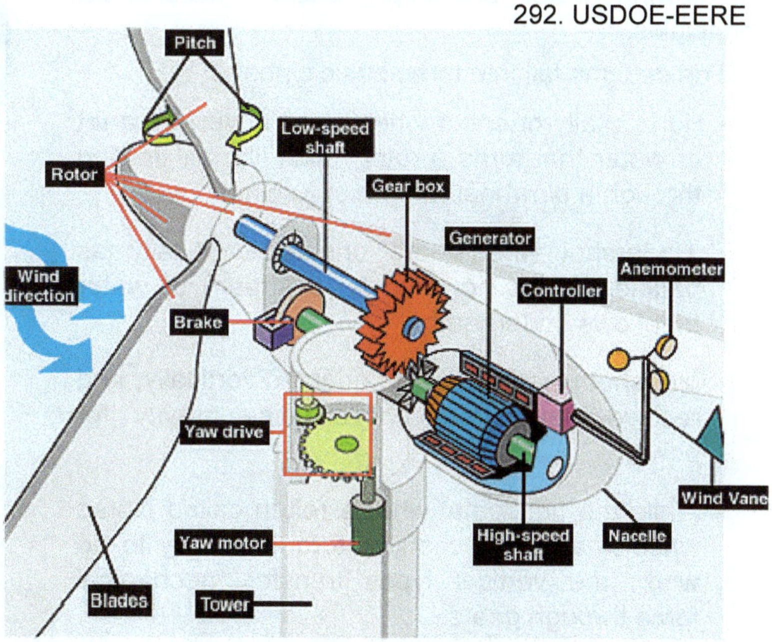

Internal Wind Turbine Components

Hydropower harnesses mechanical energy from moving water produced by dams, rivers, ocean tides and waves to generate electricity for local and commercial uses.

- The methods used to capture the energy for transfer to electrical power generators fall into two general categories:
  - Water turbines use the gravitational kinetic force of falling water or the kinetic energy of flowing currents to drive electrical generators.
  - Power take-off systems convert kinetic energy from wave or tidal motions into electricity using hydraulic pumps, air turbines and other devices to drive generators.

293. Larry Bell

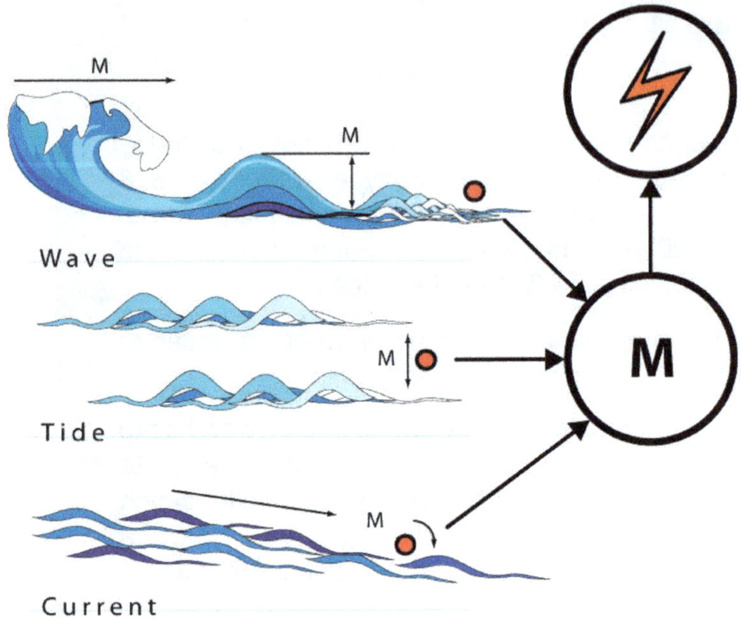

Hydro-mechanical Sources

Contemporary hydroeletric systems that use falling or flowing water still apply ancient waterwheel principles.

- The designs fall into three basic types:
  - Horizontally-oriented wheels are driven by a jet of water that turns a rotor much like air flowing through a pinwheel (a form of turbine).
  - Undershot wheels are oriented vertically as paddle wheels powered by a stream of water that flows underneath.
  - Overshot wheels are also oriented vertically, and receive water from above using gravity for power.
  - While the horizontal wheels (often called Norse wheels) apply mechanical torque directly to do work, the vertical types transfer mechanical force through gears.

294. Water History

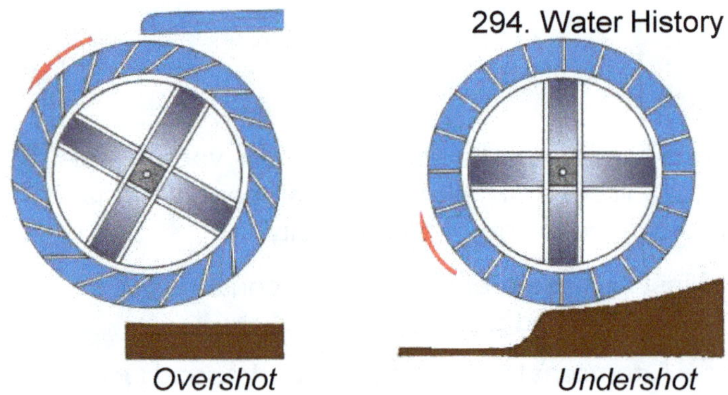

__Overshot__: use an elevated stream and require gears.

__Undershot__: are driven from below and require gears.

__Norse__: apply power directly

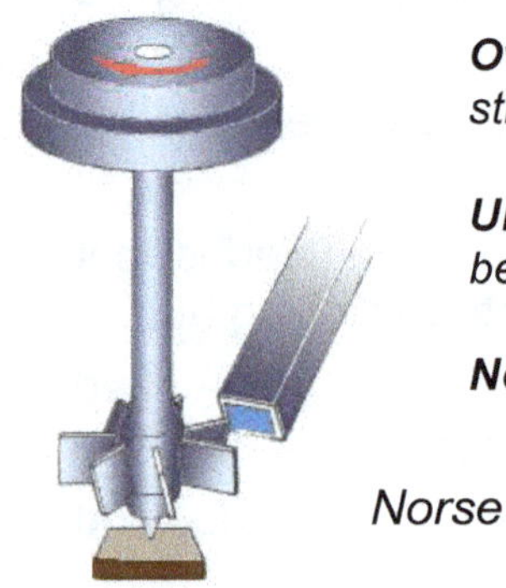

Waterwheel Types

As with other power sources, hydroelectric resources can be measured according to the amount of available power (or energy) that they can produce in a unit of time.

- In the case of energy produced by falling water, power is a function of "hydraulic head" and rate of flow:
  - For a reservoir, the hydraulic head is the height of the water level relative to the height after the water is evacuated.
  - Each unit of water can produce a quantity of work equal to its weight times the head, where:

The amount of energy (E) released by lowering an "object" of mass (M) by a height (h) in a gravitational field is:

E= mgh (where g is acceleration due to gravity)

295. Energy Encyclopedia

Hydraulic Head in a Reservoir

Modern turbines are similar to waterwheels, but accelerate by water pressure force that is directed onto blades of a spinning "runner" to transfer energy that drives a generator.

- The main difference between water turbines and waterwheels is the swirl component of water which passes energy to a spinning rotor:
  - The evolution from waterwheels to modern turbines took about 100 years, and rapidly advanced during the Industrial Revolution.
  - Their principle advantage over waterwheels is the fact that they can be much smaller, yet produce substantial power.
  - Since they turn much faster than waterwheels, they can harness greater "heads" to process water more effectively.
  - Very small turbines are ideal to provide residential and supplementary power from streams at low cost.

296. Army Corps of Engineers

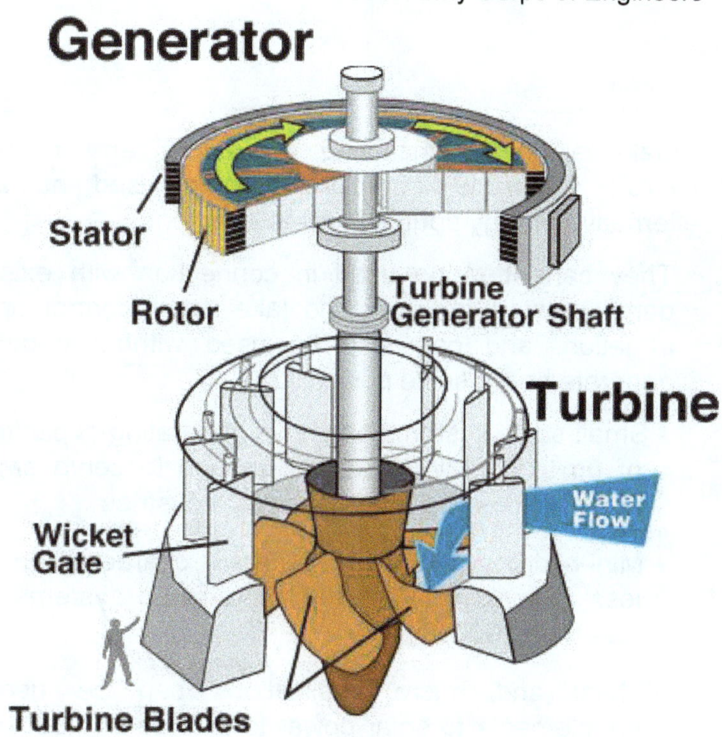

Commercial-Scale Water Turbine-Generator

Very small turbines can provide residential and supplementary power from streams at low cost.

- Francis micro-hydro turbines can operate with as little as 10ft of water head:
  - Water circulates through the center (inward flow) and provides energy through combined radial and axial flow.
  - Water enters through a spiral-shaped inlet and acts upon vanes (or a wicket gate) causing the runner to spin.
- Banki micro-hydro turbines operate much like overshot waterwheels:
  - The middle is left open, causing the water to pass through and accelerate by a venturi effect.
  - Very inexpensive home-made Banki turbines can produce up to 2 amps at 12vDC with only 3ft (1m) of water head.

297. Voith Siemens; European Small Hydropower Association

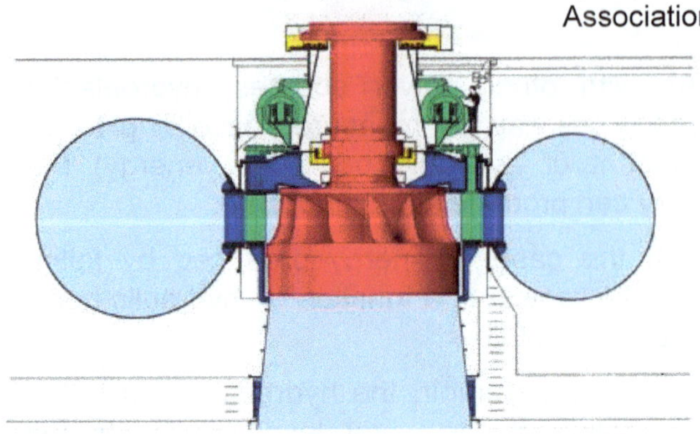

Francis Turbine

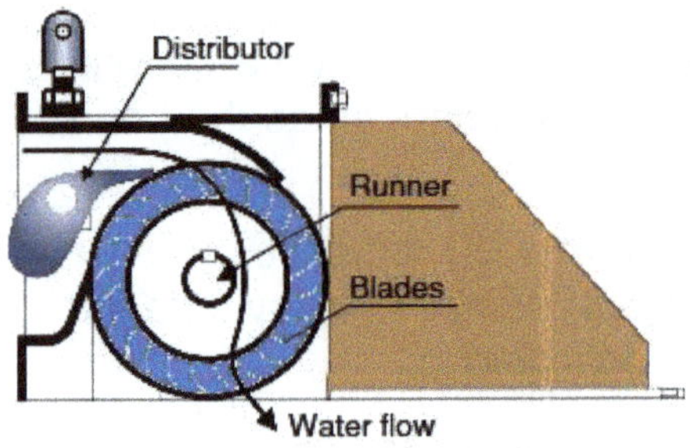

Although the efficiency of a Banki turbine is lower, it enjoys a niche market for low-cost and home-made applications.

Banki Turbine

Small-scale hydro (including "mini-hydro" and "micro-hydro") power is increasingly being used as an alternative energy source in remote regions.

- They can often be used in connection with existing dams created for river and lake water control or for irrigation, and can also be used without dams to generate home-scale power:
  - Small-scale systems can have generating capacities of up to 10 MW or more, enough to serve small communities and medium-sized industries.
  - Mini-hydro systems are generally characterized as less than 1,000 kW, and micro-hydro systems as less than 100 kW.
  - Mini and micro applications can be useful supplements to solar power to provide 24 hour/day capabilities.

298. Forcefield

This home-made squirrel cage waterwheel generator in a small stream produces a steady 1-2 amps of power 24 hours/day.

Small Scale Hydro Applications

299. World Ocean Review

Relatively new technology uses tidal stream generators that draw energy from underwater currents in a manner similar to generators powered by wind. "Seagen" in Northern Ireland uses sets of rotor turbines rated at about 1 MW that can be raised above the surface for maintenance.

The Seagen Tidal Generator

300. Engineering Business (EB)

The Stingray uses the flow of the tidal stream over a hydroplane to create an oscillating motion that operates hydraulic cylinders that drive motor generators. The devices are mounted to the seabed at depths of up to 330 ft (100m).

The Stingray Tidal Flow Generator

301. SMD Hydrovision

SMD Hydrovision's TidEL device consists of a pair of contra-rotating 500 kW turbines mounted together on a crossbeam. Mooring chains that tether the buoyant assembly to the seabed, allow the turbines to align to the prevailing tidal flow.

The TidEL Tidal Flow Generator

302. Dicander & Co.

Wave Power

Harnessing power from wave motion may eventually yield more energy than tides.

- New technologies are being developed involving generators that can be either coupled to floating devices or turned by air displaced by waves:
  - Although sometimes co-mingled with tidal power, wave power is more continuous since it is produced by ocean currents.
  - Wave power is determined by wave height, wave speed, wavelength and water density.
  - Wave size is determined by wind speed, "fetch" (distance over which wind excites the water), and seafloor depth and topography.
  - A given wind speed has a practical limit over which time or distance will not produce larger waves (a limit called a "fully developed sea").

303. Larry Bell

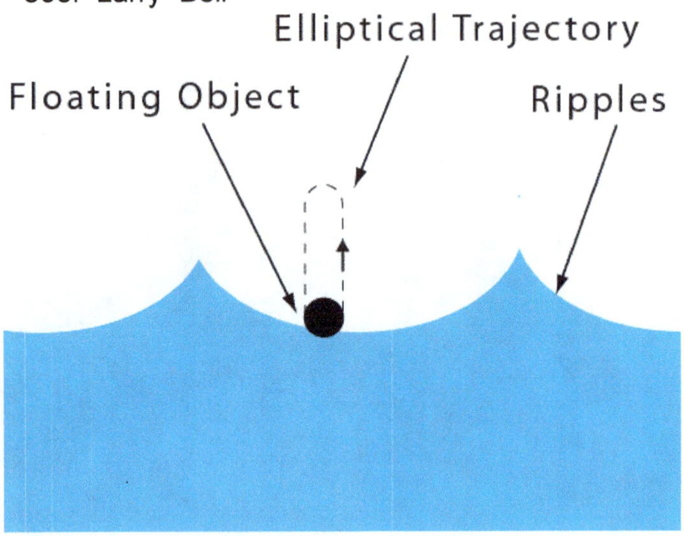

When an object bobs up and down on the surface of water it experiences an elliptical trajectory. The energy transferred to the object from the water's movement can be harnessed to do useful work (such as to drive an electrical turbine generator).

Harnessing Wave Energy

North and south temperature zones offer the best wave power locations, with prevailing westerlies in winter blowing strongest.

- Wave motion is highest at the surface, and diminishes exponentially with depth, although wave energy is also present as pressure waves in deep water:
  - The potential energy of a set of waves is proportional to wave height squared times wave period (time between wave crests), and is equal to kinetic energy.
  - Wave power is expressed as kW/m, according to the formula:

    Power=$KH^2T \approx 0.5H^2T$, where:

    K=constant, **H**=height (m), and **T**=period (sec)

Although wave power generation is primarily in experimental stages of development and implementation, it can be accomplished by a variety of methods.

- Some representative system types are characterized according to the ways they capture wave energy, by their locations of use, and by the way energy is converted to power:
  - Wave energy capture methods include wave power point absorbers (small area devices), wave power attenuators (aligned parallel to wave propagation), and wave power terminators (aligned perpendicular to propagation).
  - Locations include shoreline, offshore and deep water applications.
  - Power conversion includes hydraulic rams, elastomeric hose pumps, pump-to-shore, and hydroelectric or air turbines.

304. BWEA Marine
305. Wavegen

The Pelamis Wave Energy Converter is a point absorber that is flexibly moored near shore to swing head-on into incoming waves.

Wavegen's Land Installed Marine Powered Energy Transformer captures wave energy along a shoreline in Scotland.

Shoreline Wave Energy Converters

306. Ocean Power Delivery

The Pelamis is a semi-submerged, articulated structure composed of cylindrical sections linked by hinged joints. Wave-induced motions of the joints are resisted by hydraulic rams that pump high-pressure oil through hydraulic motors that drive generators to produce electricity. Many sections can be tethered to the shore by a single cable.

The Pelamis Wave Energy Converter

307. Wave Dragon

Wave Dragon consists of collectors that channel waves towards a double-curved ramp structure to create a water storage reservoir. A set of propeller turbines convert water head in the reservoir into electricity. The design is unique among wave energy converters because it produces electricity through a one-step system using turbines.

The Wave Dragon

308. University of Oregon

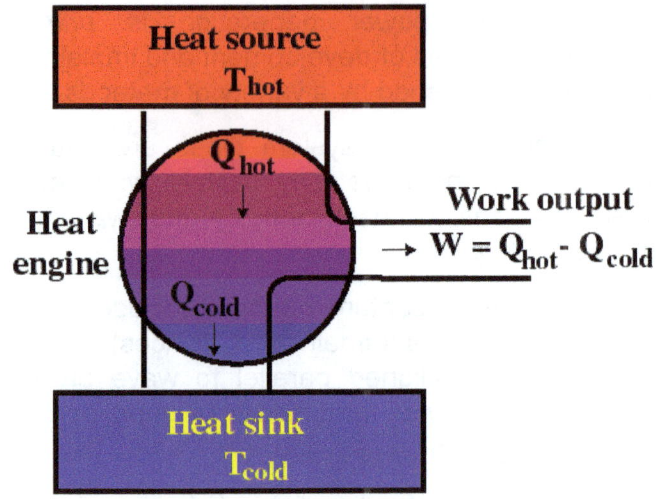

*No known engine can convert all input heat energy into work. Efficiency is measured according to work output (**W**) = temperature gradient (**Q_{hot} – Q_{cold}**), where:*

Efficiency = W/Q_{hot} = $(Q_{hot} - Q_{cold})/Q_{hot}$

Heat Engine Efficiencies

Heat engines, including steam, gasoline and diesel types, extract mechanical energy to do work from a temperature difference (or gradient) between a hot source and cold body (heat sink).

- As decreed by the Second Law of Thermodynamics, no heat engine can be 100% efficient, because not all of the source heat can be converted to do work:
    - In general, larger temperature gradients between the source and cold side produce greater potential engine efficiencies.
    - On Earth, the cold side is typically close to the ambient temperature of the environment, so most efforts to improve thermodynamic efficiency focus upon increasing the source temperature.

309. C.R. Nave

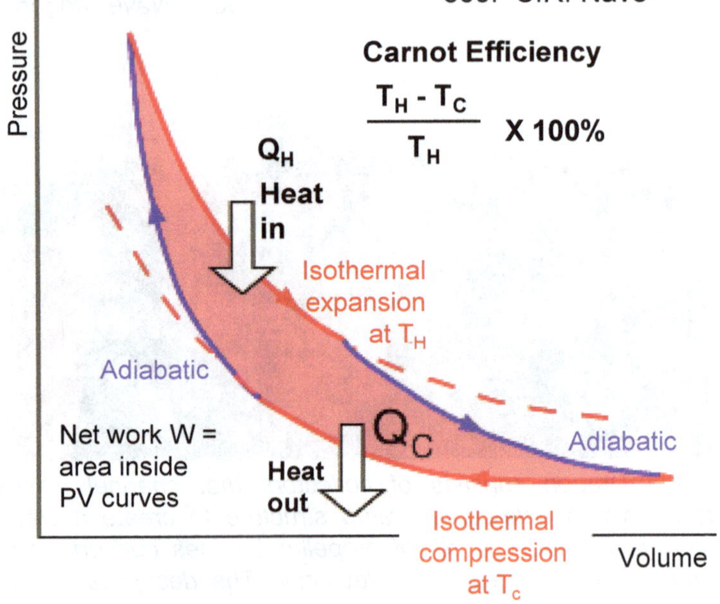

A theoretical Carnot engine is an enclosed gas piston consisting of two basic reversable processes:

- *Adiabatic Compression (T_c – T_H)*
- *Isothermic expansion of the gas in contact with heat from T_h.*

The Carnot Cycle

Heat engines operate on thermodynamic cycles that occur each time heat is converted to mechanical energy and back.

- This same general principle is applied to heat pumps such as refrigerators, but in reverse, where mechanical energy can be transferred from a colder body to a warmer one:
    - No heat engine can operate reversibly as a heat pump without putting more energy back into the system.
    - This rule can be demonstrated mathematically by an imaginary "Carnot engine" (a perfect but impossible device) that could only operate if a heat engine cycle was completely reversible.

Steam Turbines

Steam turbines are a form of heat engine that operate somewhat like water turbines using thermodynamic force rather than water pressure to provide mechanical output that is converted to electrical power.

- These turbines range greatly in size, from 1 hp (0.75 kW) units to more than 2,000 hp (1,500 kW) for utility-scale electricity:
  - Steam turbines create rotary mechanical motion that is particularly useful for driving generators, produce more than 85% of all world electricity, and generate more than three-fourths of all US electricity.
  - Thermal conversion efficiencies range from 30%-50% depending upon types and applications, usually offering excellent power-to-weight ratios.

310. TXU Energy

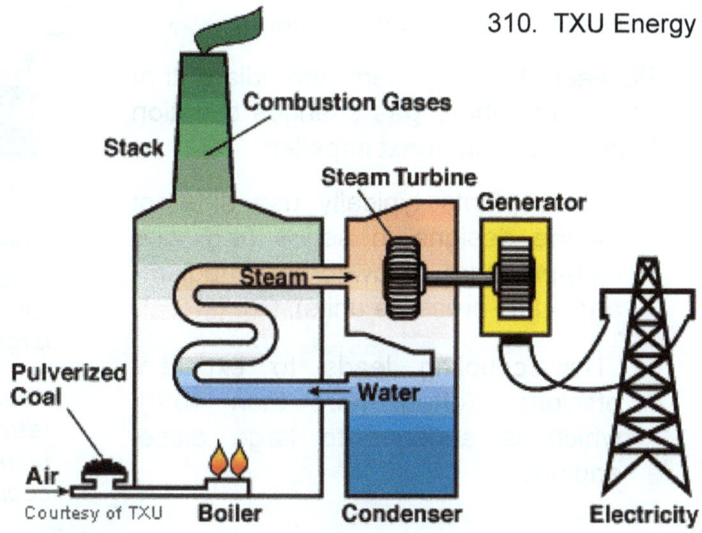

Steam turbines have replaced steam engines to drive generators that produce the majority of all US and world electricity.

Combined Cycle Heat and Power Turbines

311. Consumer Energy Council of America

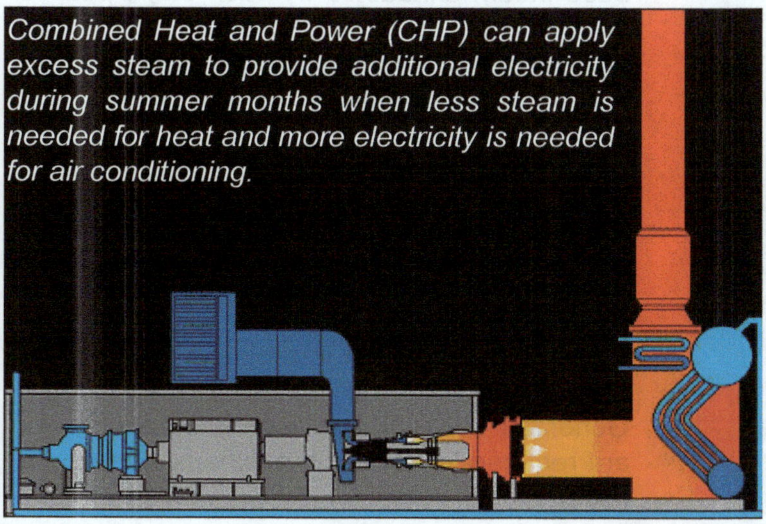

Combined Heat and Power (CHP) can apply excess steam to provide additional electricity during summer months when less steam is needed for heat and more electricity is needed for air conditioning.

About 80% of all bulk utility-scale fossil fuel plants, and all nuclear plants, use turbines to produce electricity and apply waste exhaust heat for supplementary power, thermal processes and other applications.

Steam turbines operate on a "Brayton" cycle, and are one of the oldest and most versatile technologies used to drive electrical generators and produce process heat for other applications.

- Steam turbines use stationary sets of blades (called nozzles), and moving sets of adjacent blades ("buckets" or "rotor blades") comprising impellers that are mounted on shafts to produce kinetic energy:
  - A high pressure maintained in front of an impeller, and a low temperature behind, provides a pressure and thermal gradient.
  - As steam shoots through the turbine to the rear, its energy is converted to blade velocity.
  - Exhaust heat can be recycled back to the burner to produce more power, or can be redirected for separate purposes.

Modern steam plants use multiple impellers configured in series that are referred to as "multi-level" systems.

- Between the impellers are idlers that don't turn, where gas changes direction to be used in the next impeller:
  - Power plants typically use different turbine designs in series (e.g. one high-pressure, followed by 2 medium- and 4 low-pressure units).
  - This coupling leads to excellent efficiency (often more than 40%), which is superior to large diesel engines.

312. Helmut Huetten

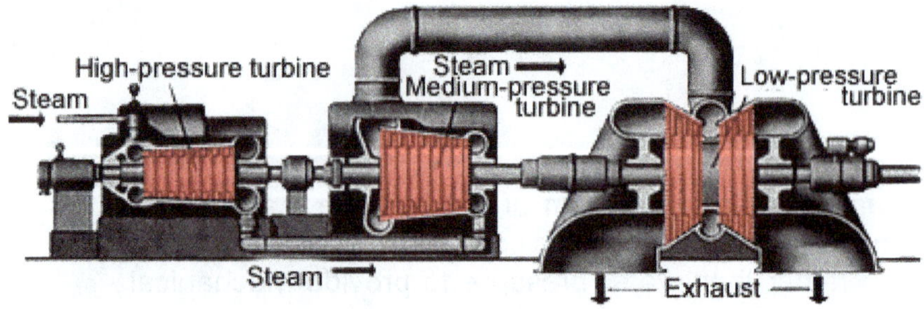

Steam cools and expands as it passes through each impeller, requiring larger, conical-shaped impellers in sequence. Coupled with a generator and fired by a nuclear reactor, they can produce large amounts of electricity, exceeding 1,000 MW.

After passage through the system, condensed water can be returned to a boiler by a pump (a special advantage in nuclear plants where radioactive thermal water is clearly separated from cooling water).

Coupling of Steam Turbines

313. Siemens
314. Dal Engineering Group

Condensing Steam Turbines

Steam turbines come in a variety of technical forms.

- Most that are used in large power plant applications are condensing-type turbines for primary power:
  - Used steam exhausted by each stage passes through a series of condensers to ensure that an optimum amount can be extracted to provide the most efficient thermal cycle.
  - Operators controlling the process can allow more steam to be used as desired, either for thermal applications, or to produce more electricity during periods of lower heat demands for other purposes.
  - The CHP waste heat is produced at no cost, although a separate Heat Recovery Steam Generator (HRSG) and associated equipment are required to convert hot exhaust gases into useful process steam.

Non-condensing (or back pressure) turbines are used in combination with condensing types or as separate units in refineries, distinct heating units, pulp and paper plants and desalination facilities where large amounts of low-pressure steam is available.

- These systems are typically smaller and less efficient than condensing types (about 20%-40% efficiency), but are also less expensive:
  - Higher pressure steam is used to drive the turbines, leaving lower pressure steam to be used for process or space heating.
  - They are often used in industrial and institutional settings which have limited power and heating needs and budgets.

315. Global Power Resources

Back pressure steam turbines are often used in connection with industrial processes that have a need for low- or medium-pressure steam.

Non-Condensing Steam Turbines

316. Ansaldo Energia

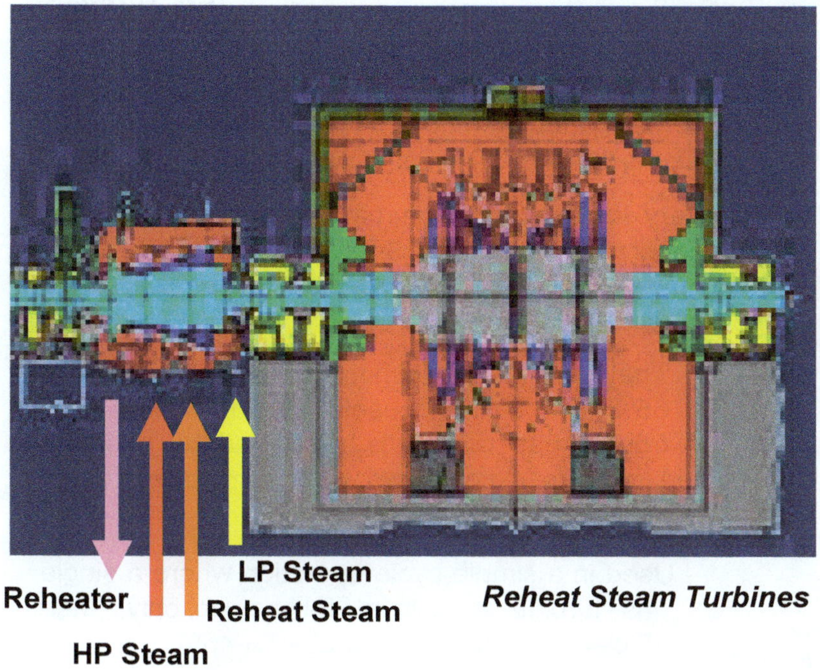

Reheat Steam Turbines

Turbine energy conversion is most efficient when the gradient between high and low pressures from input through output is as large as possible, an advantage for nuclear installations, and a frequent disadvantage for geothermal heat sources.

- In many utility-scale applications, the low-pressure steam at the outlet of a high-pressure chamber is returned to the boiler where additional superheat is added, which then goes to an intermediate pressure section of the turbine to continue its expansion:
  - Reheat is a form of energy recycling to provide higher efficiencies.
  - Reheat can also take advantage of afterburners to provide supplementary heating that greatly increases power, but markedly decreases efficiency.

A gas turbine, also called a combustion turbine, is a rotary heat engine that extracts energy from a flow of gas.

- Gases (typically air) are compressed, heated, and then expanded to drive the turbine, with excess energy over that consumed by the compressor used for power generation:
  - Similar to a steam turbine, gases directed over turbine blades cause it to spin and power the compressor.
  - Gases then pass through a nozzle to generate additional thrust by accelerating hot exhaust through expansion back to atmospheric pressure.
  - Energy is extracted in the form of shaft power, compressed air and thrust, or any combination of these to power electric generators, aircraft, trains, ships, and even tanks.

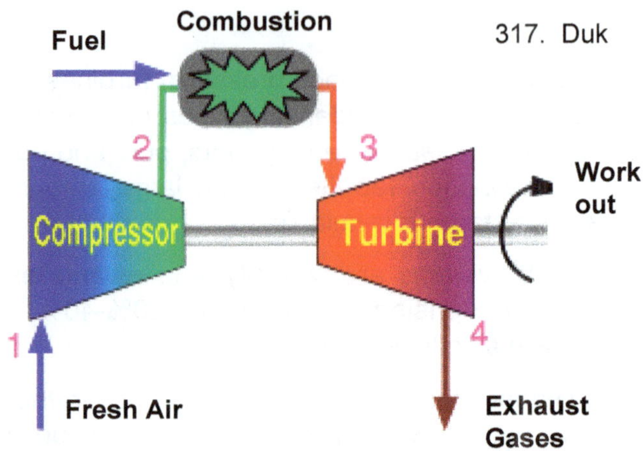

317. Duk

Gas turbines operate using a constant-pressure Brayton (or Joule) cycle and have three types of components:

- *Compressors intake and pressurize ambient air.*
- *Burners (or combustion chambers) use fuel to heat air to a constant pressure.*
- *Expansion turbines extract heat energy to produce work output and exhaust (or recycle) excess heat.*

Gas Turbines

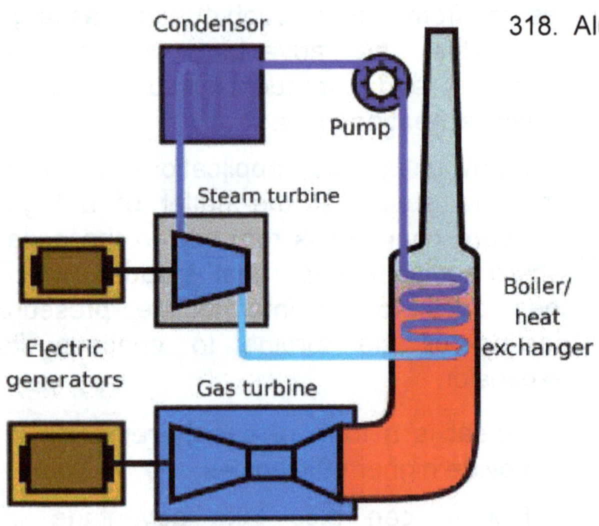

318. Alureiter

Gas turbines are often used in combined cycle operations with high-pressure steam turbines to recycle exhaust heat. This improves thermal power efficiencies and provides cogeneration of heat and electricity.

Combined Cycle Gas/Steam Turbines

Industrial gas turbines range in size from truck-mounted mobile plants to enormous complex systems for bulk utility-scale plant operations.

- Very large industrial turbines for electrical power generation operate at 3,000 rpm or 3,600 rpm to maintain AC grid frequency, and can produce large outputs:
  - They can be particularly efficient (up to 60%-65% electrical and useful thermal) when heat from a gas turbine is recovered by a steam turbine in a combined heat and power (CHP) installation.
  - Used in a simple-cycle operation where a single gas turbine is utilized for power only, their efficiencies approach about 40%-45%.

319. Austrian Energy Agency

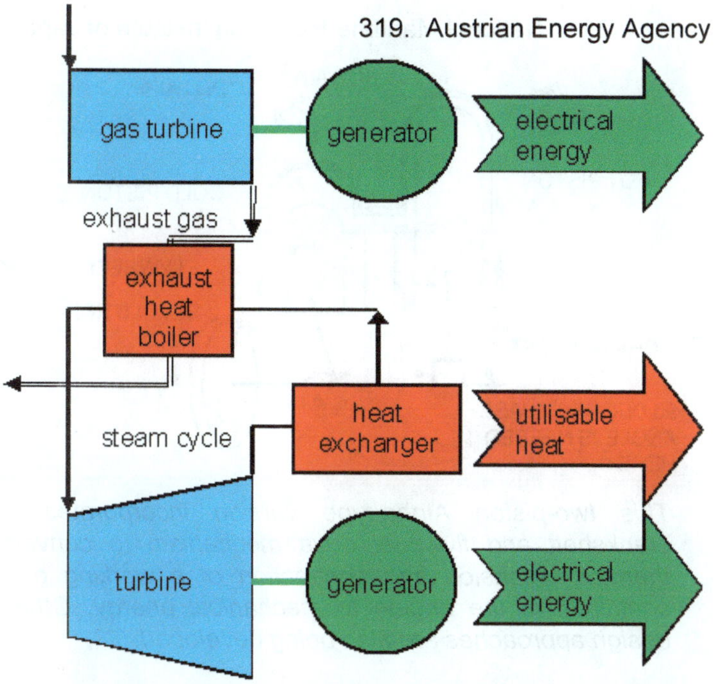

Gas turbines can provide power and heat in stand-alone applications, or can be combined with steam turbines for greater power and efficiencies.

As with all heat engines, the overall efficiencies and pollutant emissions produced are influenced by the fuel sources used and their particular applications.

- Each type offers special advantages for specific circumstances:
  - Operations using a single simple-cycle gas turbine for power only require smaller capital investments than CHP systems that recover exhausts for auxiliary steam and hot water, and are most useful when small amounts of power are needed.
  - Combined-cycle gas turbines using single stand-alone units can be installed much more rapidly than more expensive and complex gas-steam turbine configurations.
  - Combination gas-steam turbine systems are most efficient and applicable for large-scale bulk utility cogeneration plants.

Gas Turbine Applications

The portability and relative simplicity of stand-alone simple-cycle gas turbines offers versatility for a broad range of industrial, institutional and commercial uses.

- Larger units can be turned on and off in minutes to adjust to power demand loads with efficiencies approaching 40%, and small ones can be easily installed to support site-specific needs:
  - Steel industries use gas turbines to drive air compressors for blast furnaces, and process industries use them for chemical, paper and other material production.
  - Large commercial and institutional organizations use gas turbines for on-site CHP, including space/water heating and to drive absorption chillers for cooling and refrigeration.
  - Small turbines drive compressors for long-distance natural gas pipelines and petrochemical refinery processes.

320. Energy and Environmental Analysis, Inc.

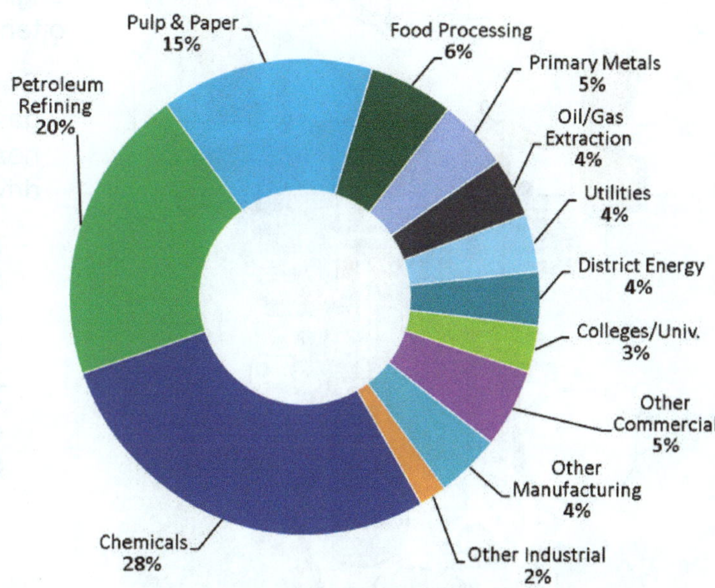

Existing Simple Cycle Gas CHP Capacities

321. National Maritime Research Institute of Japan

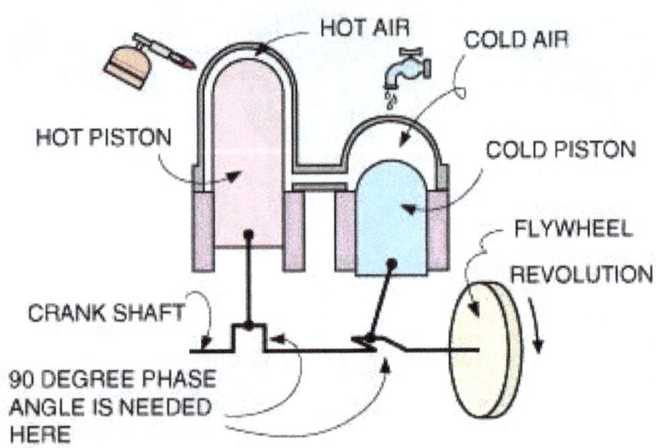

This two-piston Alpha-type version incorporates a crankshaft and flywheel drive mechanism to convert thermal expansion and contraction of a working gas contained in the engine to mechanical energy. Other design approaches are also being developed.

Stirling Cycle Engines

Sterling cycle engines are piston-operated devices that can offer high thermal efficiencies using temperature gradients to produce relatively low electrical power levels (up to a few kilowatts).

- Although they are not suitable for industrial applications and are not yet commercially available, they are gaining interest for residential and small business-scale applications for special reasons:
  - Sterling engines can run on chemical processes that produce cold as well as heat, such as sublimation of dry ice or boiling of liquid nitrogen.
  - They can also run on virtually any means to create heat, including fossil materials, biomass, waste gases and solar thermal.

322. PES Network

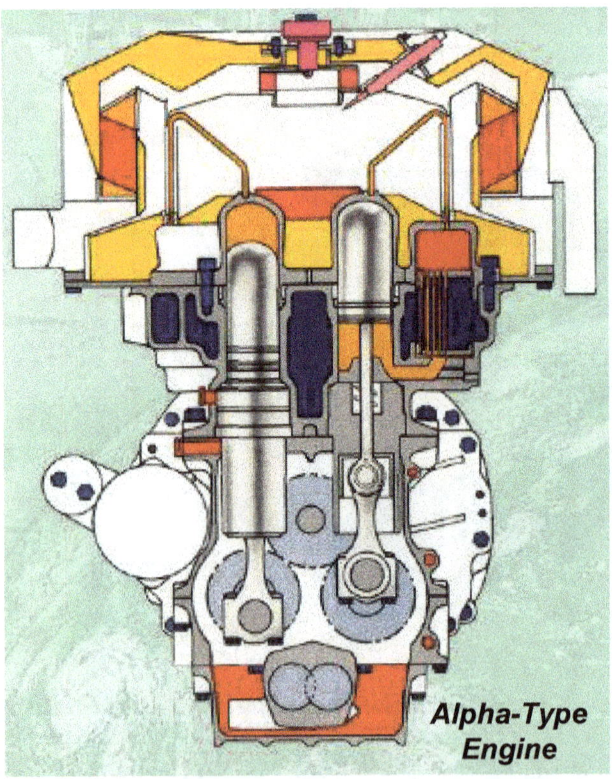

Stirling Cycle Engines

Sterling engine designs represent a variety of configurations, engineering concepts and application sizes that are most often in experimental development stages.

- Common features provide for use of a contained working gas (typically helium or hydrogen) which is repeatedly heated and cooled, causing expansion and contraction to drive one or more pistons:
  - A "regenerator", consisting of a wire mesh positioned between cold and hot regions, stores much of the thermal energy contained in the working gas for use in the next cycle after each expansion stage.
  - The same basic processes can be applied in reverse to convert mechanical energy into heat or cold using Sterling heat pumps or refrigerators.

323. Karls Ruhe University

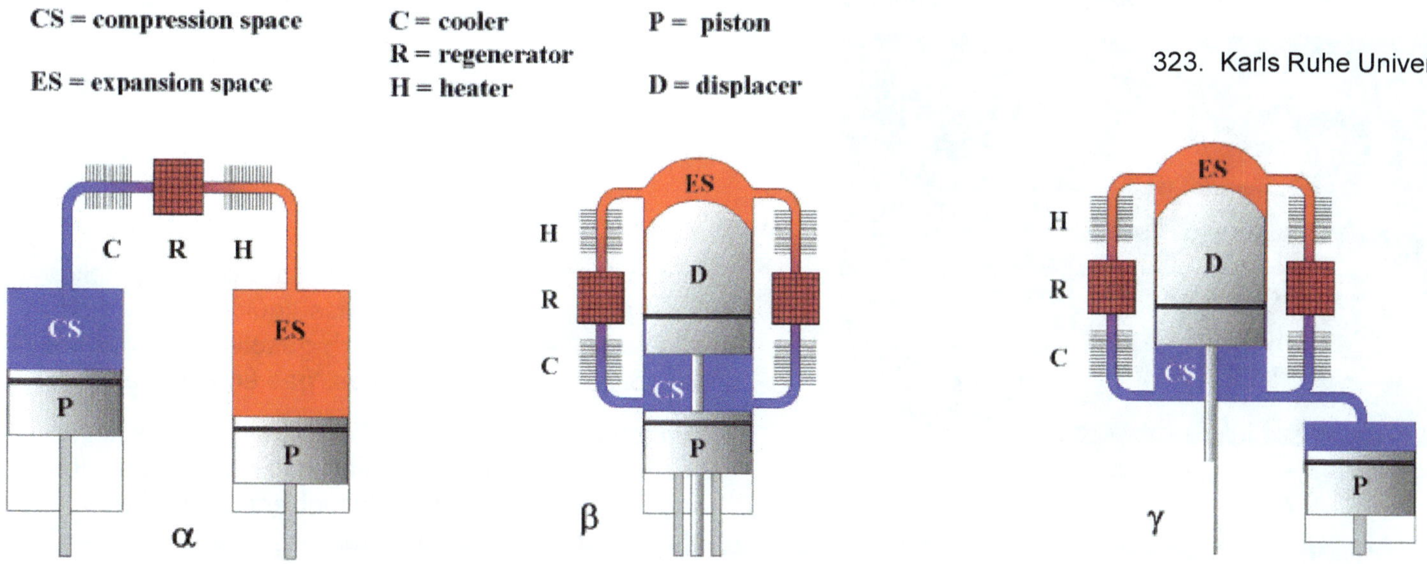

The Alpha engine contains two pistons in separate cylinders (one hot and one cold), each driven by different heat exchangers. It has a very high power-to-volume ratio, but presents some technical problems associated with the hot piston seals.

The Beta engine has a single piston within the same cylinder and on a common shaft with a displacer piston that functions only to move working gas between hot and cold heat exchangers. It is usually combined with a flywheel, and avoids a need for moving seals.

The Gamma engine has a single power piston and displacer piston mounted in separate adjacent cylinders connected to the same flywheel. Gas flows freely between the two cylinders within a single body. This configuration is the simplest, but has a lower compression.

Representative Stirling Engine Types

Although the term "internal combustion engine" can be applied to gas turbines, it is usually used in reference to reciprocating piston engines.

- Such engines are widely used for mobile applications due to relatively high power-to-weight ratios with good fuel energy density:
  - The most common fuels used today are derived from petroleum hydrocarbons, including diesel, gasoline and liquefied petroleum gas (LPG), although liquid and gaseous biofuels are gaining popularity.
  - Some engines can run on hydrogen, but this requires modifications to the conventional engine's cylinder block, head and head gasket to seal in the explosive gas (which can still leak out undetected).

324. Larry Bell

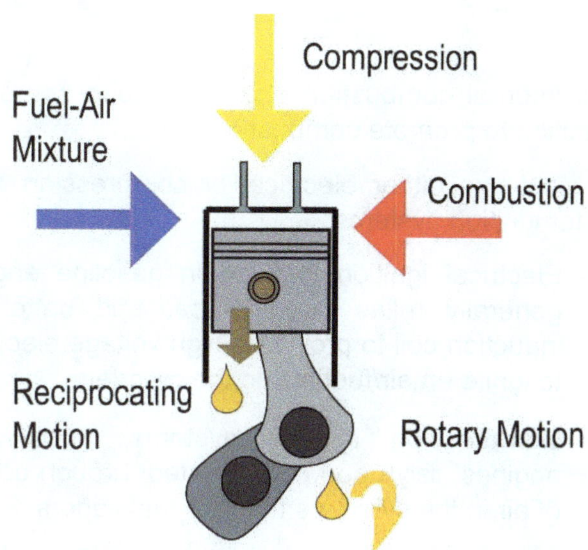

Internal Combustion Engines

325. Matthia Sebulke
326. Madgasters Photos

Engines that use gases for fuel are properly called gas engines, and those that use liquid hydrocarbons are called oil engines, although gasoline engines are often colloquially referred to as “gas engines” also.

- Limitations on fuels used are their abilities to be easily transported through an engine’s combustion chamber and release sufficient energy in the form of heat for practical benefits:
  - Diesel fuel engines are generally heavier, noisier and more powerful than gasoline, and are also more fuel-efficient in many instances.
  - Sophisticated, fuel-efficient diesel engine cars have become prevalent in Europe since the 1990s, and currently constitute about 14 percent of the market.

All internal combustion engines require some means of ignition to promote combustion.

- Most use either electrical or compression heating for ignition systems:
  - Electrical ignition is used in gasoline engines, and generally relies upon a lead-acid battery and an induction coil to provide a high voltage electrical spark to ignite an air/fuel mix in the cylinders.
  - Compression ignition systems, such as diesel engines, rely upon heat created through compression of air in the cylinders to ignite fuel vapors.
  - Homogeneous charge compression ignition (HCCI) is an emerging new technology that combines features of electrical and compression types, where ignition occurs in several places at one time to cause the fuel/air mixture to burn nearly simultaneously.

327. Popular Mechanics

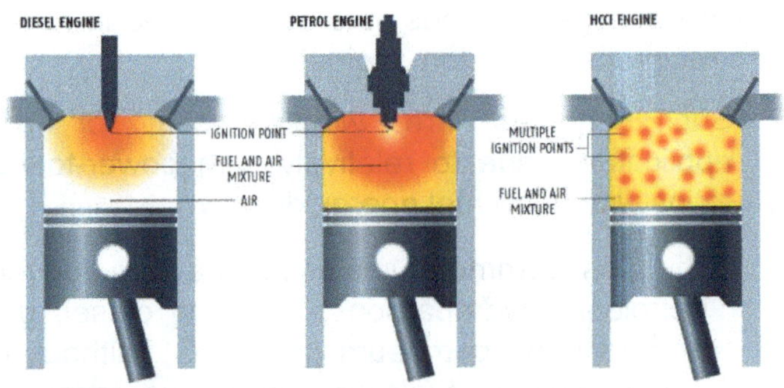

In HCCI and petrol engines, fuel and air are mixed before combustion, preventing soot emissions associated with diesel engines. Only HCCI engines have multiple ignition points throughout the chamber which enable lean combustion at low temperatures to offer higher efficiency and avoid formation of nitrogen oxides (NOX). Ignition relies upon heat from compression rather than a spark.

Engine Ignition Systems

Most internal combustion engines use expanded gas in closed cylinders to drive pistons.

- Conventional diesel and gasoline engines used in modern vehicles operate on a four-stroke ("Otto") cycle that has one power stroke per cycle:

1. During the initial stroke, a valve opens to take air into the cylinder (in a diesel engine), or a mixture of air and fuel (gasoline engine) as the piston moves down.
2. Next, the piston moves up to compress the air (diesel) or air/fuel mixture (gasoline) to increase the pressure and temperature. (In a diesel, fuel is then sprayed into the cylinder.)
3. When the piston returns to the top of the cylinder, an electric spark (gasoline) or heat of the mixture (diesel) ignites the gases, forcing the piston down (power stroke).
4. At the bottom of the piston's travel, unburned gases are forced out through a port that then closes, and the cycle repeats.

328. NCERT India

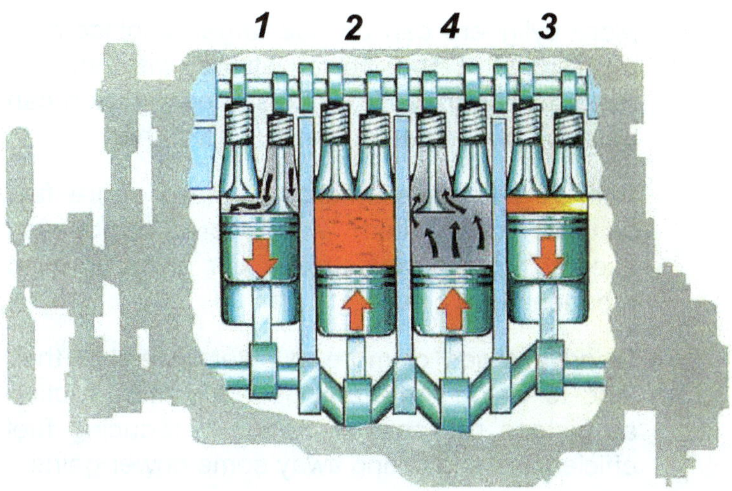

Diesel engines are similar to gasoline engines except that air alone enters on the intake stroke and the fuel is injected or sprayed into the cylinder at the end of the compression stroke to be ignited by heat rather than a spark.

Diesel Engines

329. Eric Pierce

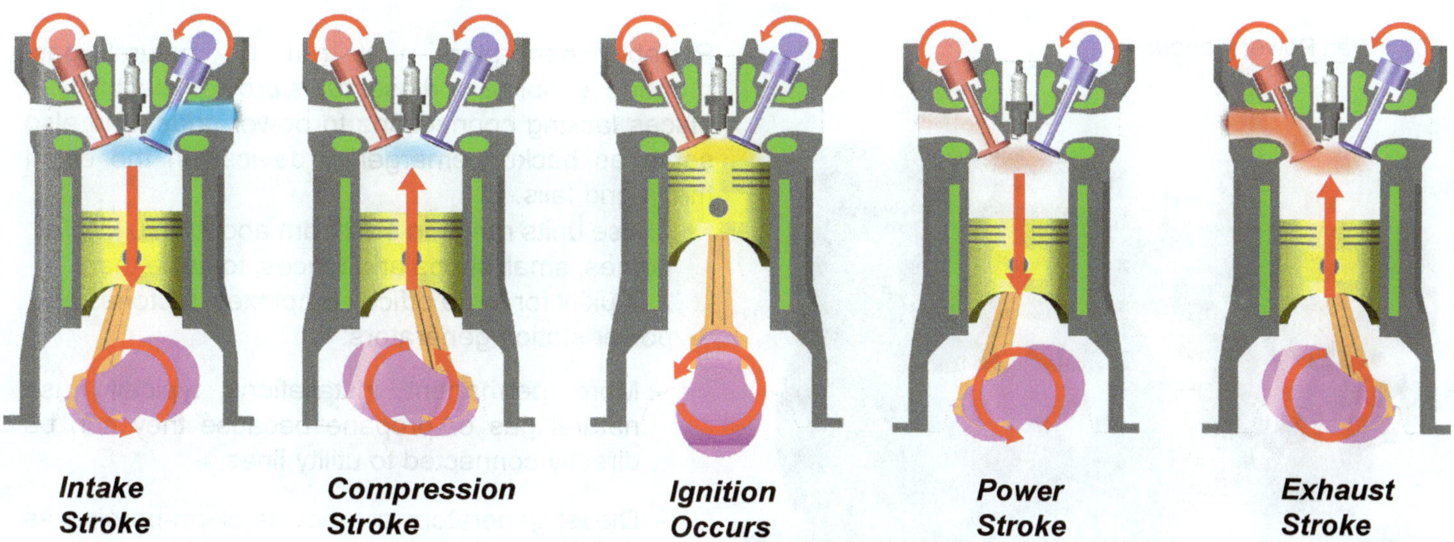

Sequences in a Four-Stroke Gasoline Engine

Gasoline Engines

- Having more cylinders yields potential benefits along with disadvantages:
  - More cylinders can provide larger displacement of pistons in the power stroke with smaller reciprocating mass (the mass of each piston can be less), making an engine run smoother.
  - With larger displacement vs. mass, more fuel can be combusted, with more combustion events (more power strokes per given time) to produce more power.
  - Disadvantages of more cylinders are that they tend to weigh more and generate more friction as pistons rub against cylinders, reducing fuel efficiencies and taking away some power gains.
  - Engines can lose about 30% of power through exhaust heat, 6% through friction, and 36% through water heating (leaving about 20% to do work).

330. Hawaii Community College

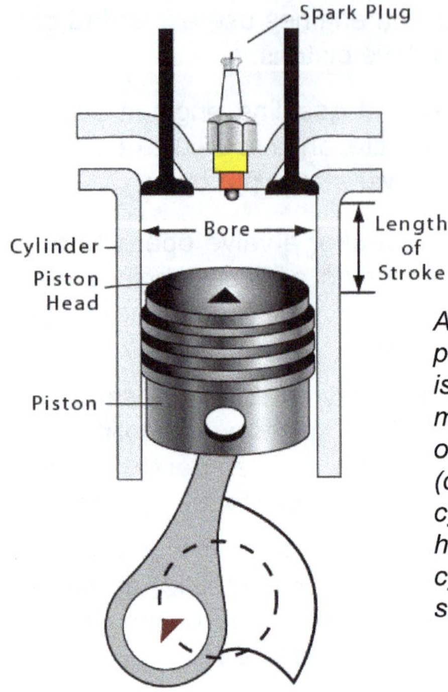

An engine's total piston displacement is determined by multiplying the area of the piston head (diameter of the cylinder bore) by the height of the cylinder (length of stroke).

Total engine displacement (size) is the number of cylinders times the displacement of one of the pistons measured in liters or cubic inches.

Engine Capacity / Volume

331. Watt Power Canada

Diesel Generator

Portable generators that run on natural gas, propane, gasoline or diesel fuels provide electricity in places lacking connections to power grids, and also serve as backup emergency devices in the event that a grid fails.

- These units range in size from about 1-30 kW for homes, small shops and offices, to more than 2,500kW for large office complexes, factories and power station generators:
  - More permanent installations typically use natural gas or propane because they can be directly connected to utility lines.
  - Diesel generators are not as clean-burning as gas or gasoline units and require large holding tanks, but are currently 20%-30% less expensive to operate due to lower fuel costs than other types.

Various types of engine and fuel combinations present different environmental impacts.

- Gasoline and diesel engines produce significant CO_2, sulfur and nitrogen oxide emissions due to incomplete combustion of hydrocarbon fuels:
  - Diesel engines produce a wide range of pollutants, including aerosols of many small particles that can lodge deep in lungs and present health hazards.
  - LPG burns much more cleanly and completely than gasoline and diesel fuels, and doesn't contain sulfur.
  - Hydrogen engines primarily release water, but when air is used as the oxidizer, nitrogen oxides are also produced.

332. EIA

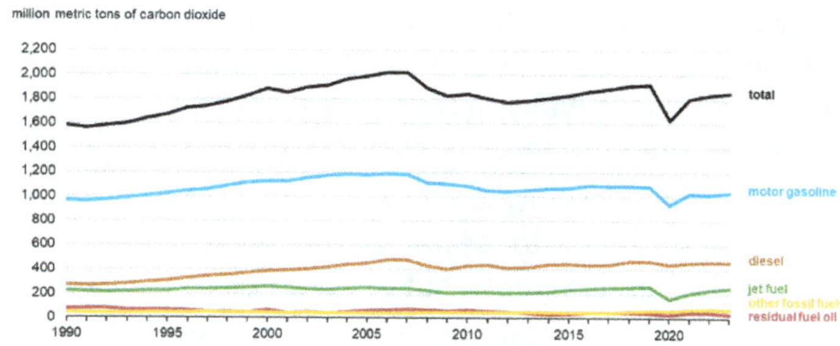

CO2 emission by fuel source, 1990-2023

333. Honda

Fuel Cell Vehicles (FCVs)

334. Honda

Honda Vehicle

A fuel cell is an electrochemical energy conversion device that uses a simple chemical reaction to combine hydrogen and oxygen into water, producing electrical current in the process.

- Fuel cells work much like electrolysis, which is used to separate water into hydrogen and oxygen, but in reverse:
  - At the anode, hydrogen molecules give up electrons to form hydrogen ions (using a platinum catalyst).
  - In the most common type, a proton electrolyte membrane (PEM) allows protons (but not electrons) to pass through, causing hydrogen ions to flow to the cathode, while electrons flow through an external circuit.
  - No pollution is produced, and the only byproducts are water and heat.

335. Romary

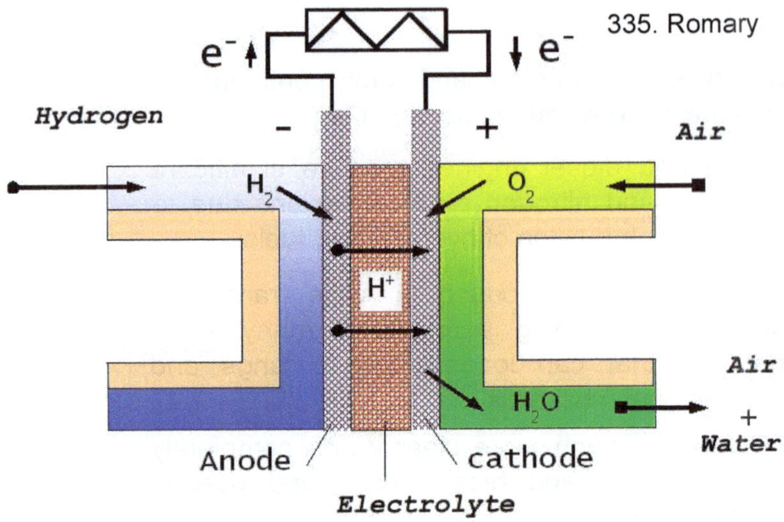

$H_2 \rightarrow 2H^+ + 2e^-$

On the anode side: hydrogen diffuses to the anode catalyst to dissociate into protons and electrons. Protons move through a membrane to the cathode, and electrons are released.

$O_2 + 4H^+ + 4e^- \rightarrow 2H_2O$

On the cathode side: oxygen molecules react with electrons (which have traveled through the external circuit) and with protons to form water vapor and/or liquid water.

Fuel Cell Principle

336. USDOE-EERE

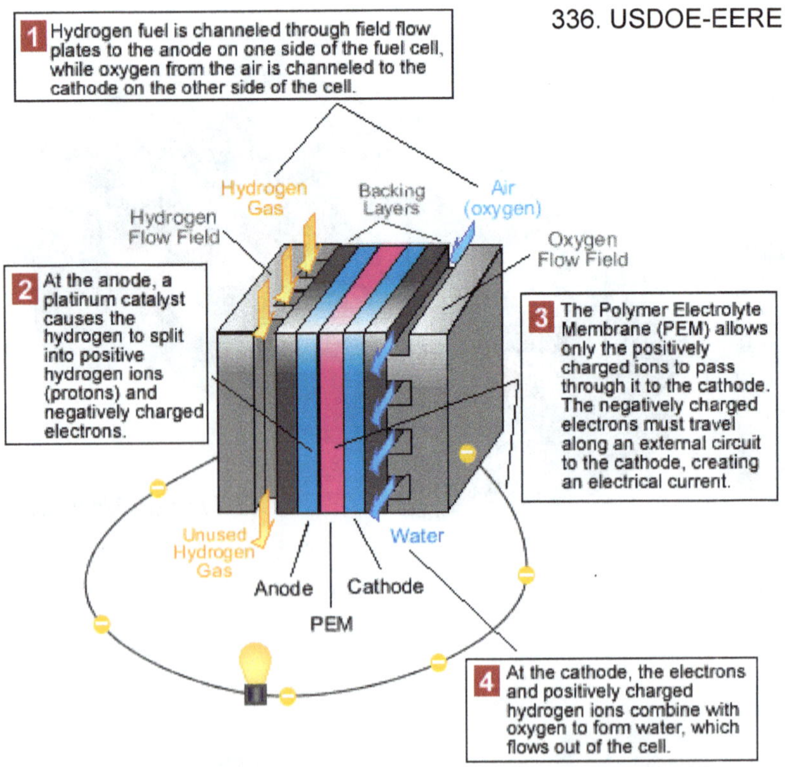

Typical Fuel Cell Stack

A typical fuel cell produces about 0.86 volts.

- To create enough voltage, cells are layered and combined in series and in parallel into a “fuel cell stack” usually containing more than 45 cells:
  - Different types of fuel cells vary in terms of materials used, influencing cost.
  - Platinum-coated anode catalyst which cause hydrogen to split into positive hydrogen ions (protons) and negatively charged electrons are very expensive.
  - New means are being developed to reduce platinum requirements without reducing performance.

337. Staxon

Fuel Cell with Water Cooling

Standard proton electrolyte membrane fuel cells require that the membrane must be constantly hydrated with water evaporation occurring precisely at the rate it is produced.

- If water evaporates too quickly causing the membrane to dry, resistance across it will increase and eventually cause it to crack and produce heat damage:
  - Flow control is essential to maintain a constant ratio between the reactant and oxygen.
  - Proper temperature must be maintained throughout the cell to prevent destruction through thermal loading.
  - Automotive fuel cells require about a 5,000 hour lifespan (equivalent to about 150,000 miles [240,000 km]) under extreme temperatures.

338. European Commission

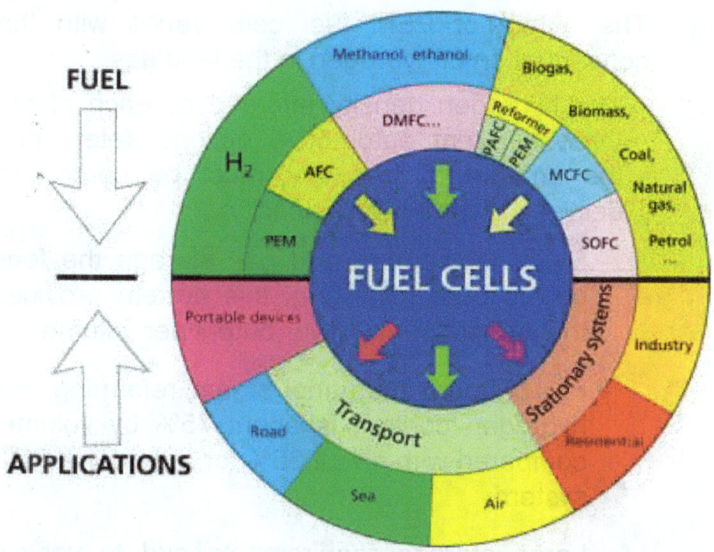

Fuel Cell Types and Applications

Fuel cells are available in a variety of types for diverse uses:
AFC: Alkaline (Space Applications)
PEMFC: Polymer Electrolyte Membrane Carbonate
DMFC: Direct Membrane
PAFC: Phosphoric Acid
MCFC: Molten
SOFC: Solid Oxide

Fuel cells are not constrained by the maximum Carnot cycle as combustion engines are (they do not operate with a thermal cycle).

- As a result, they can have very high efficiencies in converting chemical energy to electricity:
  - Typical fuel cells have an efficiency of about 50%, but this is dependent upon current through the cell (more current, less efficiency).
  - Hydrogen cell efficiency is about equal to cell voltage divided by 1.23 volts operating under standard conditions.
  - Voltage varies with fuel used and quality and temperature of the cell.
  - If 50% of the “available energy” content of hydrogen is converted into electrical energy, the remaining 50% will be converted into heat.
  - Fuel cell vehicles running on compressed hydrogen may have power plant-wheel efficiency of 22% (17% with liquid hydrogen).

Various fuel cell systems use hydrogen that is stored in gaseous or liquid forms.

- Automotive fuel cells generally fall into three categories: those that use compressed hydrogen gas storage; those that reform methanol to obtain hydrogen onboard the vehicle; and those that use onboard systems that apply partial oxidation (POX) of hydrocarbons obtained from crude oil derivatives (gasoline, diesel fuel or middle distillates):
  - Generally speaking, the higher the hydrogen content by volume of the fuel source, the better the fuel cell performance due to greater power density.
  - Systems that use direct hydrogen storage tend to be the most simple in design, but entail more complexity in terms of fuel transport infrastructure requirements.

339. JM Ogden et.al., Journal of Power Sources

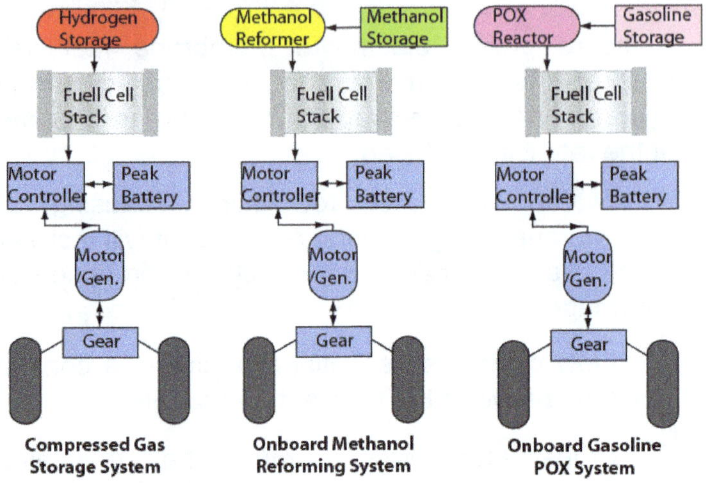

Various automotive fuel cell systems store hydrogen in the form of compressed gas, or convert methanol or petroleum fuels to hydrogen onboard the vehicles.

Types of Automotive Fuel Cell Systems

340. JM Ogden et al., Journal of Power Sources

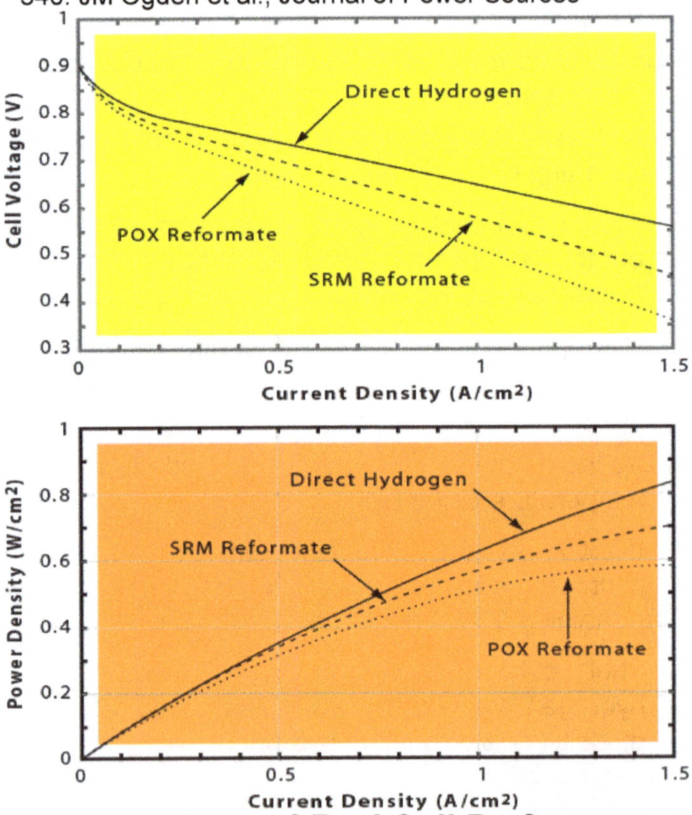

Comparison of Fuel Cell Performances

The output of PEM fuel cells varies with the concentration of hydrogen in the feed gas.

- The hydrogen density afforded by each of the three different automotive options determines the amount of fuel that must be carried to deliver a given power capacity:
  - Since for compressed gas storage the feed gas is pure hydrogen, this system provides the highest peak power output per volume.
  - For onboard methanol steam reforming, the hydrogen content is about 75% by volume, compared with about 35% for a gasoline POX system.
  - Large storage requirements add to vehicle weight and energy inefficiencies, both for fuel and for increased structure to carry the fuel.
  - In general, onboard methanol fuel cell cars may weigh about 10% more than compressed gas vehicles, and gasoline POX vehicles may weigh about 19% more.

Fuel cell economies of different options are significantly impacted by energy conversion characteristics of fuel system types.

- Compressed hydrogen systems are about one-third more efficient in converting energy than those that convert hydrogen from fuels onboard a vehicle:
  - Approximately 15%-25% of energy is lost in converting methanol or gasoline/diesel to hydrogen.
  - Although methanol and gasoline fuel cell cars offer about the same fuel economy, about twice as much methanol is needed due to a lower energy density, requiring larger storage tanks and weight.
  - Onboard steam reforming response times are typically 40%-50% slower than onboard POX systems, requiring more backup energy to be rerouted through a battery with resulting charging/discharging power losses.

342. JM Ogden et al., Journal of Power Sources

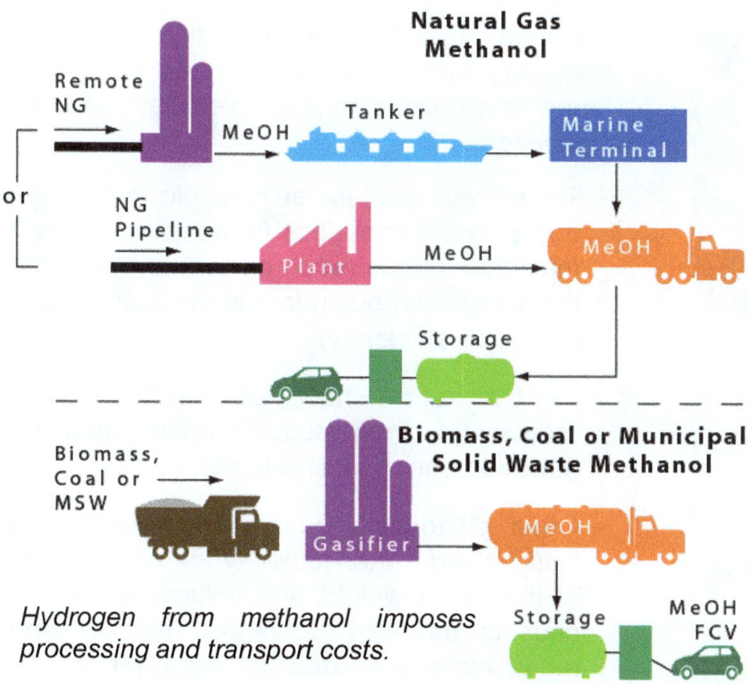

Hydrogen from methanol imposes processing and transport costs.

Fuel Cell Hydrogen from Methanol

341. JM Ogden et al., Journal of Power Sources

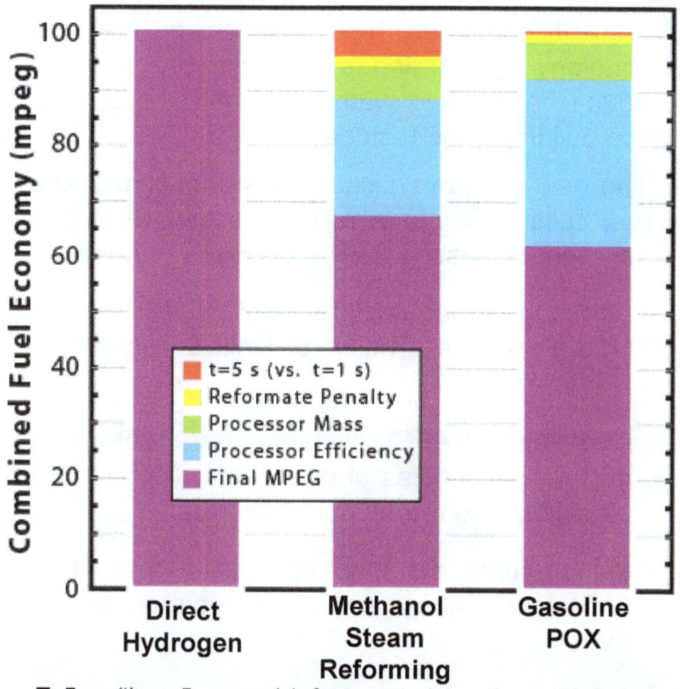

T=5 s *(time=5 seconds) factors a delay time allowed for fuel processing by steam reformers and POX systems to meet peak demands, requiring batteries to fill energy gaps. This adds recharge penalties.*

Cumulative Fuel Economy Losses for Different Fuel Cell Types

Assessments of alternative fuel cell options for automotive applications should also take broader energy conservation and environmental considerations into account.

- The sources of hydrogen fuel used by different system types is a key factor influencing relative benefits vs. downsides:
  - If compressed or liquefied hydrogen is obtained from natural gas/methane reforming, the depletion of these resources, combined with energy used for compression and transport, add real costs.
  - Hydrogen obtained from water electrolysis adds substantial costs for electricity that competes with other uses, and may produce greenhouse gases.
  - Hydrogen derived from methanol or gasoline/diesel consumes energy for processing and transportation, although transport can use conventional roadway and distribution infrastructures that reduce costs.

A recent combined heat and power fuel cell technology is emerging to generate electricity and produce hot air and water for family residences, office buildings and factories.

- The new system typically uses phosphoric acid fuel cells (PAFCs) which comprise the largest segment of existing CHP products worldwide:
  - The largest manufacturer of PAFC fuel cells is UTC Power, a division of United Technology Corporation.
  - Combined PAFC systems can potentially provide efficiencies of nearly 80% (45% - 50% electricity, the remainder as thermal).
  - Molten-carbonate fuel cells are also used for CHP applications, and solid-oxide prototypes exist.

343. Wisconsin Distributed Resources Collaborative

Fuel cells generate efficient heat and power continually without use of generators. CHP applications emit significantly less CO_2 than reciprocating engines per unit output, and operate at lower noise levels that enable plants to be located nearer to users to reduce electricity transmission losses. Presently high implementation costs are a disadvantage.

Combined Heat and Power Fuel Cell Use

344. Whole Building Design Guide

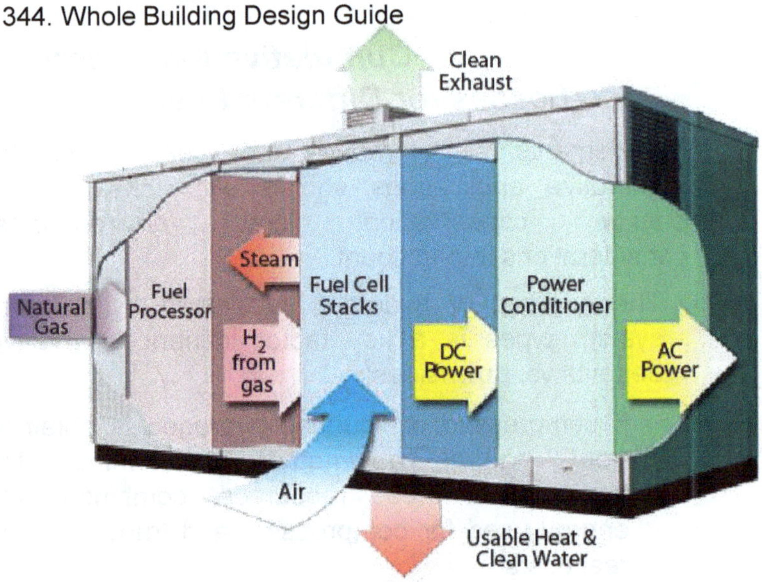

Fuel cell power plants are typically comprised of three primary system elements:

1. ***A fuel processor*** *(or reformer) extracts hydrogen from natural gas or other fuel.*
2. ***A power section*** *containing fuel cell stacks produces electricity and releases water and heat byproducts.*
3. ***A power conditioner*** *converts DC to AC current.*

Typical Fuel Cell Power Plant

Phosphoric acid and other fuel cells are similar to proton electrolyte membrane types except for the electrolyte transfer chemistry.

- Although PAFCs are typically used for stationary power generation, they have also found applications in large vehicles such as city buses:
  - These cells are less susceptible to damage due to some impurities from reformed fossil fuels than PEMs (such as carbon monoxide that binds with platinum catalysts at anodes to reduce efficiency).
  - PAFCs are, however, sensitive to "poisoning" from sulfur, which must be removed from gasoline fuels.
  - Other disadvantages are that PAFCs are less powerful than other types of fuel cells of the same weight and volume, generally causing them to be larger, heavier and more expensive due to platinum anode catalyst costs.

Photovoltaic (PV) power systems produce electricity through solar electric conversion.

- When light photons of sufficient energy strike a solar cell, they knock electrons free in the silicon crystal structure, forcing them through an external circuit (battery or DC load) to then return to the cell and repeat the process.
  - The voltage output from a single cell is about 0.5 V with an average directly proportional to the surface area.
  - Typically about 30-36 cells are wired in series (+ to -) in each solar module.
  - A solar module with a 12 V nominal output (~17 V at peak power) can then be wired in series and/ or in parallel with other modules to form a complete solar array to charge a 12, 24 or 48 volt battery bank.

345. US DOE-EERE

Solar Cell

Solar Electric Conversion

346. NJ Solar Power

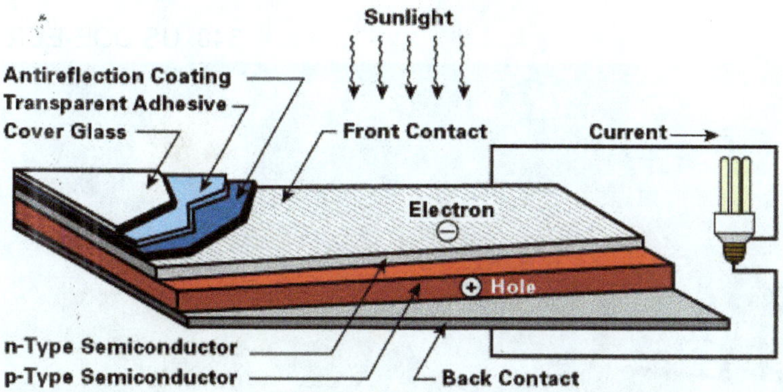

A typical solar cell consists of a glass or plastic cover or other encapsulant, an antireflective layer, a front contact to allow electrons to enter a circuit, and the semiconductor layers that create a current.

Solar Electric Conversion

Semiconductor layers comprise the most important parts of a solar cell because this is where electrons are freed and electric current is created.

•Several different semiconductor materials are used to make layers in different types of solar cells, each with particular benefits and drawbacks:

- Manufacturing cost and complexity are influenced by these materials and device structures based upon factors which include:

… deposition in a vacuum environment

… amount and type of materials

… the number of steps involved

… special processing conditions and steps

The most common PV array design uses flat plate modules (or panels) that can be either fixed in place or allowed to track the Sun.

- Fixed arrays are simplest because they lack moving parts and are relatively lightweight:
  - They are suitable for many locations, including most residential roofs.
  - While less efficient than tracking systems, they are also substantially less expensive
- Tracking arrays are useful when high efficiency is desired using smaller installations:
  - Optimal orientation to the Sun enables more energy to be captured with less surface area.

347. US DOE-EERE

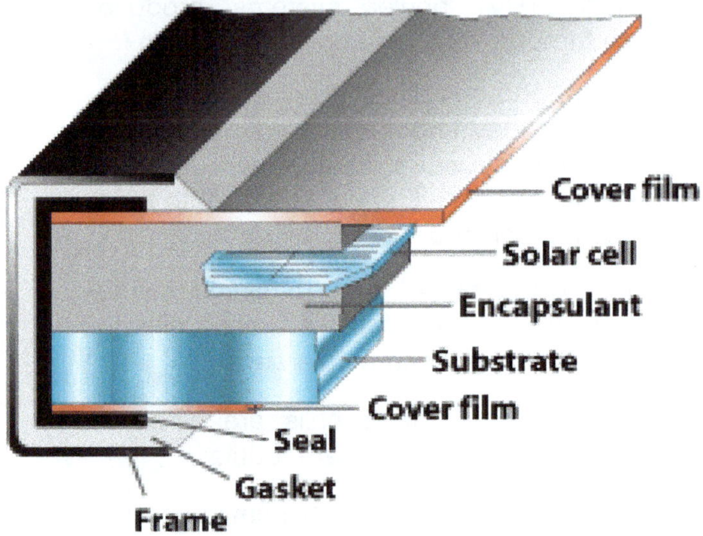

Typical flat plate module design uses a substrate of metal, glass or plastic to provide structural support on the back, an encapsulant material to protect the cells, and a transparent glass or plastic cover.

Flat Plate PV Design

Flat plate collectors draw a substantial amount of energy from diffuse sunlight.

- These systems containing a large number of cells mounted on a rigid, flat surface, and can use both direct sunlight and diffuse sunlight from clouds, the ground, and nearby objects.
  - In clear skies, the diffuse component of sunlight accounts for 10%-20% of the total solar radiation on a horizontal surface.
  - On partially sunny days, up to 50% of radiation is diffuse.
  - On cloudy days, 100% of the radiation is diffuse.

348. US DOE-EERE

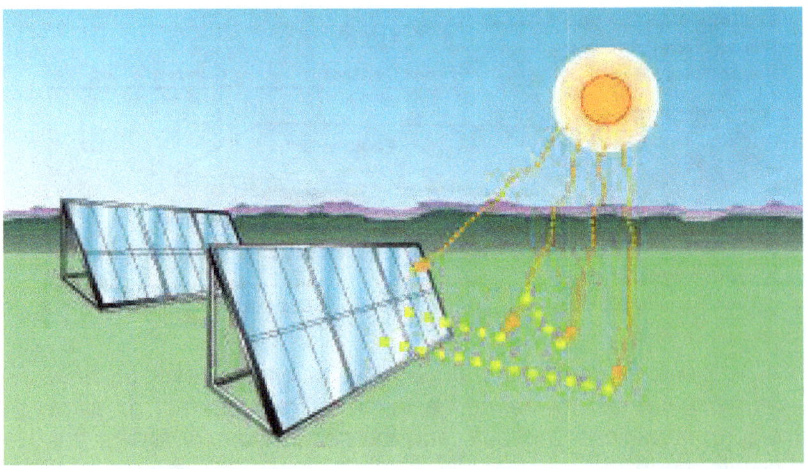

PV arrays must be mounted on stable, durable structures that can withstand wind, rain, hail and other adverse conditions. Much solar energy is captured from diffuse radiation.

Diffuse Energy Collection

349. US DOE-EERE

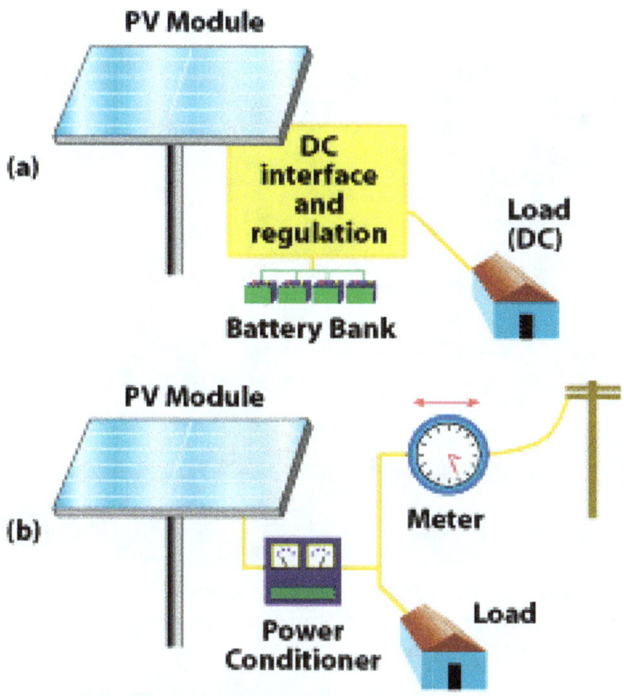

a: Stand-alone system
b: Grid-connected system

PV Systems and Subsystems

PV systems are selected and designed to meet particular user applications.

- Complete PV energy systems are essentially comprised of three major subsystems:
  - On the power generation side, a subsystem of PV devices (cells, modules and arrays) converts sunlight to DC electricity.
  - On the power use side, a "load" subsystem (such as a home) applies the PV electricity.
  - A "balance of system" (BOS) subsystem regulates and controls how the generated power is "conditioned" to meet load requirements.
- A stand-alone PV system typically depends upon battery storage when sunlight isn't available, while a utility grid-connected system may not require energy storage::
  - The BOS subsystem may include storage devices, along with DC-AC inverters.

350. NJ Solar Power

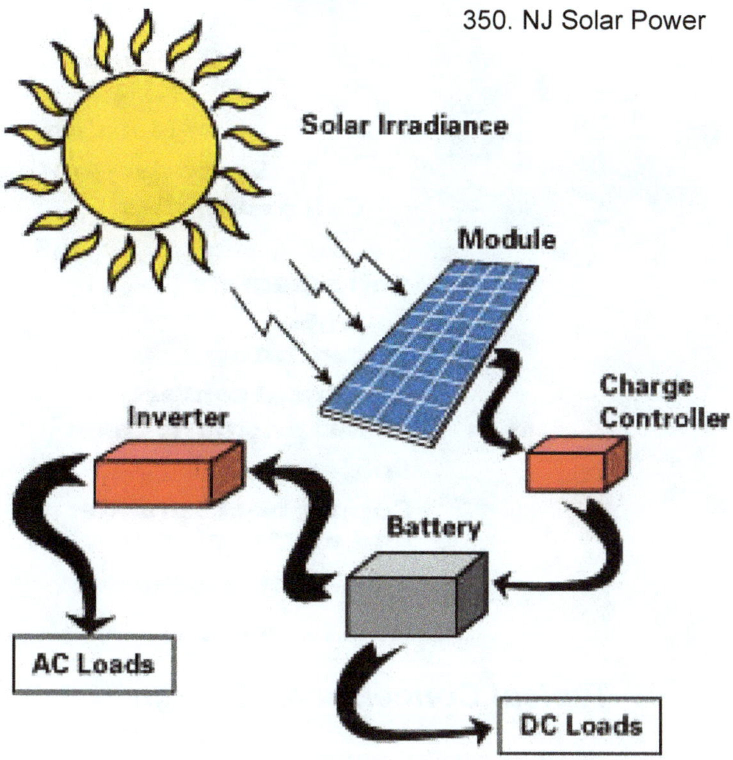

PV System Components

A typical small-scale PV installation consists of the following:

- **PV array**- a series of 9-54 solar panels (residential) that generate DC electricity. Typically covers an area of about 125-750 ft^2 (12-70 m^2), either roof- or ground-mounted.
- **Inverter**- an electronic device that converts the DC to standard AC electricity compatible with the electrical grid.
- **Rails**- anodized aluminum rails are mounted to a roof or ground installation to hold PV panels in place.
- **Battery backup**- charged by the solar system to provide power at night and cover outages. (This is optional since most systems are "grid-tied" and utilize the electrical grid for storage.)

Although not suitable for small projects, solar concentrator PVs make use of relatively inexpensive materials such as plastic lenses and special metal housings to focus more energy onto cells, thereby increasing output.

- These devices can be very effective for large-scale power generation, offering several advantages over flat plate approaches:
  - Fewer of the expensive solar cells are needed to accomplish power objectives.
  - Solar cells are more efficient using concentrated light.
  - Concentrators can be made of small individual cells which are easier/cheaper to produce than large-area, high-efficiency cells.

351. US DOE-EERE

This high-performance solar power generator has been developed for the Arizona Public Service Company's Solar Test and Research (STAR) center.

Concentrator PV System

A typical basic concentrator unit consists of a lens to focus sunlight, cell assemblies, a secondary concentrator to reflect off-center light, various electrical contacts, and support housings.

- These devices can be very effective for large-scale power generation, offering several advantages over flat plate approaches:
  - Both reflectors and lenses can be used in these units to concentrate light.
  - Fresnel lenses use a miniature "sawtooth" design to focus light.
  - Even the best lenses can transmit only 90%-95% of the direct incoming light, and none can focus diffuse sunlight (20% of solar radiation available on clear days).

352. US DOE-EERE

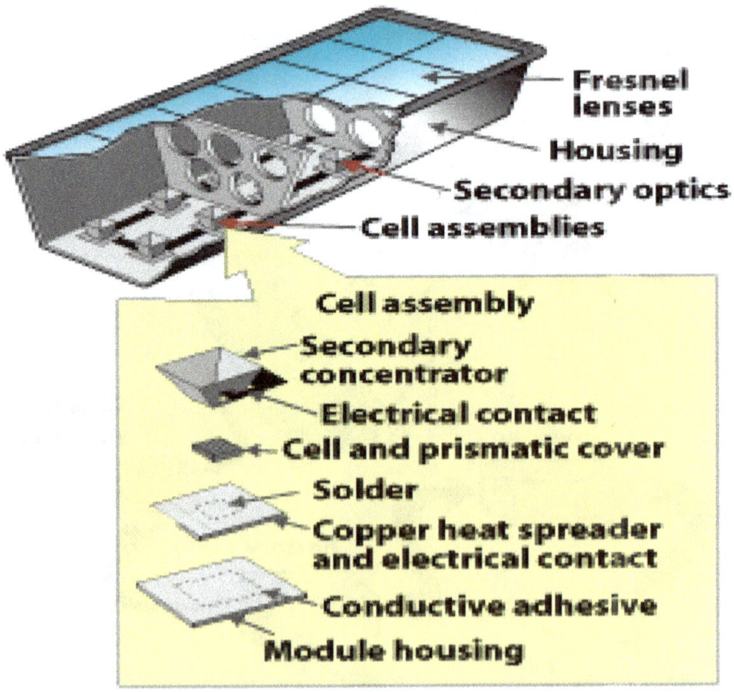

Typical Concentrator Design

353. Kelvin Hulsey Illustration, Inc.

Energy can be stored in a variety of forms including electrochemically in batteries, electrostatically in supercapacitors, mechanically using flywheels, and chemically in the form of hydrogen and other fuels.

- Batteries play very diverse roles in low and moderate energy storage applications, and improved technologies are especially vital to advance electric vehicle (EV) and hybrid electric vehicle (HEV) performance:
  - Electric vehicles currently have very constrained power and range due to limited battery capacities and added weight they impose.
  - Hybrid electric vehicles have much smaller electrical storage needs than EVs, but require batteries with higher power capacities and abilities to withstand many more charge/discharge cycles.

Toyota Electric Vehicle

Batteries

354. Effpower Sweden

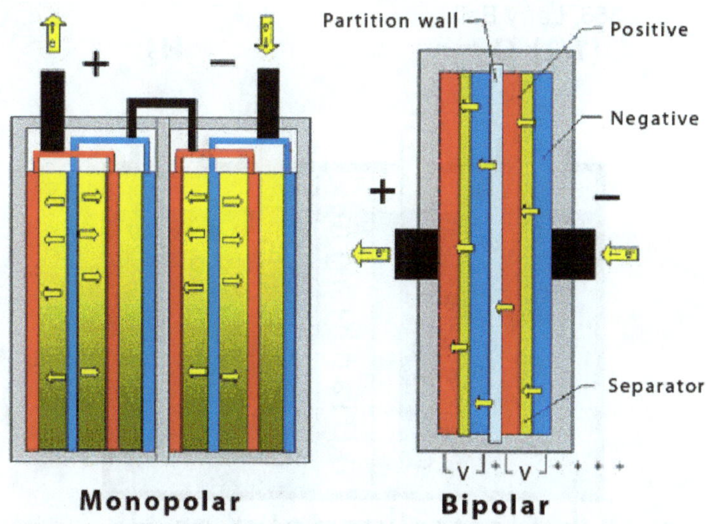

Standard monopolar batteries introduce current in one post at an upper end and have it exit at a positive post on the opposite upper side, producing unevenly distributed higher currents in the upper section near the posts. Bipolar batters use a stack of serially-coupled bipolar electrodes that distribute current more evenly and apply conducting materials more efficiently, creating less resistance and providing more compact high energy.

Many advanced battery technology developments are being driven by needs of the military, electronics industries and automotive EV and HEV markets for lighter, more powerful, smaller and less expensive portable and rechargeable units.

- Conventional monopolar lead-acid batteries which continue to be the standard for today's vehicles may eventually be replaced by bipolar types using other materials that provide better performance features:
  - The US Congress appropriated more than $1 million for a DOD contract to Electro Energy, Inc. in Danbury, Connecticut to support continued development of a patented bipolar water-cell nickel metal hydride (NiMH) battery for military aircraft.
  - Costs for these and other improved devices can be expected to fall due to mass production economies and expiration of exclusive patents.

Battery Technologies

Growing concerns about fuel costs and greenhouse gas emissions are causing vehicle manufacturers and consumers to seek better battery alternatives.

- Among the many varieties of available and emerging battery technologies, three types appear to offer special promise for EV and HEV applications:
  - Lead-acid (Pb-acid) batteries (both monopolar and bipolar) are least expensive owing to a long history of use and lowest material costs.
  - Nickel-metal hydride (NiMH) batteries currently provide high specific power (power per weight) which is particularly important for EVs which depend entirely upon batteries for power and range (less important for HEVs which carry most of their energy in the form of liquid or gaseous fuel).
  - Nickel-cadmium (NiCad) and NiMH batteries last much longer than Pb-acid, but may lose some life cycle advantage as bipolar Pb-acid batteries are advanced.

355. Effpower LIC™ Bipolar Concept

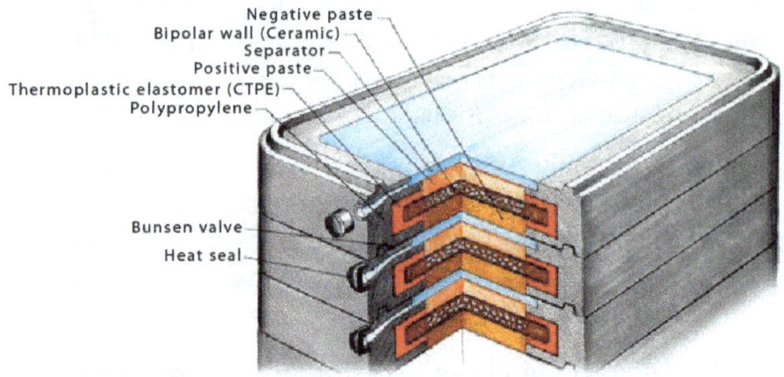

This bi-polar lead-acid battery design uses lead-infiltrated-ceramic (LIC™) plates as partitioning walls between cells which are connected in series (2 volts per stacked cell)

Lead-Acid Bi-Polar Battery

356. Larry Bell

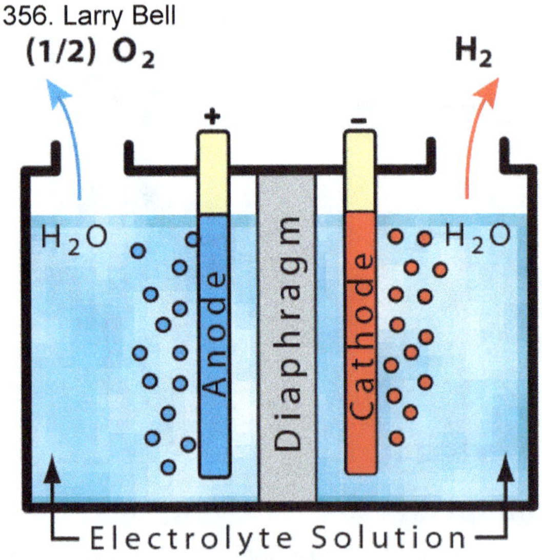

Since pure water conducts electricity very poorly, a water-soluble electrolyte (acid, base or salt) is added to close the circuit. Each electrode attracts ions of the opposite charge. Platinum is typically used because other metals tend to react with oxygen at the anode instead of being released as a gas.

Electrolysis to Produce Hydrogen

Hydrogen can be produced by applying a direct electric current to water with a soluble electrolyte in a manner that separates H_2O into the diatomic molecules H_2 and O_2.

- This electrolysis process is the opposite reverse reaction to what occurs in fuel cells, and in both cases, the energy conversion efficiency is typically less than 50%:
  - Electrolysis is more efficient at high temperatures than at room temperature because some of the energy to drive the reaction is supplied by heat (which is less expensive than electricity).
  - Only about 6% of hydrogen world-wide is produced using electrolysis, most often for on-site purposes.
  - Platinum electrodes generally used in the process add substantial costs to energy conversion penalties.

Supercapacitors (sometimes also called "ultracapacitors" or "electric double-layer capacitors") store energy in an electrostatic field rather than in a chemical state as batteries do.

- These devices have low energy densities (typically less than 15 Wh/kg), but can have very high power densities (4,000 W/kg or more):
    - Low energy densities render supercapacitors unsuitable as primary energy sources for EV and HEV applications, but they can offer excellent means to capture and store energy from regenerative braking and to boost power for sudden demands.
    - Used in conjunction with batteries for regenerative charging, they can avoid wasteful dumping of excess charges that batteries can't accommodate in order to increase vehicle ranges.

357. Axeon Power

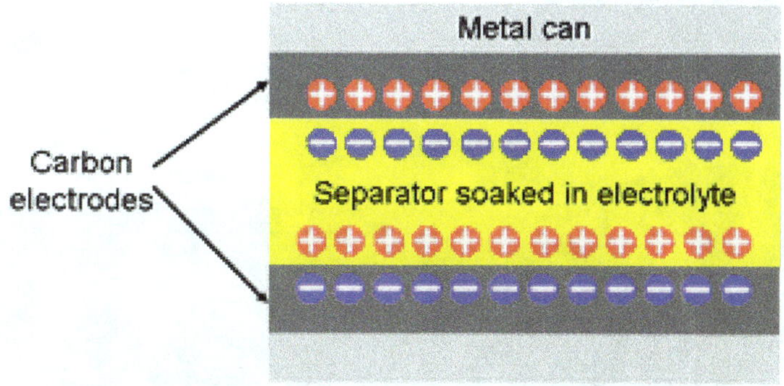

Supercapacitors look much like batteries, but do not use chemical storage. They consist of two electrode plates of opposite polarity separated by an electrolyte. The capacitor is charged by applying voltage across terminals, causing current to migrate to the surface of the electrode of opposite polarity.

Capacitor / Battery Comparison

Device	Energy density Wh/L	Power density W/L	Cycle life Cycles	Discharge time Seconds
Batteries	50-250	150	1 - 10^3	> 1000
Capacitors	0.05 - 5	10^5 - 10^8	10^5 - 10^6	<1

Supercapacitors

358. NEDO Japan

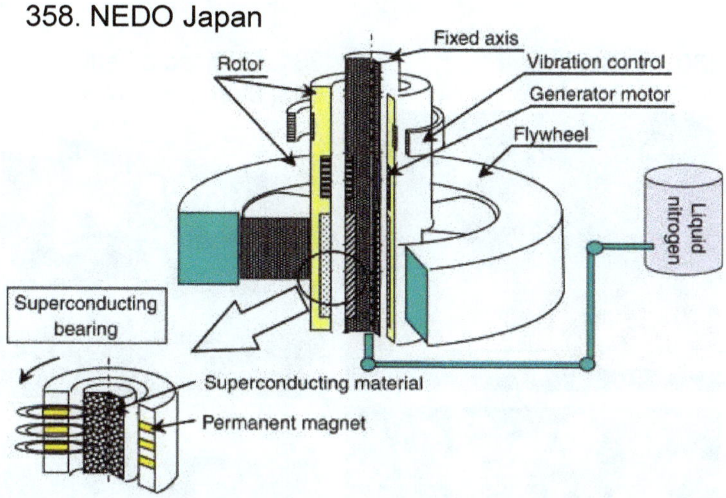

Super flywheels must be very strong to avoid being torn apart by rotational forces at speeds of 100,000 rpm or more. A 1-foot diameter flywheel rotating at that speed would attain a rim surface speed of about 3,570 mph (4.8 times the speed of sound), with a centrifugal force on rim particles equal to 1.7 million Gs. Very strong materials such as fused silica are used along with extremely low-friction bearings.

Super Flywheels

Modern super flywheels store kinetic energy in high-speed rotating drums that form rotors of generators which provide short periods of electricity to fill power gaps in large power grids.

- When surplus energy is available it is used to add momentum to a drum/rotor for times when it is needed:
    - A super flywheel spinning at 100,000 rpm might deliver 100 kW for 15 seconds, then recharge at the same rate until backup generators can be brought back on line during a power outage (usually within 5-20 seconds).
    - Super flywheels have been proposed as power boosters for EVs, but this idea raises concerns about containment of high-speed rotors in case of accidents or mechanical failures, requiring massive and heavy enclosures that would offset power advantages.

359. ITI Arturo Maligani

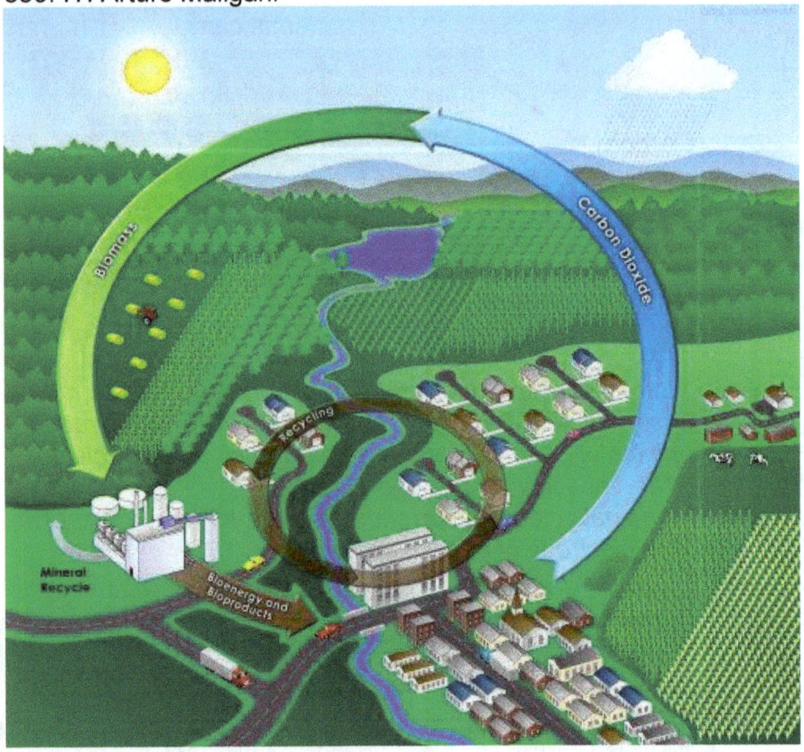

Biomass Conversion

Biomass residues from plant matter such as trees, grasses, agricultural crops, and other biological materials can be converted into useful gaseous hydrogen as well as liquid fuel forms.

- Government and industry organizations in the US, Europe, and many developing countries are actively cultivating energy crops specifically for biomass fuel production:
  - As fossil-based fuels become scarcer and more expensive, biomass conversion becomes more attractive and economical.
  - Bioenergy already ranks second to hydropower as a renewable US energy source, and accounts for 3% of primary US energy production.

Useful biomass products can be obtained from many sources, including dedicated energy crops and trees, agricultural food and feed crops, crop wastes and residues, aquatic plants, and animal and municipal wastes.

- Conventional technologies convert those materials into diverse energy-related products:
  - Biopower technologies presently produce more than 10 gigawatts of US electricity.
  - Biofuel research and development is advancing production and product quality.
  - Biomass-derived materials are gradually replacing non-renewable traditional petrochemical-derived products, including chemicals, plastics, fibers, and structural substances.

360. Ian Smith
362. EESI

361. Minnesota Legislature
363. Marcello Movra

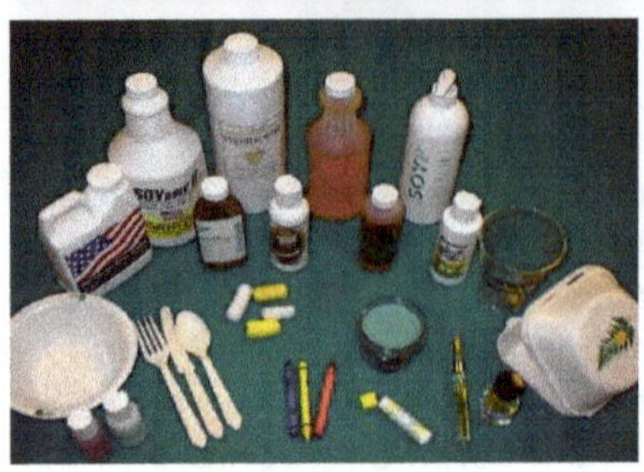

Biomass Products

Humans have used biomass energy (or "bioenergy") ever since people began to burn wood to cook food and keep warm.

- Wood is still the largest bioenergy resource today, but it is only one among a great many others, including:
  - Food crops, such as corn stover, potatos, vegetable oils (rapseed, corn, safflower), and brewery waste.
  - Animal crops, such as switchgrass, wheat straw, and other grassy plants.
  - Residue from forestry and sawmills (sawdust, woodchips, and pallets/crates).
  - Animal and landfill wastes (methane in particular).

364. Seeds of Change
366. Flagstaffotos
365. Fairfax Co. Pub. Schools
367. US Fish and Wildlife Service

Bio-Energy Resources

368. USDA-ARS
369. Stan Shebs

Bio-Energy Combustion

Burning biomass releases about the same amount of carbon dioxide as burning fossil fuels produced by photosynthesis millions of years ago.

- Advocates argue that biomass burning releases carbon dioxide that is largely balanced by carbon dioxide captured by plant growth ("new" greenhouse gas vs. "old" greenhouse gas):
  - As with all fuels, true accounting of greenhouse emissions must also factor in pollutants released through fuel production and processing, including crop-growing and harvesting.

Most biomass-fired power plants burn feedstocks directly for steam-driven turbines.

•Some biomass industries use spent steam from these plants to heat buildings as well as for manufacturing processes:

- As with other power generators, CHP can greatly increase overall energy efficiency.
- Paper mills are the largest current producers of biomass power, and typically use the generated electricity and process heat to support recovery of pulping chemicals.
- Co-firing, which mixes biomass with fossil fuels such as coal, can significantly reduce process emissions (especially sulfur dioxide).

370. Council of Industrial Boiler Owners

Direct Use of Biomass Fuel

371. Biomass Energy

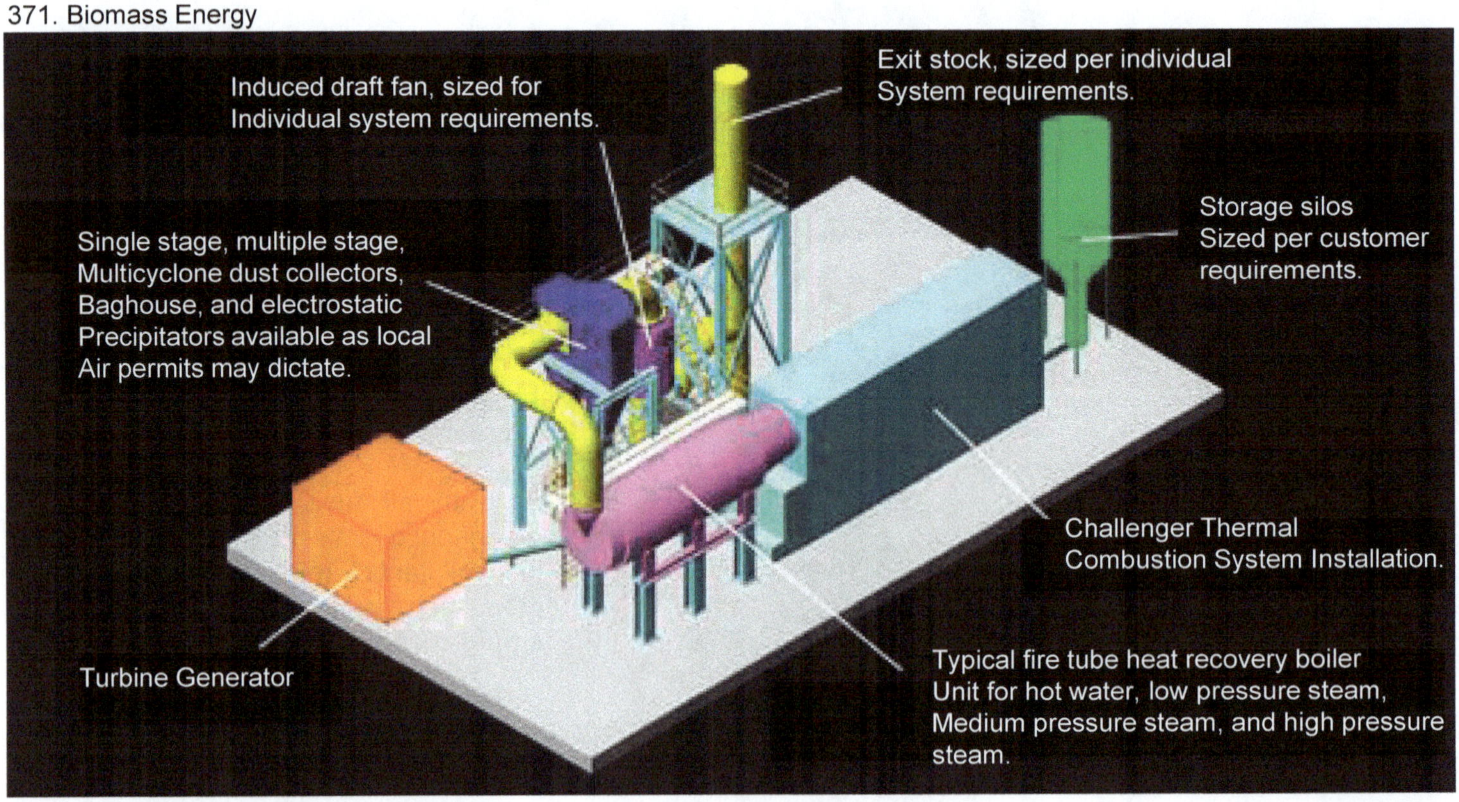

Typical Direct-Fired System

In addition to use as a direct fuel, biomass is converted to bio-oils, biogas and petroleum additives through various chemical means:

- Key among these are biochemical, chemical and thermochemical processes:
  - Biochemical (or organic) processes use natural decay methods (anaerobic digestion and fermentation) to produce methane and liquid fuels and feedstock.
  - Chemical processes mix biomass and other fuel products together into combined forms using catalysts to create liquid fuels.
  - Thermochemical processes (pyrolysis and gasification) use heat to convert biomass and fossil fuels into hydrogen, biogas and other energy and feedstock materials.

372. Larry Bell

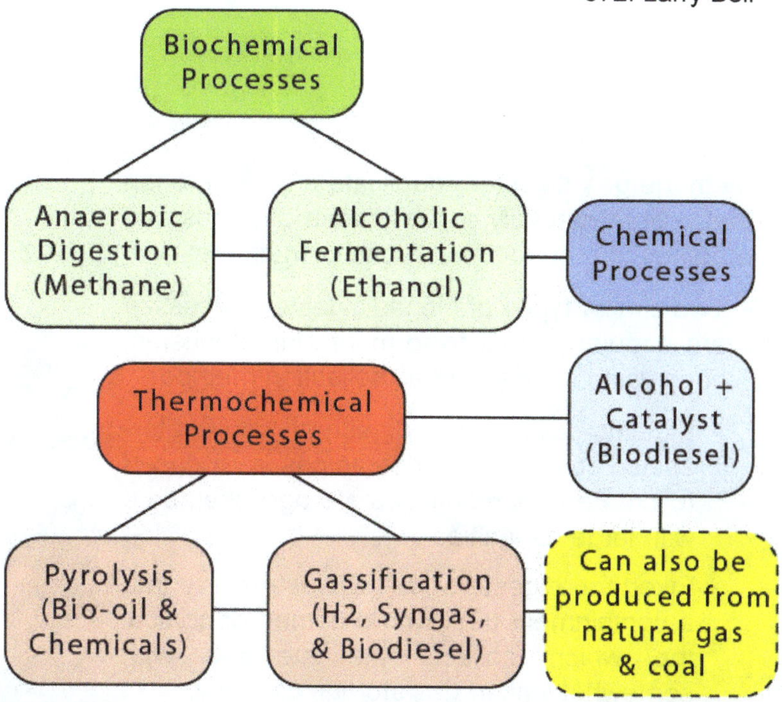

Biofuel Conversion Methods

373. INRA Montpellier

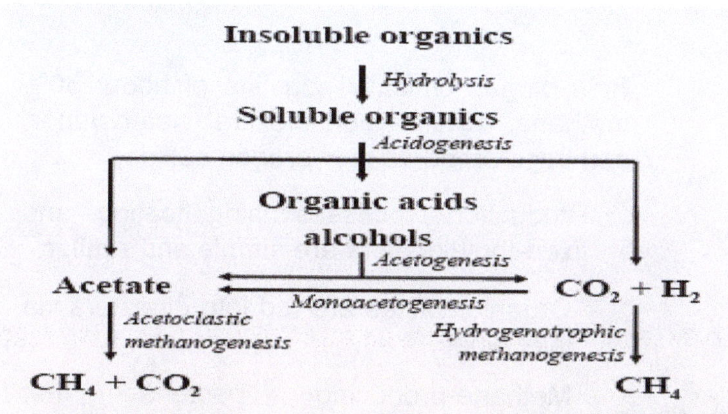

Anaerobic digestion transforms organic materials under oxygen-free conditions into biogas, nutrients, salts and cell matter. The process involves an acidification (acid) phase, and a methanogenic (methane) phase.

Anaerobic digestion is used for treatment of municipal and industrial sludge, human and animal wastes, and municipal waste waters. Benefits include energy production/ conservation, waste/ odor removal, pathogen control, and greenhouse gas reduction.

Anaerobic Digestion

Methane gas is produced through anaerobic digestion which breaks down organic matter by microbiological activity in the absence of air.

- Two common man-made technologies are used:
  - Fermentation of human and/or animal wastes in specially-designed digesters kills bacteria in the manure and produces methane.
  - Wells drilled into landfills release methane from decaying organic matter which is piped to central points where it is filtered and cleaned before burning.
  - Anaerobic digestion is used for treatment of municipal and industrial sludge, human and animal wastes, and municipal wastewaters. Benefits include energy production / conservation, waste / odor removal, pathogen control, and reduction of greenhouse gases.

374. Larry Bell

Methane gas digesters duplicate a phenomenon that occurs naturally at the bottom of ponds and marshes to produce marsh gas or methane.

- Two similar types of biogas digester processes are popularly used to ferment animal wastes, each using a different gas collection method:
  - Floating-cover type digesters use water-sealed covers which rise as gas is produced to create expanding gas storage chambers with large capacities.
  - Fixed-top digesters feed a waste-slurry into a containment tank with a domed or conical top which collects the methane with somewhat limited gas storage capacity.

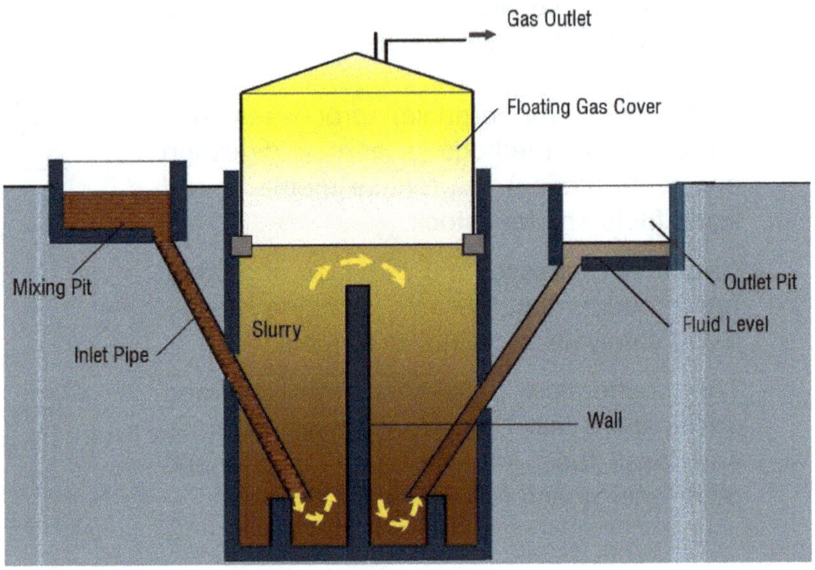

Floating-Cover Anaerobic Digester

375. Clarke Energy

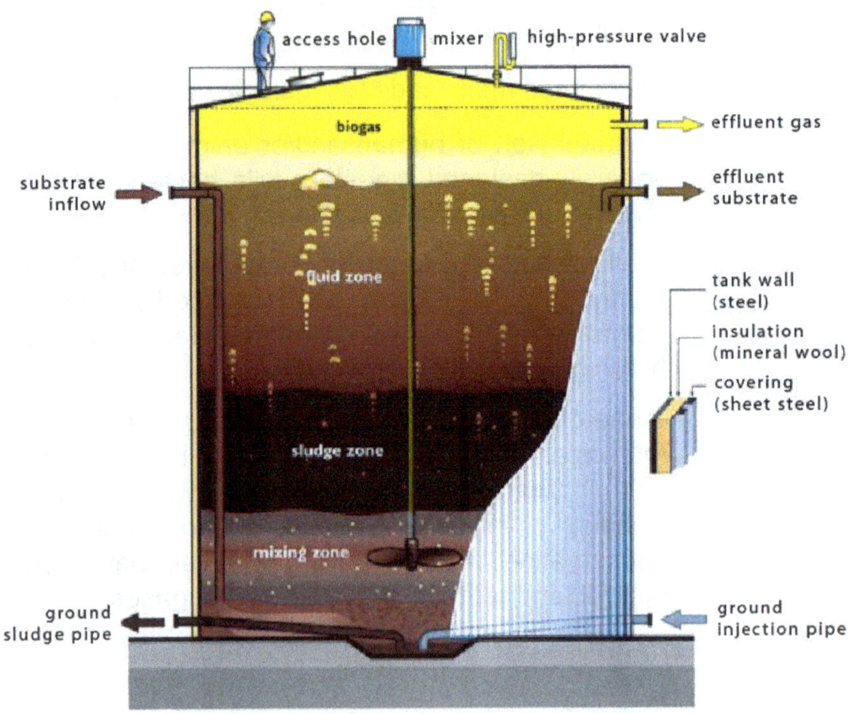

Fixed-Top Anaerobic Digester

Raw biogas typically consists of about 60% methane, 40% carbon dioxide, water vapor, and trace amounts of hydrogen sulfide.

- Production processes using floating- and fixed-top digesters are simple and similar:
  - Organic wastes are fed into digesters via inlet pipes.
  - Methane-producing bacteria that transform the wastes into component parts operate most efficiently at temperatures between 86°F - 172°F (30°C - 60° C) (requiring heat to be added in colder climates).
  - Digestion time ranges from weeks to months, depending upon feedstock and temperature.
  - Residual slurry removed at outlets can be used as a fertilizer.

Ethanol (also called ethyl alcohol, grain alcohol or simply alcohol) is produced both biologically by fermentation of sugars with yeast, and through hydration of petrochemical ethylene.

- Glucose for organic ethanol can be obtained either from biomass cellulosic wastes or from grain crops such as corn, switchgrass and sugarcane:

- Ethanol produced from cellulose requires hydrolyzing enzymes to digest the materials, adding substantial costs.
- Ethanol obtained from corn and other grains requires energy to grow the plants and convert them into alcohol.
- Closed-loop agricultural ethanol production can reduce energy costs by using fermenting energy from manure produced by livestock that are fed distillation byproducts, and by applying leftover manure for fertilizer to grow more grains.

376. Larry Bell

1. Milling:

The corn (or other grain) is passed through hammer mills that grind it into a fine power called “meal”.

2. Liquefaction:

The meal is mixed with water and alpha-amylase and passed through cookers to reduce bacteria levels.

3. Saccharification:

The mash from the cookers is cooled, and a second enzyme (gluco-amylase) is added to convert the liquefied starch to sugars.

4. Fermentation

Yeast is added to ferment sugars into ethanol and CO_2. Using several fermenters, the process continues until complete.

5. Distillation:

The fermented mash (or “beer”) containing about 10% alcohol is pumped through a multi-column distillation system where alcohol is removed from unfermented solids, leaving about 96% strength.

6. Dehydration:

Alcohol passes through a dehydration column where water is removed. Most plants use a molecular sieve to capture any remaining water, and the anhydrous ethanol is now about 200 proof.

Typical Ethanol Production Processes

377. NCP

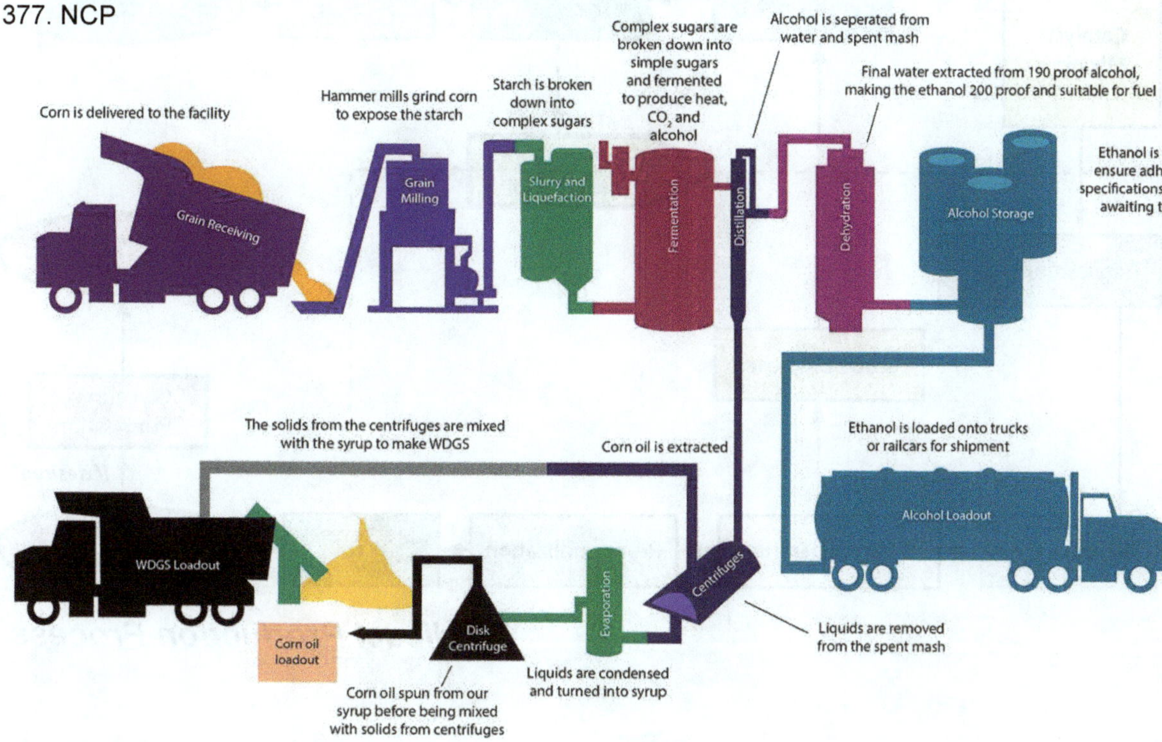

Distillous grain recovered from ethanol production is a high-protein livestock feed nutrient, and “syrup” containing other solids can be combined with it or sold separately.

Ethanol Production and Products

Biodiesel is produced through a process in which organically-derived oils are combined with alcohol (ethanol or methanol) in the presence of a catalyst to form ethyl or methyl ester.

- Biomass-derived ethyl or methyl esters can be blended with conventional diesel fuel or used in a 100% form:
  - Biodiesel can be made from a variety of fats and oils, including soybean or canola, animal fats, vegetables and microalgae.
  - The fats/oils are first degummed, and acid is removed.
  - Methanol and a catalyst are then added for esterification prior to final washing and drying.

378. Risoe

Biodiesel Production

379. Fleet Challenge Canada

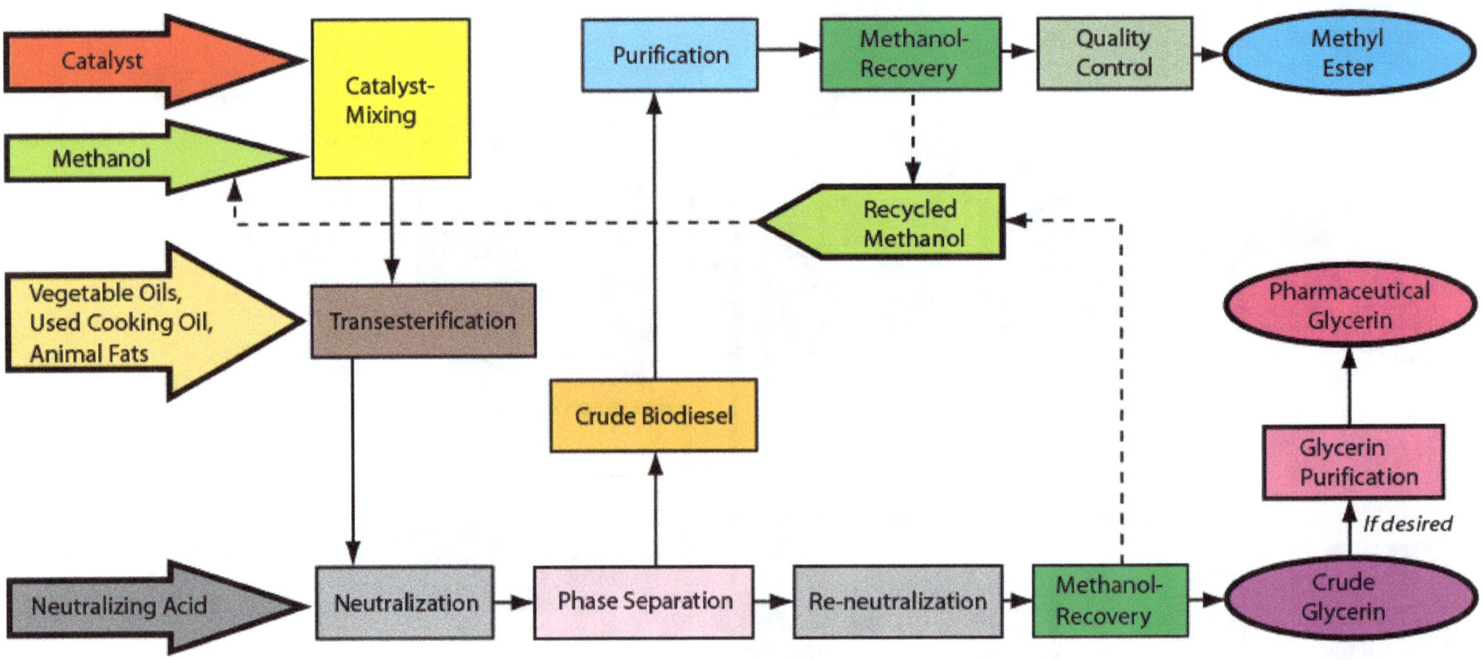

Biodiesel Production Process

Methanol (methyl alcohol) can be blended with other liquid fuels to reduce greenhouse gas emissions in internal combustion engines, and can serve as an ideally simple one-carbon molecule hydrogen carrier for fuel cells.

- Methanol (CH_3OH) can be produced from syngas originating from several sources:
  - Syngas from anaerobic digesters or landfills can be reacted with CO_2 to create methanol.
  - Syngas from biomass pyrolysis can also be reacted with CO_2 (one ton of biomass yields about 45 gallons [720 liters] of methanol).
  - Carbon monoxide and hydrogen from coal gasification can be reacted with steam (currently a major source).
  - Hydrogen from electrolysis or methane steam reduction can be reacted with CO_2 to produce the fuel.

380. Larry Bell

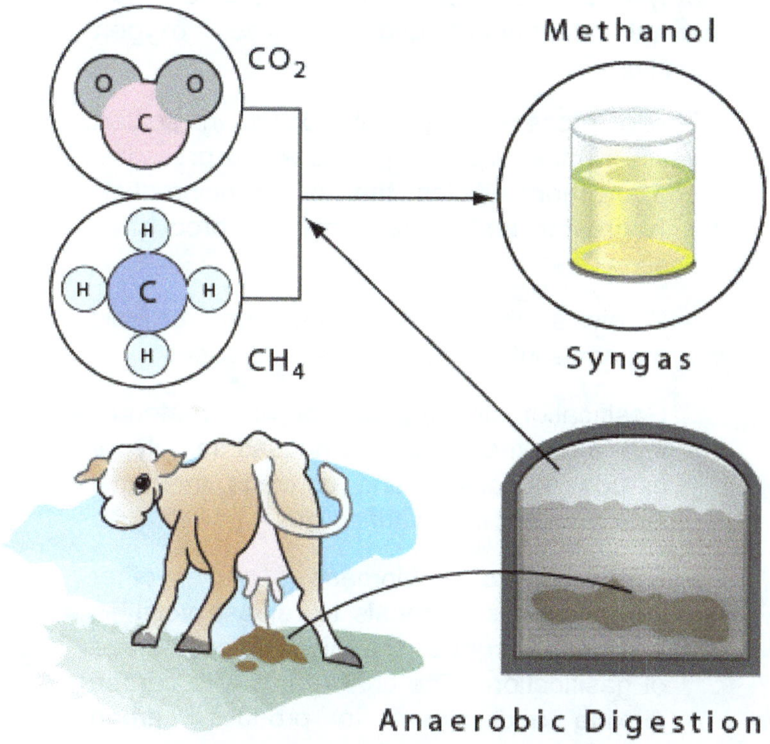

Methanol from Biodigesters

Many research organizations are pursuing ways to develop efficient and inexpensive ways to produce methanol from CO_2 using various enzyme reactions.

- One method being studied reduces CO_2 to methanol by coupling reactions of three enzymes using NADH from photosynthesis in linked conversion stages:
  - The enzymes are encapsulated into a porous silica sol-gen solution that solidifies into a gel within a buffer solution containing small amounts of NADH.
  - The NADH dehydrogenase serves as an electron donor for each of the three enzyme conversion stages.
  - After about two days to allow time for the NADH to diffuse into the gel, CO_2 is added, producing methanol through enzymatic reaction with a nearly 44% yield.

381. Larry Bell

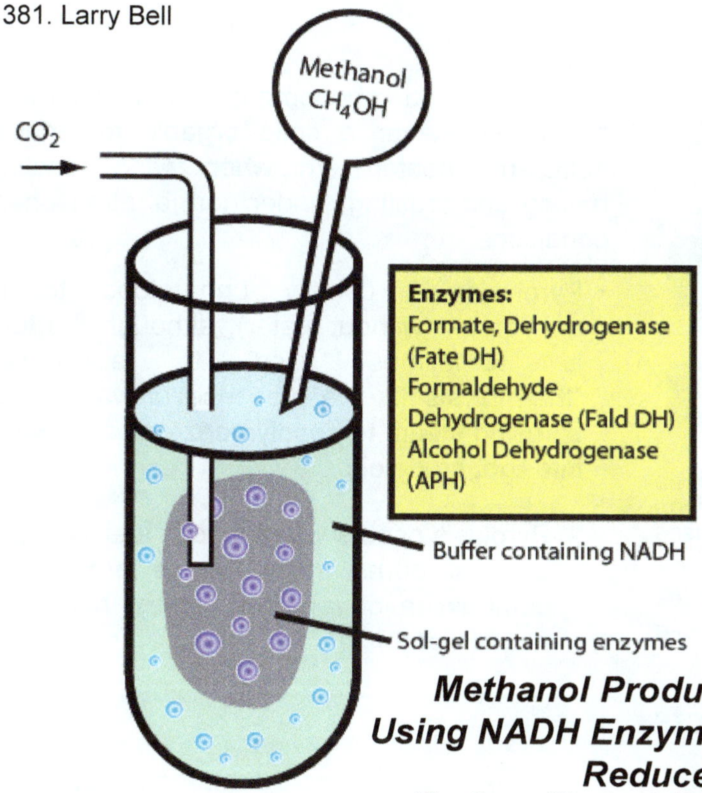

Methanol Production Using NADH Enzymes to Reduce CO_2
(Southern Illinois U. Process)

382. Claus Hindsgaul

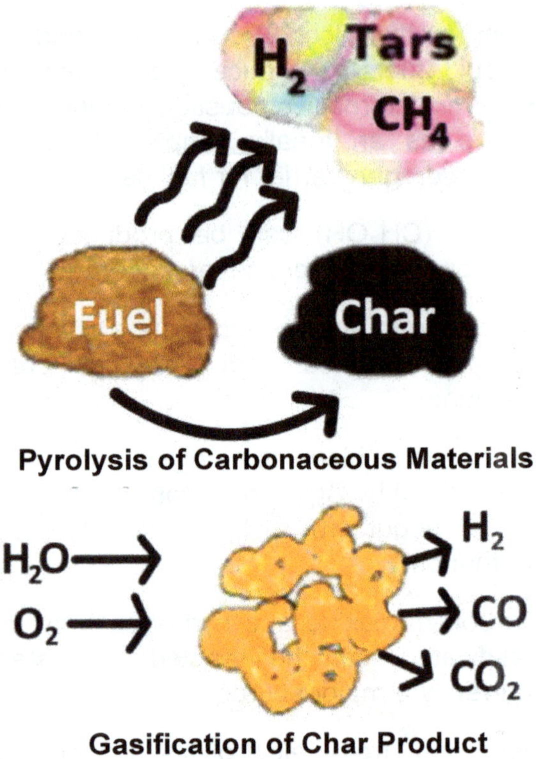

Pyrolysis and gasification are related forms of thermal treatment under limited oxygen conditions.

- These processes typically occur in sealed vessels under high temperature and pressure, and are more efficient than direct incineration with greater energy recovery for productive uses:
  - Pyrolysis chemically decomposes organic materials into solids, liquids, and gases.
  - Gasification decomposes organic materials into a synthetic gas composed largely of carbon monoxide, and converts inorganic waste into an inert vitrified gas.
  - Pyrolysis of biomass and other carbonaceous materials to release volatiles and char occurs at the initial process stage of gasification. (The char reacts with carbon dioxide and steam to produce carbon monoxide and hydrogen.)

Pyrolysis and Gasification

383. Alcyon Engineering

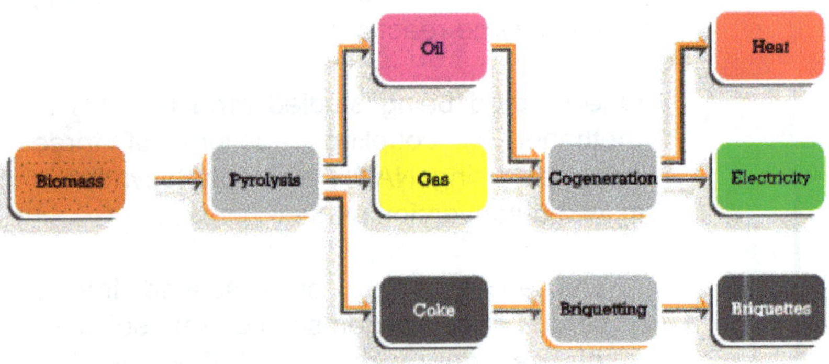

When biomass decomposes at elevated temperatures, three primary products are formed: gas, bio-oil and char. At high temperatures the bio-oil decomposes again into secondary gases and polymetric tar.

Ancient use of anhydrous pyrolysis produced charcoal, and the process is currently used on a massive scale to turn coal into coke for metallurgy (particularly steelmaking).

Pyrolysis is a phenomenon that commonly occurs whenever a solid organic material is sufficiently heated (e.g. when frying, roasting, baking and toasting) under normal atmospheric conditions.

- Pyrolysis is usually understood to be anhydrous (without water), although the term is sometimes used to encompass “thermolysis” in the presence of water (such as using steam to depolymerize organic waste into light crude oil):
  - Pyrolysis occurs when a solid fuel like wood or coal burns. (The flames are due to combustion of released gases, not of the actual wood.)

Pyrolysis Products

Biomass pyrolysis produces oil that can be used directly as fuel, and also as a feedstock for valuable chemicals and materials.

- Pyrolysis oil (or "bio-oil") is a renewable liquid fuel that can be easily stored and transported:
  - "Fast pyrolysis" has achieved commercial success as a method to produce oil for engines, turbines, and boilers.
  - Bio-oils can be upgraded to high-quality hydrocarbon fuels, although presently only at high energetic and financial costs.
  - The yields and products of bio-oil depend upon the feedstock, process type and conditions, and product collection efficiency.

384. Madur

Pyrolysis Bio-Oil Production

385. European Commission, DG Research

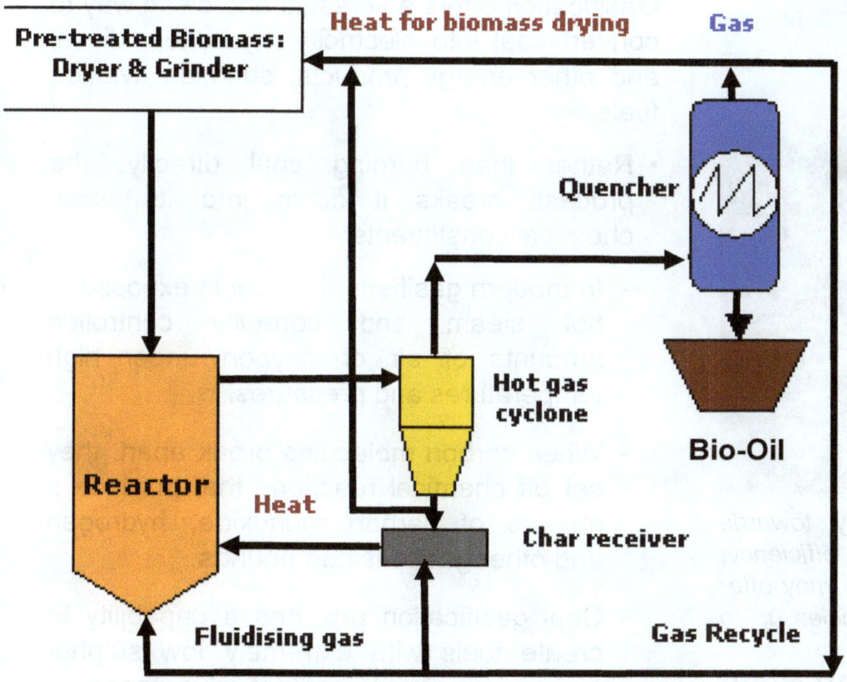

Fast Pyrolysis Processing

Fast pyrolysis of biomass can yield liquid energy fuel and chemical products with yields up to 75% based upon starting biomass weight.

- Fast heating of biomass greatly reduces the amount of char – forming reactions to optimize liquid fuel product:
  - This process rapidly heats the biomass stock to temperatures of 750°F - 1110°F (400°-600°C).
  - Product yields are temperature-dependent. (Maximum oil yields are obtained at about 930°F [500°C].)

Biomass can be gasified to produce a synthesis gas composed primarily of hydrogen and carbon monoxide (also called "syngas" or "biosyngas").

- Gasification uses a thermochemical process similar to pyrolysis which exposes the stock material to high temperatures in the absence of oxygen, causing it to decompose:
  - Syngas can also be converted using a "Fischer-Tropsch" process to create Fisher-Tropsch diesel fuel which is similar to petroleum diesel.
  - These same fuels can be produced from natural gas using a similar process.

386. Larry Bell

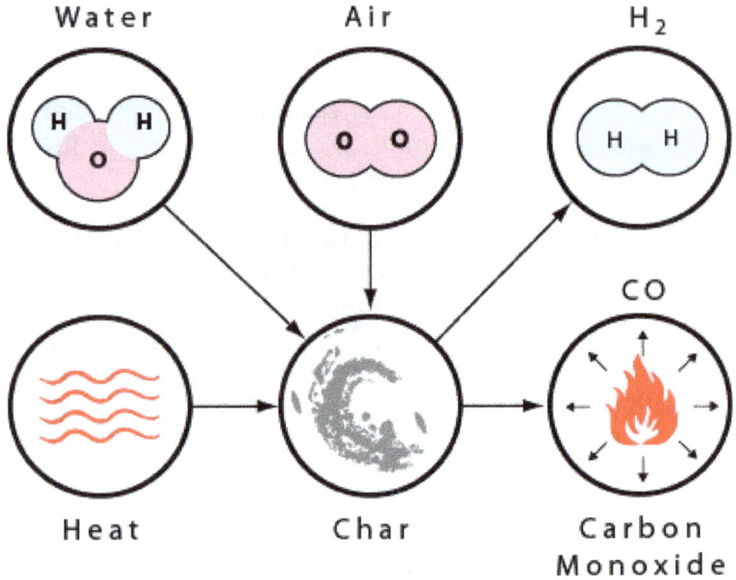

Gasification of Char

387. University of ND

The clean coal technology field is moving rapidly towards gasification with CO_2 capture, offering 73% thermal efficiency. Using hydrogen produced in gas turbines for electricity may offer long-term potentials to achieve overall system efficiencies up to 60%

Coal Gasification Plant

Gasification offers a versatile and clean way to convert coal into electricity, hydrogen, diesel and other energy products, such as aviation fuels.

- Rather than burning coal directly, the process breaks it down into its basic chemical constituents:
  - In modern gasifiers, the coal is exposed to hot steam and carefully controlled amounts of air or oxygen under high temperatures and pressures.
  - When carbon molecules break apart, they set off chemical reactions that produce a mixture of carbon monoxide, hydrogen and other gaseous compounds.
  - Coal gasification provides a capability to create fuels with extremely low sulphur and carbon dioxide content advantages.

The gasification process was originally developed in the 1800s to produce gas for lighting and cooking, and has been utilized for production of synthetic chemicals and fuel since the 1920s.

- Wood gasifiers producing "gasogene" provided fuel used to power vehicles in Europe during World War II, and the process has subsequently gained much broader applications:
  - Integrated gasification combined cycles (IGCC) produces electricity, ammonia, and liquid fuels, and is capable of producing hydrogen for fuel cells.
  - Recent gasification technologies have been developed that use plastic-rich waste as feedstock to create methanol.

388. US DOE Fossil Energy Office

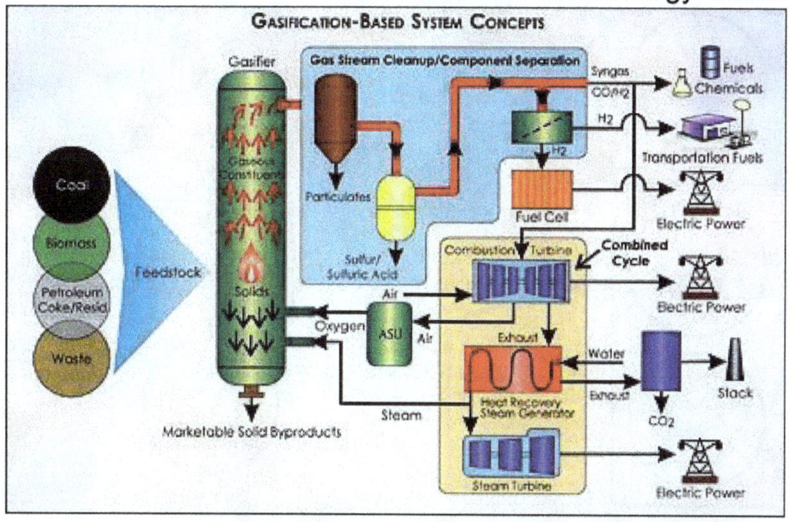

Gasification Stages and Products

389. Missoula News

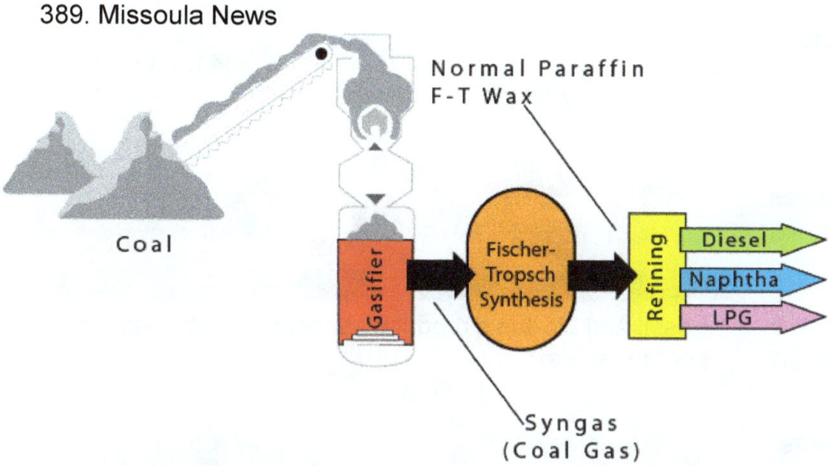

The original Fischer-Tropsch process to convert coal into liquid fuel was developed by German researchers during the 1920s and is still used to produce diesel fuel today in South Africa and China. Although currently more expensive than diesel from crude oil, this approach is likely to become more cost-competitive as petroleum sources are depleted and prices rise.

Liquid Fuels from Coal

Researchers at the Penn State University Energy Institute have demonstrated that jet fuel for high-performance military aircraft can be produced from coal, with the possibility that similar methods might also produce fuels for conventional jet engines in commercial aircraft.

- Two alternative methods have been successfully tested:
  - One method is a variant of a process used in petroleum refining where bituminous coal is combined with a byproduct petroleum oil and heated until the mixture becomes fluid and the liquid portion distills off as jet fuel (leaving solid coke as a useful secondary product).
  - The second process mixes refined coal-derived chemical oil (a byproduct of the coke industry), with a light petroleum oil, and adds hydrogen to produce jet fuel as a distillate.

390. Larry Bell

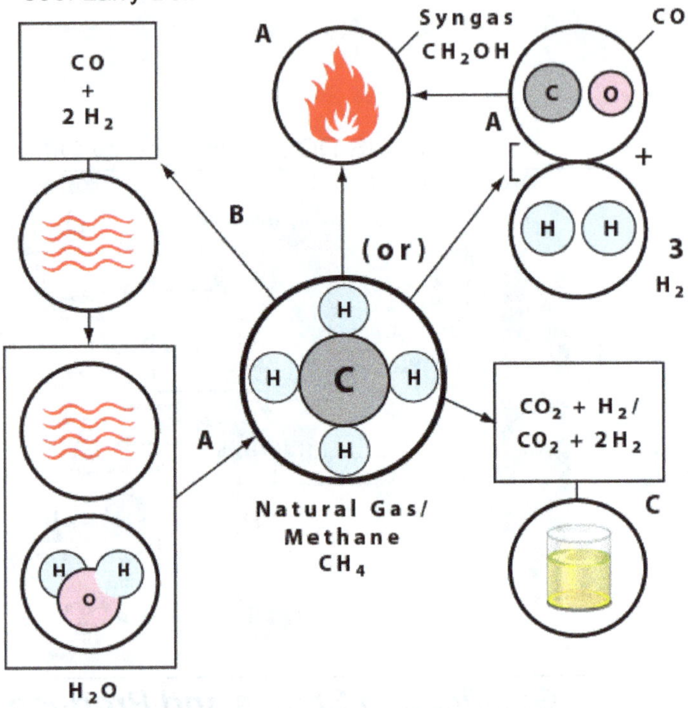

Hydrogen, Methanol and Syngas Production Using Steam Methane Reforming (SMR)

Steam methane reforming using natural gas is the most widely used method to produce hydrogen, and can also be used to produce methanol and syngas.

- SMR uses moderate pressure (2 MPa), high temperature 1560°F (850ºC) steam, and a nickel catalyst:
  - **Option A**: Methane reacts endothermically with steam to create carbon monoxide and hydrogen ($CH_4 + H_2O \rightarrow CO + 3H_2$) or syngas ($CH_2OH$).
  - **Option B**: Methane undergoes partial oxidation with oxygen to release heat through an exothermic reaction which is used to drive the SMR process in-situ through "autothermal reforming" as a precursor step to conserve energy in producing methanol ($CH_4 + 0.5O_2 \rightarrow CO + 2H_2$).
  - **Option C**: The ration of CO and H_2 from option B is adjusted through a water-gas shift reaction ($CO + H_2O \leftrightarrow CO_2 + H_2$), and CO and H_2 react with a second catalyst (copper, zinc oxide and alumina) to produce methanol ($CO + 2H_2 \rightarrow CH_3OH$).

Available power production options that can supply substantial percentages of US requirements are few, each presenting disadvantages that must be reconciled within comprehensive energy strategies.

- Only three general types of primary energy sources are likely supply more than a few percent of national electricity, heating, and transportation needs*:
  - All fossil sources are non-renewable and release CO_2 and other atmospheric pollutants when combusted.
  - All biofuels also release combustion pollutants, and although often cleaner, only marginally extend rather than replace fossil fuels; require energy for production and processing that may release pollutants; and some compete with land use essential for livestock/food crops.
  - Nuclear power releases no atmospheric pollutants, but presents problematic safety, security and waste storage concerns.

*Hydrogen (fuel cells) are net power consumers, not producers.

341. Larry Bell

Fossil and bio-fuels produce atmospheric combustion contaminates

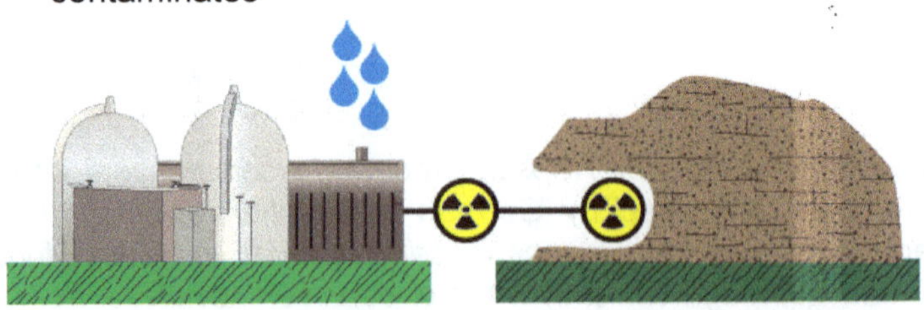

Nuclear plants emit only water, but produce wastes that must be sequestered.

Pollution Tradeoffs

Other general power production options offer important opportunities to reduce energy consumption from fossil, biofuel and nuclear sources, but are principally linked to local and off-grid applications.

- All of these "secondary options" are governed by geographic and site-specific conditions that constrain expansion capacities and utilization:
  - Mechanical energy from wind and hydropower is "free" and renewable, but is limited to locations with satisfactory seasonal climate and prevailing weather conditions.
  - Wind and solar are intermittent and somewhat unpredictable.
  - Hydroelectric and geothermal energy expansion is severely limited by a scarcity of practical and unexploited sites, along with environmental restrictions on new development.

392. SKF Group 393. EV World
394. How Stuff Works 395. ACRE

Mechanical and solar power is limited to sites with favorable weather conditions.

Hydroelectric and geothermal expansion is constrained by a scarcity of sites and environmental restrictions.

Expansion Limitations

396. Larry Bell

Energy Sources	Long-Term Supply	Utility Capacity	Energy Efficiency	Environment Issues
Fossil Fuels: • Natural Gas • Petroleum • Coal	Dwindling natural gas & petroleum reserves. Longer coal reserves.	Primary sources of electricity, heat, vehicle fuels and chemicals.	Relatively high compared with other fuel options (use-dependent).	CO_2 & other combustion emissions are greenhouse & health liabilities.
Biofuels: • Biomass burning • Bioenergy crops	Renewable energy sources, but may consume energy for production and transport.	Depends upon croplands & energy for acquisition, processing & distribution.	Depends upon biomass source, yields and process/ distribution costs.	May reduce greenhouse emissions, depending on production/ processing.
Nuclear Electric Plants	Can offer long-term potential using breeder reactors to process fuels.	Currently provides19% of US electricity with large growth possible.	Comparable to fossil fuels in production and plant costs.	Clean emissions but requires storage/ protection of spent fuels.

Primary Sources

Energy Sources	Long-Term Supply	Utility Capacity	Energy Efficiency	Environment Issues
Mechanical Energy • Wind Turbines • Hydropower	Renewable sources with limited site applications for substantial benefits	Depends upon local site features & seasonal/ daily weather conditions	Can be very efficient, primarily for power grid balancing.	Wind is clean. Hydro can impact marine ecosystems.
Geothermal Power Plants	Renewable energy source, but very limited growth possibility	Depends upon limited sites with suitably high/ accessible temperatures.	Depends upon size/ temperature of heat reservoirs for power plants.	Clean, but can have potential adverse subterranean impacts.
Solar Power: • PV systems • Utility-scale concentrators	Renewable, but primarily a back-up or power-balancing supplement.	Depends upon geographic climate & daily site weather.	Influenced by atmosphere, weather & size/type of system used.	Clean energy with excellent conservation benefits.

Secondary Sources

Since power production sources are limited, it is important to use all of these assets resourcefully through efficient conversion methods.

- Various systems offer special advantages and disadvantages that relate to application needs:

  - Internal combustion engines offer relatively high weight-to-power ratios that are particularly useful for vehicular applications, but produce pollutants and efficiency penalties due to incomplete fuel combustion.

  - Electric turbines and generators can offer high efficiencies, particularly in combined cycle CHP applications, but often consume non-renewable fuels and emit greenhouse gases.

  - Fuel cells burn cleanly, but are expensive and consume more energy than they yield for hydrogen production, transfer and conversion.

  - Photovoltaics offer valuable backup and off-grid applications, but cost-recovery can require many years despite rapidly growing use.

Energy storage between periods of excess production and demands can impose significant efficiency penalties.

- In addition to storage of potential energy using hydrokinetic water columns or in the form of liquid and gaseous fuels, a few key system devices are useful for general applications.

  - Batteries are presently the most practical way to store electricity for relatively small-scale purposes, including electric and hybrid electric vehicles, and while very efficient, impose weight penalties that reduce overall mileage economies.

  - Supercapacitors provide very high power densities for utility-scale power balancing, and can capture and store energy from hybrid vehicle regenerative braking to extend ranges.

  - Super flywheels are used to fill short power gaps in electrical power plants, but present mass penalties along with safety risks for vehicles in the event of accidents and structural high-speed rotor failures.

397. Larry Bell

Systems	Applications	Advantages	Disadvantages
Internal Combustion Engines • Diesel • Gasoline	Stationary & mobile mechanical & electrical power for vehicles and small-scale local uses.	Relatively good weight-to-power ratios using a variety of liquid & gaseous fuels.	Incomplete combustion emits pollutants and relatively low power-to-wheel efficiencies.
Electrical Turbines/ Generators: • Steam • Gas	Utility and industrial-scale electricity & heating often using combined heating & power.	Provide more than ¾ of all US electricity with high CHP efficiencies using various fuels.	Systems using fossil sources release CO_2 & other greenhouse gasses & consume resources.
Fuel Cells: • Fixed arrays • Tracking arrays	Used to power vehicles & to provide heat & electricity for local facilities.	Transportable energy transfer devices that can use hydrogen obtained by various means.	Fuel requires more energy to obtain & transfer than it releases at about 50% efficiency.
Photovoltaics • Fixed arrays • Tracking arrays	Principally used for small-scale water heating & off-grid backup site power.	Rapidly growing market driven by conservation & environmental interests is reducing costs.	Effectiveness depends upon sites & cost recovery can require several years.

Energy Production Systems

398. Larry Bell

Systems	Applications	Advantages	Disadvantages
Batteries • Monopolar • Bipolar	Have an extremely broad range applications & are a pacing technology for electric & hybrid vehicles.	Are commercially available in many types and sizes, & can offer very high energy conversion efficiencies.	Tend to be relatively heavy & large, adding weight that compromises vehicle economy.
Supercapacitors	Can be useful to capture & store energy from hybrid vehicle regenerative braking to optimize economy.	Have very high power densities (4,000 W/kg or more) contained within simple lightweight devices.	Have very low energy densities (usually less than 15 Wh/kg) which limits broader uses.
Super Flywheels	Often used to fill gaps in utility-scale electrical plants for short-term power balancing.	Have very high power densities with rapid re-charge capacities to cover brief power outages.	High speed rotors require massive enclosures & present safety risks for vehicle applications.

Energy Storage Systems

Power Production Options

A variety of methods are used to convert organic fossil and biomass materials into relatively clean gaseous and liquid fuels, nutrient fertilizers and valuable chemical feedstocks.

- Some of these methods apply natural biological processes and conventional waste treatment and industrial production technologies:

- Methane alternatives to natural gas and associated byproducts can be produced using biochemical anaerobic digestion and fermentation of biowastes.

- Chemical catalysts can convert plant wastes and grains into biodiesel fuel and other energy products.

- Thermochemical pyrolysis and gasification are used in combination to release bio-oils, hydrogen and syngas from carbonaceous materials and realize “clean coal” benefits.

- Steam methane reforming is used to produce hydrogen, methanol and syngas from natural gas, and can do the same from biodigested methane.

399. Larry Bell

Systems	Applications	Advantages	Disadvantages
Biochemical: • Anaerobic Digestion • Alcoholic Fermentation	Produces methane, liquid fuels, & chemical feedstocks through natural bacterial & chemical processes.	Can be applied using organic, municipal & agricultural wastes for resource recovery & bio-remediation.	Must be applied on a practical scale to realize practical development & operation benefits.
Chemical	Reacts bio-chemical, thermo-chemical & other materials to produce fuels and derivatives.	Can convert alcohols & sugars from organic sources into biodiesel and other energy products.	May require expensive catalysts that impose costs that reduce advantages.
Thermochemical: • Pyrolysis • Gasification	Use conventional industry processes to produce bio-oil, hydrogen, sugars & liquid fuels.	Extensively used to create "clean coal" (gasification) & may be a source of diesel and jet fuel.	Use of a natural gas as the heat source consumes a dwindling natural resource.
Steam Methane Reforming (SMR)	Is the most widely used process to produce hydrogen from natural gas/ methane	Is also used to produce methanol & syngas for local plant & commercial applications	Use of natural gas as a feedstock consumes a valuable & dwindling natural resource.

Processing Methods

400. Larry Bell

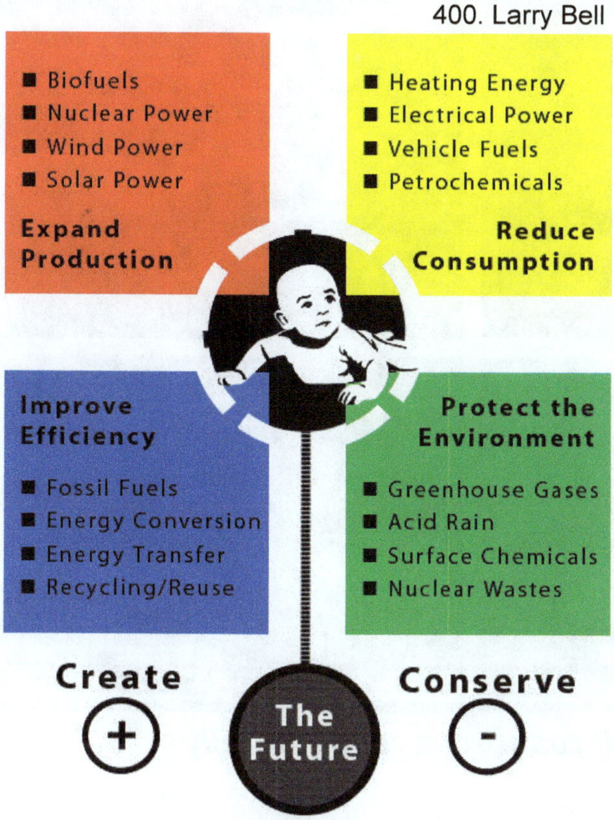

The Only Real Options

Regrettably, no simple or singular solution will supply future resource needs at expanding or current consumption rates; no prospects exist for technology breakthroughs that will provide cheap or limitless energy; and no known strategy will eliminate all impacts of our present industries and lifestyles upon regional and global ecosystems.

- Our only real options demand combined efforts to develop and expand more environmentally-friendly alternatives to fossil fuels, to make more effective use of all resources, and above all, to apply aggressive conservation practices at all levels of scale that reduce wastes:

  - Global volatility and environmental vulnerability require that we enact decisive interventions now that allow time for transitions essential to prevent devastating crises later.

  - Lives of today’s children and those who will follow will be greatly affected by our actions.

Utilizing All Opportunities

401. Greenpeace
402. University of Georgia
403. Minnesota Legislature
404. Sci/Tech
405. About Spas
406. Green Car Journal
407. Ministry of Technology
408. Tomorrow Bulletin
409. US DOE-EERE
410. Umpqua Watersheds
411. Rise
412. US DOE-NREL

We have no choice but to utilize all resources resourcefully

413. Texas A&M
414. US DOE-EERE
415. CA Fuel Cell Partnership
416. Celilo Group Media
417. G-man
418. Touchstone Hme. Ener.
419. Montana DEQ
420. Gunder-friend Productions
421. Eau Claire Eco. Dev.
422. Energy Xchange
423. Project GHB
424. Tewkesbury Borough
425. Sysmac Electronic
426. David Darling
427. Yelm, Farm
428. How Stuff Works

PART III: WORKING APPLICATIONS

429. EIA

Fossil fuels provide about 90% of all world energy.

- Oil has been the largest single source, followed by coal and natural gas, but a consumption trend is shifting towards coal, particularly in North America:
  - Coal remains much more abundant than known deposits of oil and natural gas, with an estimated 250 year US reserve supply.
  - Most known oil reserves are projected to be exhausted within about 40-60 years, depending upon the rate of growth in global consumption, along with possible extraction from sandy tars that may provide some added time.
  - Most known natural gas reserves are also projected to be depleted within about 60 years.

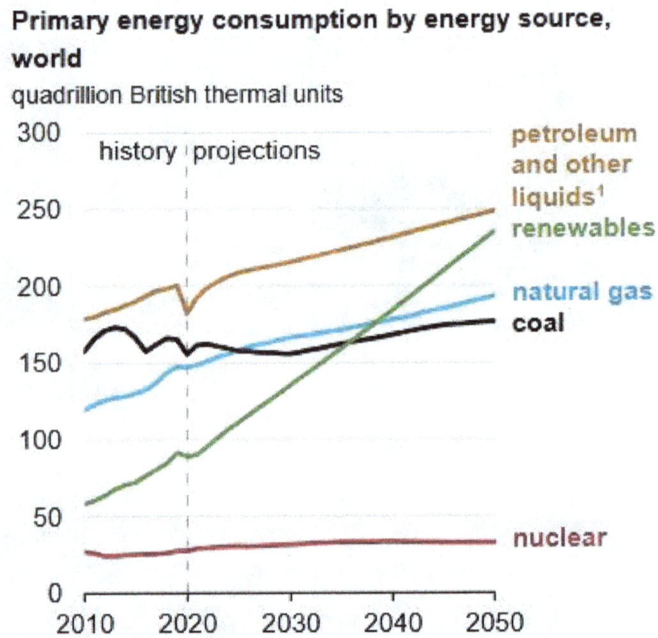

World Energy Demand Trends

430. U of Rochester
431. LA Review Journal
432. Clipmarks
433. Age Concern Cymru

International Trade

Travel and Tourism

Industries and Jobs

Health and Hardships

Future Impacts

Based upon best estimates, it is realistic to expect that global oil and natural gas reservoirs will be substantially drained before children attending kindergarten today reach typical retirement ages.

- Given our current high level of dependence upon these resources, a full range of consequences is daunting to contemplate:
  - International relationships and commerce will most certainly be impacted in major ways, including shifts in trade and production, business travel and tourism, and geo-political tensions and alignments.
  - National fuel shortages will create competition between user sectors and groups, including military, manufacturing, agriculture and transportation.
  - Power and heating costs will accelerate relocations of business and households to warmer locales (particularly those with good access to energy), leaving many older and poorer residents behind.

Today, the US consumes about 16 million barrels per day of petroleum, which is approximately 20% of the world's oil. Imports constitute a significant portion of this country's balance-of-trade deficit

- More than 66.6% of US oil consumption is in the transportation sector, where energy demands are growing rapidly:
  - The US consumed about 376 million US liquid gallons of gasoline per day in 2023.
  - US consumption levels are driven by high per capita ownership of automobiles, large vehicles, private vs. public transportation preferences and relatively low gasoline costs compared with many other countries.
  - US consumers, on average, spend a smaller fraction of their incomes on gasoline now than in previous decades.
  - Only about 52.6% of these gasoline costs are for crude oil, with 14.4% for refining, 17% for taxes, and 15.7% for distribution and marketing.

434. US EIA, 2024

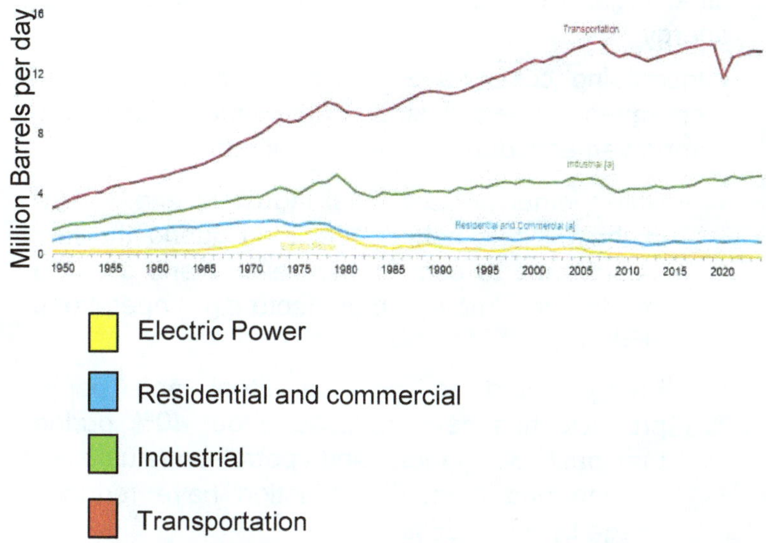

US Petroleum Consumption by Sector

435. US DOE - EIA

Total US Energy Usage, by Sector (2023)

The more wealthy nations use the most energy, and the US uses more per capita than other countries.

- About half of the US energy consumption is in transportation and residential sectors that are largely controlled by individual consumers, while the remaining industrial and commercial expenditures are mostly controlled by business, government entities and facility managers:
  - Approximately 52% of transportation energy is consumed by gasoline-powered vehicles (mostly personally-owned), 23% by diesel-powered trains, ships and trucks, and about 12% by aircraft.
  - Energy use in the residential sector varies across the country largely due to regional climate and regulation differences, with about half expended on space conditioning.
  - Space conditioning is also the biggest energy user in the commercial sector (about 30%), with lighting representing another 17%.

The industrial sector, including manufacturing, construction, agriculture, mining, waste treatment and other organizations, consumes about 33% of all US energy.

- Increasing costs have forced many of the more energy-intensive industries to implement significant improvements during recent decades:
  - Unlike other sectors, total industrial energy use in the US has actually declined during the last decade due to conservation efforts and growing trends to move manufacturing operations offshore.
  - Energy used to produce steel and paper products has been reduced about 40% during the past 30 years, and petroleum/aluminum refining and cement production have reduced usage by about 25%.
  - Energy required for fresh water treatment and distribution often consumes a large amount of electricity and natural gas.

436. Agency for Natural Resources

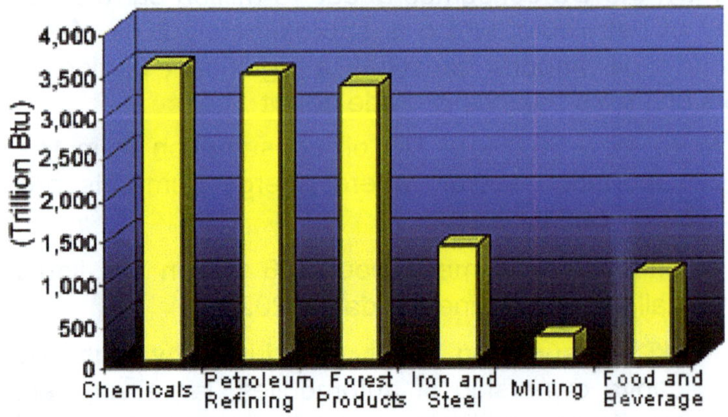

Forest product processing is the largest user of steam and combined heat-power systems; while petroleum refining is the largest user of fired heating systems (iron and steel are third). Chemicals are the largest users of electrical motors, and motors also represent about 25% of mining energy use.

Top Six US Energy System Users (Trillion BTU)

437. New Mexico's Passenger Rail Action Group

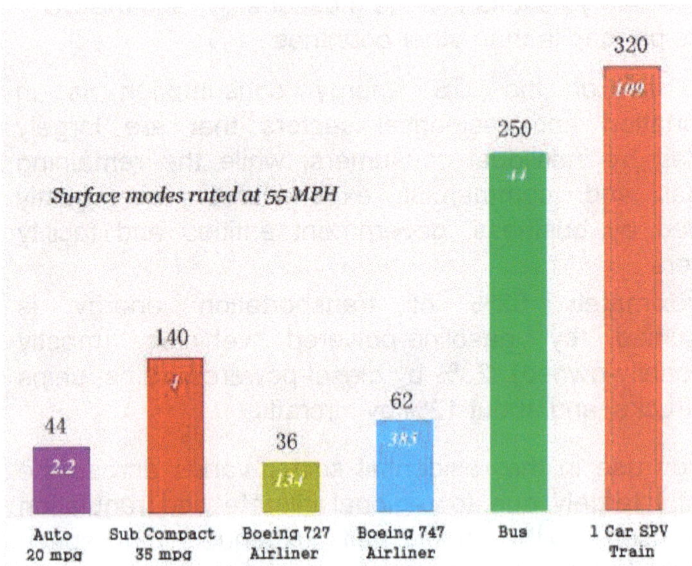

Unit = passenger miles per gallon of fuel (italicized number denotes passenger loads)

Fuel Efficiency of Various Transportation Modes

In the transportation sector, the oil supply crisis of the1970s prompted creation of a US Federal Corporate Average Fuel Economy (CAFE) program in 1975 that imposed fuel economy standards upon automobile manufactures.

- Automotive economy improvements and a "gas guzzler" tax assessed upon manufacturers since 1978 had positive effects that were offset after 1990 by the growing popularity of utility vehicles, pickup trucks and minivans that fell under more lenient truck standards:
  - Many state and local governments work to reduce vehicle energy consumption through subsidized public transportation, carpooling, "kiss-'n-ride" programs, and high-occupancy vehicle lanes.
  - Speed limits also work to improve vehicle operating economy, where a reduction from 65 mph to 55 mph can provide 15% more efficiency.

The commercial sector, including retail stores, business and government buildings, schools and other facilities, consumes a substantial amount of energy for space heating and lighting.

- Lighting, comprising about 10% of this energy, is the most wasteful component, often resulting from excessive illumination and use of incandescent sources that also add to space cooling loads:
  - Buildings with professional management and centralized control of energy conservation tend to perform much more economically than others.
  - Fluorescent lighting (about 75% more efficient than incandescent fixtures) has become a commercial standard, but produces certain adverse health effects.
  - Consistent hours of operation, programmed thermostats and lighting controls, and good solar planning can produce substantial benefits.

438. US DOE EIA

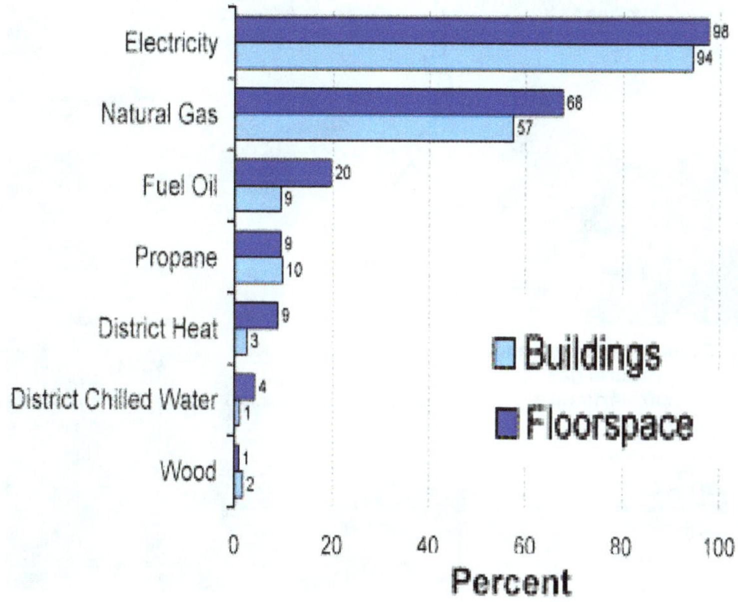

Energy Sources Used in Commercial Buildings

439. US EIA

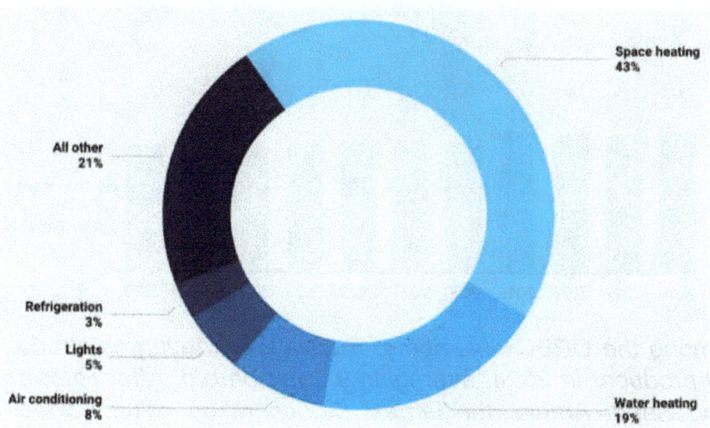

US Home Energy Consumption Averages

Although the efficiencies of furnaces and air conditioners have increased steadily since the energy crises in the 1970s, domestic energy demands continue to rise.

- A substantial portion of these growing demands is attributable to American lifestyle changes:
  - Average US home sizes have increased significantly, from about 1,500ft² (140m²) in 1970, to 1.792ft² (166m²) by 2024.
  - Single-person households have become more common.
  - The number of homes with central air conditioners increased from 23% in 1978 to 88% by 2020.
  - Lighting and water heating represent about 45% of average energy budgets, and lighting alone can consume 5% or more of the total energy in milder climates.

440. US EIA

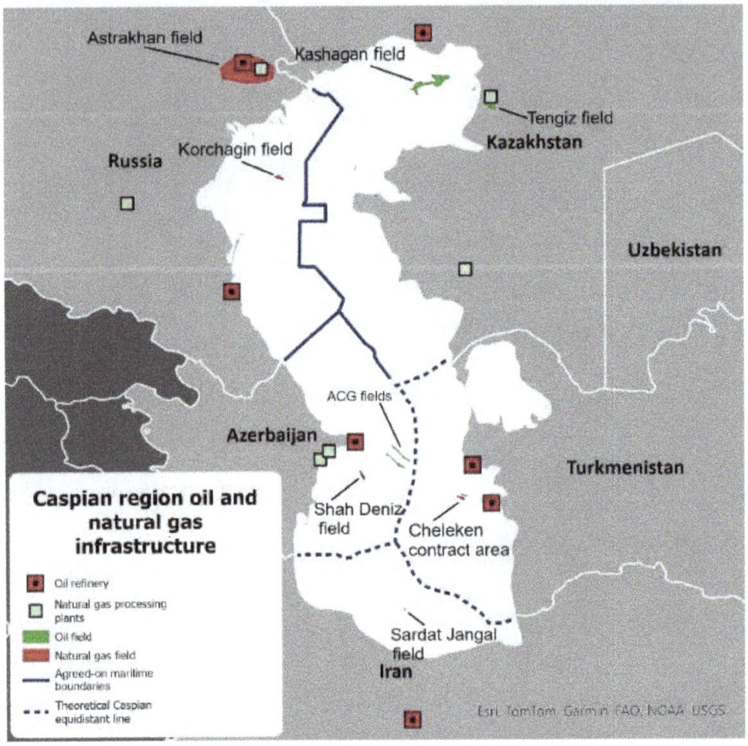

Many countries in the Caspian Sea region have long histories of political instability, and terrorism presents a growing threat to oil and gas pipeline security.

Estimated Caspian Sea Region Oil Reserves

Oil continues to be a critical fuel and chemical resource, and continued US dependence upon imported oil is becoming increasingly problematic and unrealistic.

- Although before 1950 the US was a major world oil producer, the EIA estimates that this country currently has less than 2% of known oil reserves:
  - Members of the Organization of Petroleum Exporting Countries (OPEC) produce about 34% of the world's present oil supply, and nearly 30% comes from countries in the Persian Gulf.
  - The Caspian Sea region in a geologic basin of Central Asia is another substantial oil source which may eventually produce about 4 million barrels / day, but this region is landlocked and depends upon pipelines rather than tanker ships to move the product.

441. US DOE - EIA

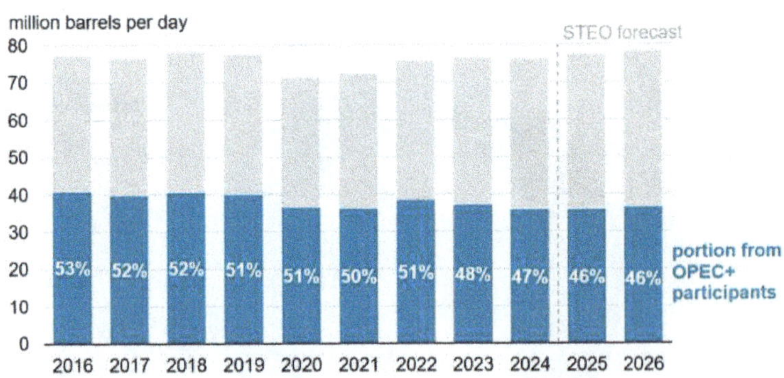

Among the OPEC+ members, Russia was the largest crude oil producer in 2024, averaging 9.2 million b/d. After Russia and Saudi Arabia, the largest producers by volume were Iraq (4.4 million b/d), the United Arab Emirates (2.9 million b/d), and Kuwait (2.5 million b/d)

World Crude Oil Production

Decisions by OPEC in 1973 to tighten oil production created shock waves of alarm, causing long queues of vehicles at gasoline stations and motivating many nations and corporations to think about new ways to lessen reliance upon petroleum products.

- The ability of OPEC to exercise extensive influence over production and prices, in combination with exponential consumption growth in China, India and other countries, are reasons for great ongoing concern:
  - Many experts believe that global oil production has already peaked, or soon will, with demand rates permanently outpacing available supplies.
  - A real problem is that we may only become broadly aware of how serious this condition is after it is too late to implement remedies.

442. Northeastern University College of Computer and Information Science

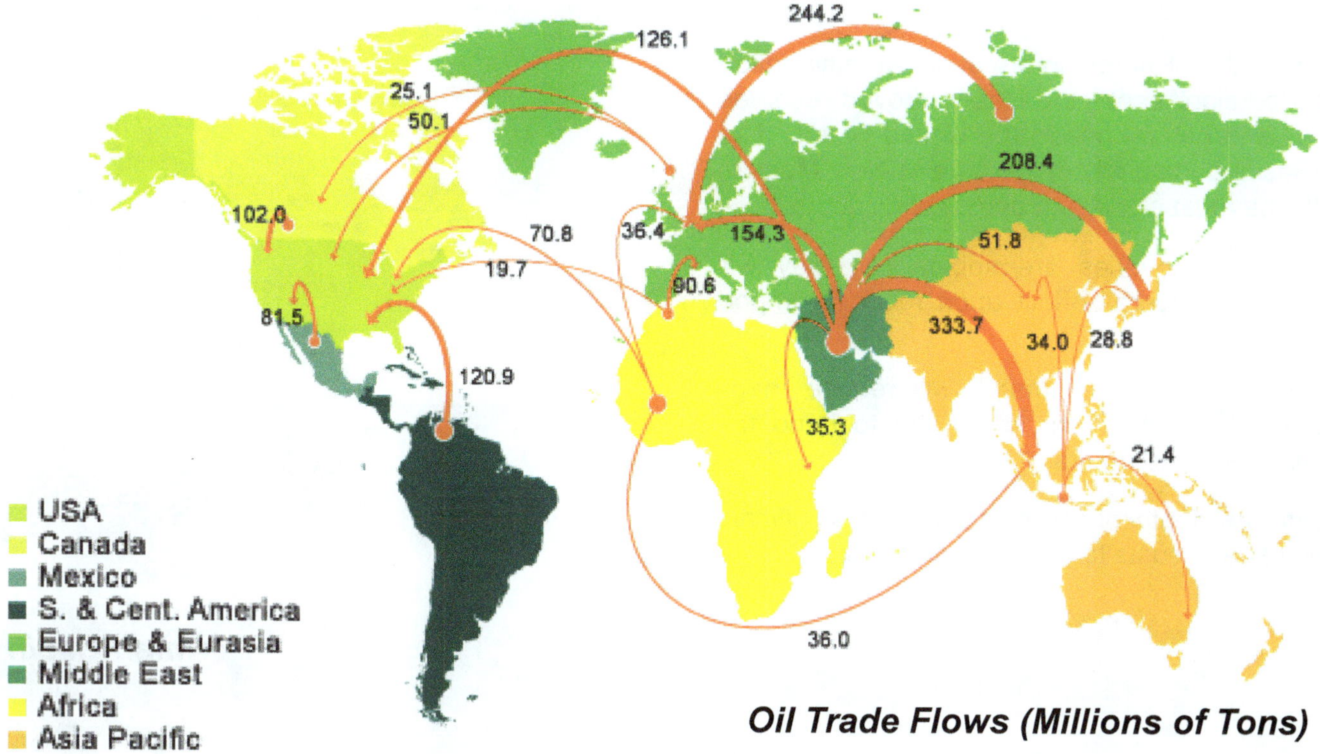

Oil Trade Flows (Millions of Tons)

443. OEC

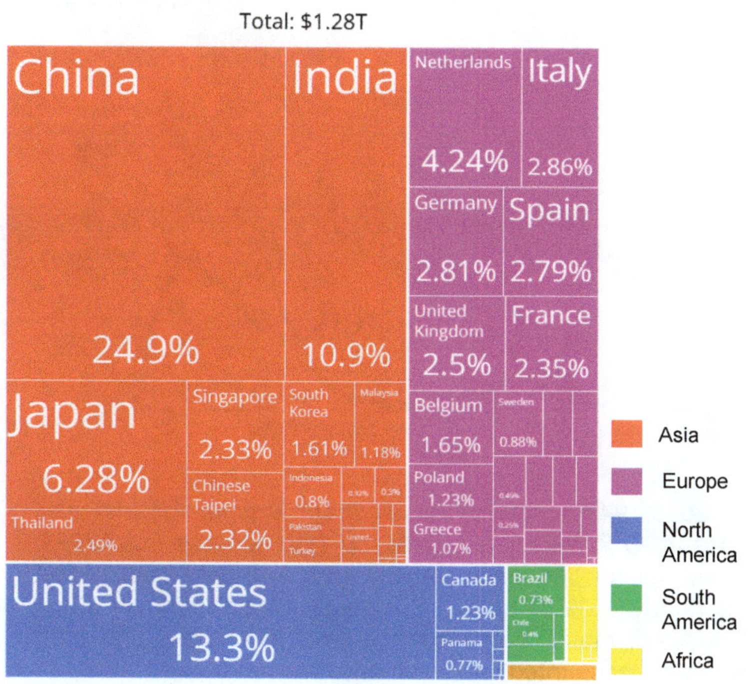

Global Importers of Crude Petroleum- 2023.

Global competition for the world's rapidly dwindling oil resources makes transitions to non-fossil fuel alternatives inevitable and urgent.

- This competition is certain to become more aggressive as developing nations continue to increase their consumption rates through industrial growth and economic prosperity:
  - According to the US DOE's Energy Information Administration (EIA), worldwide oil consumption is expected to grow by 50% by 2050, and demand in developing countries projected to increase significantly due to increasing automobile ownership.
  - The US currently imports about 6.28 million barrels/day of crude oil, and while the US has become a net exporter of petroleum products, it still relies heavily on imports to meet domestic demand.

444. Permianpartnership

In 2022, the U.S. Energy Information Administration (EIA) reported significant increases in proved reserves of crude oil and natural gas in the United States. The Permian Basin, particularly in Texas and New Mexico, is currently the most significant domestic oil deposit

- While these discoveries are substantial, they will have significant overall effect upon reducing import requirements:
  - Oil companies estimate that it will require many years and tens of billions of dollars to bring the newly-tapped oil to market.
  - The extreme depth of the oil field adds significant challenges and costs to drilling operations.

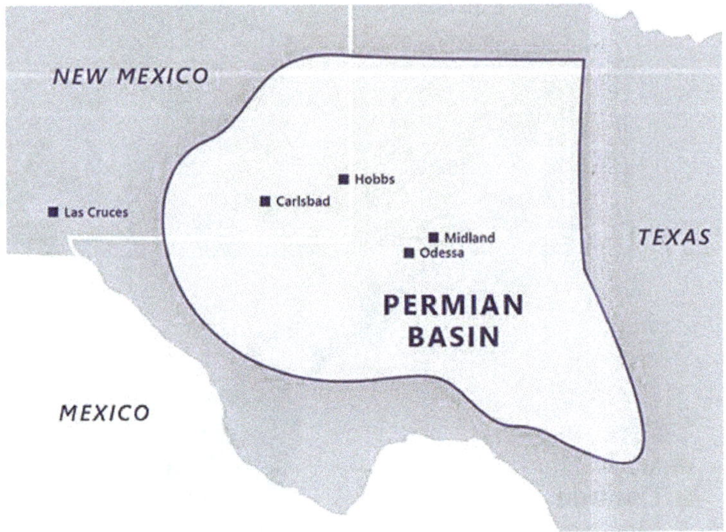

Permian Basin

445. US EIA

Global energy consumption continues to rise at a rapid rate, although this trend appears to be improving due to expanding conservation efforts in more developed countries.

- While continuing US shifts to coal and renewable fuel sources are predictable due to increasing oil and natural gas production costs and dwindling supplies, history shows that such transitions from one dominant source to another require substantial time:
  - Coal use decreased in relation to oil and gas over a 50 year period between about 1910 and 1940, and transitioning back is requiring about the same time.
  - We might imagine that transitioning to alternative renewable sources will also require many years, demanding that real progress will be needed soon to avoid economic and social turbulence.

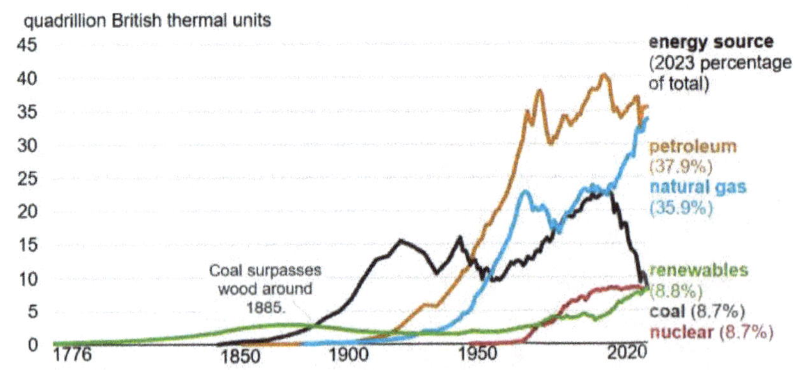

In 2023, 94 quadrillion British thermal units (quads) was consumed in the United States, a 1% decrease from 2022. Fossil fuels—petroleum, natural gas, and coal—accounted for nearly 83% of total U.S. energy consumption in 2023. Non-fossil fuel energy—from renewable sources and from nuclear—accounted for the other 17%.

Historical US Fuel Use by Percentage

446. MSCI- institute, EIA

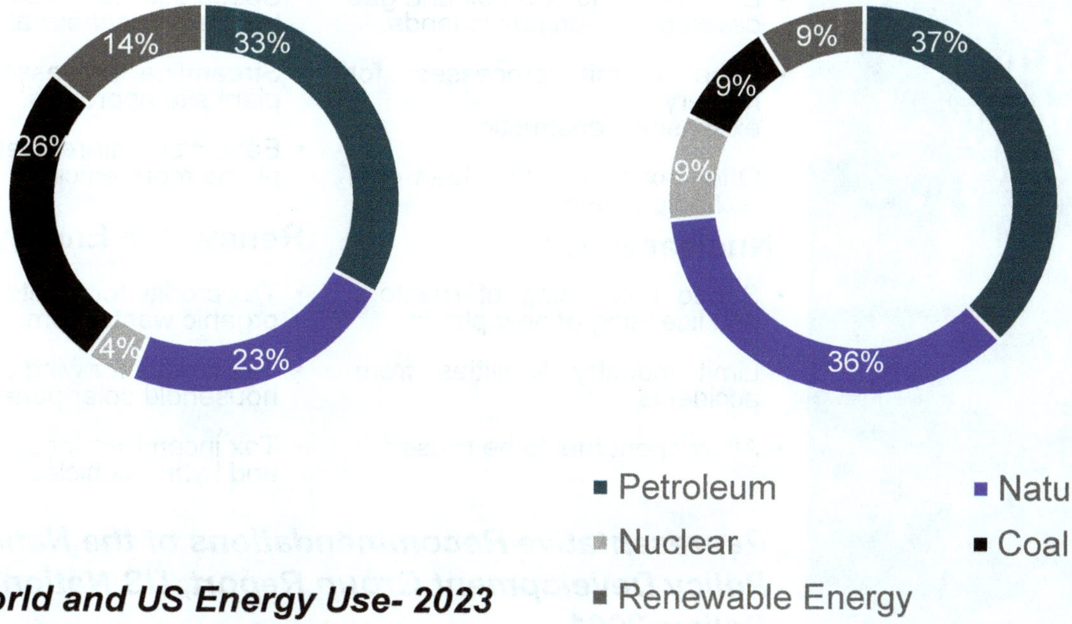

World and US Energy Use- 2023

Forward-looking energy strategies and developments will not occur without controversies and compromises.

- Growing needs and costs to fuel industry, domestic heating/cooling and power, transportation and other services are certain to prompt public policies that are not popular with everyone.
  - Nuclear power must be considered as an important option which can reduce CO_2 emissions on one hand, while also presenting legitimate safety and security concerns.
  - Relaxation of some clean air regulations may occur in the interest of providing ample and affordable fossil fuel-derived electricity for industry and domestic consumption.
  - Subsidies and tax incentives for renewable fuel source development and use will be disagreeable with many who challenge their economic viability.

447. US DOE - EIA

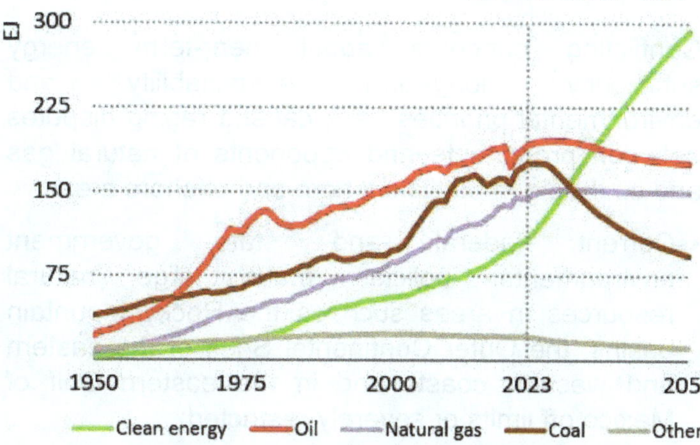

A scenario based on current policy settings, sees clean energy (includes renewables, modern bioenergy, nuclear, abated fossil fuels, low-emissions hydrogen and hydrogen-based fuels) poised for huge growth, while coal, oil and natural gas each reach a peak by 2030 and then start to decline. Other includes traditional use of biomass and non-renewable waste.

US Fuel Consumption Projected, 1950 - 2050

448. US Government

Fossil Fuels:

- Ease restrictions on oil and gas development on public lands.
- Ease permit processes for refinery expansion/construction.
- Offer tax breaks for clean coal technology use.

Nuclear Power:

- Speed relicensing of reactors and licensing of new plants.
- Limit industry liabilities from accidents.
- Allow spent fuel to be reused.

Power Plants:

- Speed license procedures for new hydroelectric/thermal plants.
- Streamline processes for power plant site approvals.
- Ease clean air regulations to make plants more efficient.

Renewable Energy:

- Tax credits for plants that use organic waste/biomass.
- Tax credits for wind energy and household solar panels.
- Tax incentives for alternative fuels and hybrid vehicles.

Representative Recommendations of the National Energy Policy Development Group Report, US National Energy Policy, 2001

Conflicting concerns about near-term energy sufficiency, long-term sustainability and environmental priorities have caused raging disputes between proponents and opponents of natural gas and oil drilling in public onshore and offshore areas.

- Current federal and state government environmental policies make large natural resources in areas such as the Rocky Mountain basins, the Outer Continental Shelf of the eastern and western coasts and in the eastern Gulf of Mexico off limits or severely restricted:
  - Drilling proponents argue that drilling can be environmentally safe, and that prohibitions and restrictions cause continued dependence on foreign sources and price escalations.
  - Environmental groups believe that offshore pipelines harm marine life and salt marshes, while onshore development poses threats to wildlife habitats, fragile soils and archaeological resources.

449. Independent Petroleum Association of America

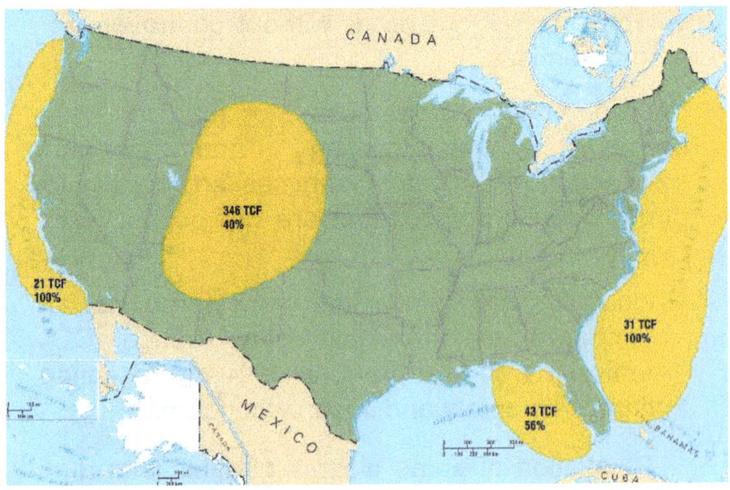

As a recent US Department of Interior study indicates that approximately 80% of the nation's untapped economically recoverable natural gas on the Outer Continental Shelf (OCS) is located in areas already open to leasing.

Natural Gas Restricted Reserves (Trillion Cubic Feet and Percentages)

450. US DOE EIA

North American Natural Gas Reserves

According to the US Department of Interior, about 85% of US natural gas comes from domestic onshore and offshore areas, and most of the rest comes from Canada.

- During 2022, total domestic land reserves were estimated to be about 691 trillion cubic feet (TCF), with significant portion owned by the Federal Government.
  - Approximately 73% of domestic reserves are concentrated in Texas, New Mexico, Wyoming, Oklahoma, Colorado and Gulf of Mexico offshore areas.
  - Proven reserves concentrated in the Gulf of Mexico off the coasts of Alabama and Louisiana total about 16.73 TCF, and off Texas, about 5.3 TCF.
  - Current economically-recoverable reserves for Federal Offshore Gulf of Mexico areas are estimated to be 100.3 TCF, mostly in deep water (more than 400 feet deep in western Gulf areas).

451. US EIA

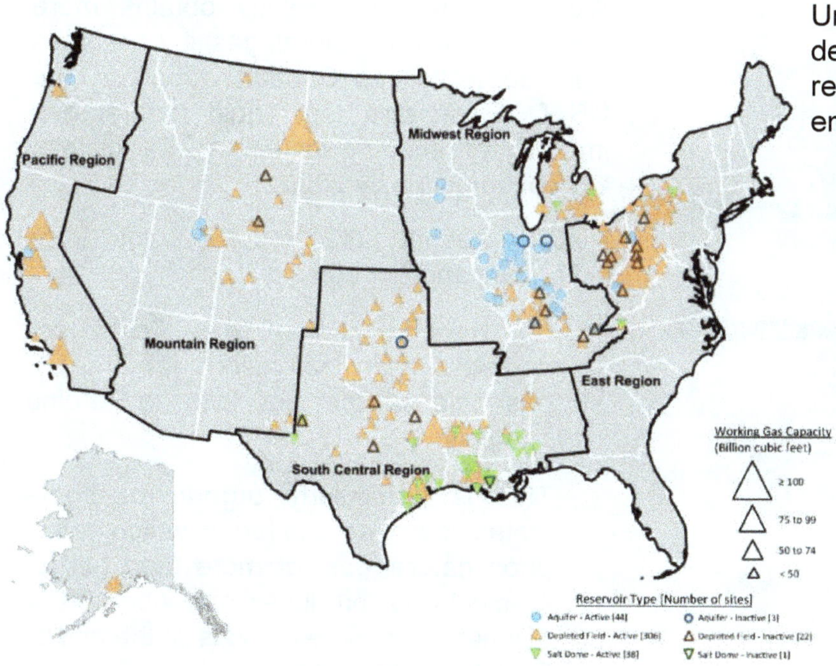

Underground Natural Gas Storage Accommodations

Unpredictable seasonal and periodic peak demands in regions remote from pipeline sources require means to store natural gas for emergencies.

These storage needs are growing in many regions as a result of population and industry growth, and will continue to expand nationally as dwindling national gas supplies create chronic shortages.

- Adequate storage is already lacking in New England and Middle Atlantic coastal areas where LNG is a critical part of the regions' heat and power supply during severe cold snaps.
- The 2025 cold snap resulted in the fourth-largest withdrawal from underground natural gas storage in the Lower 48 states that caused large energy price swings.

The US natural gas infrastructure is well developed, including production, liquefaction or regasification of LNG, storage and transport (pipelines and tankers).

- Pipelines carry natural gas in both directions between Canada and the US, and between the US and Mexico (which imports more than it exports):
  - Canadian gas flows to and from the US through several large pipelines feeding markets in the Midwest, Northeast, Pacific Northwest and California (nearly 99% of US imports).
  - A major portion of pipeline capacity within the US connects key production areas of Texas and Louisiana to markets in Western, North-eastern and Midwestern regions.

452. US EIA

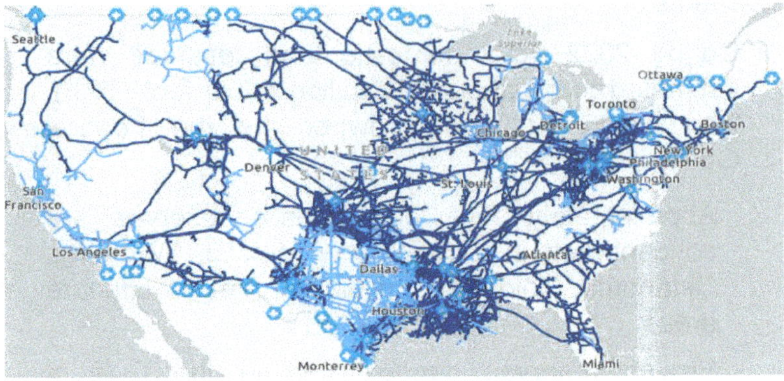

Pipelines connect US producers and markets within the US, Canada and Mexico as noted:
(1) Texas and Louisiana production areas
(2) Texas and Louisiana to the US West
(3) Texas and Louisiana to the Northeast
(4) Texas and Louisiana to the Midwest
(5) Canadian production areas

Natural Gas Pipelines

453. Lngallies

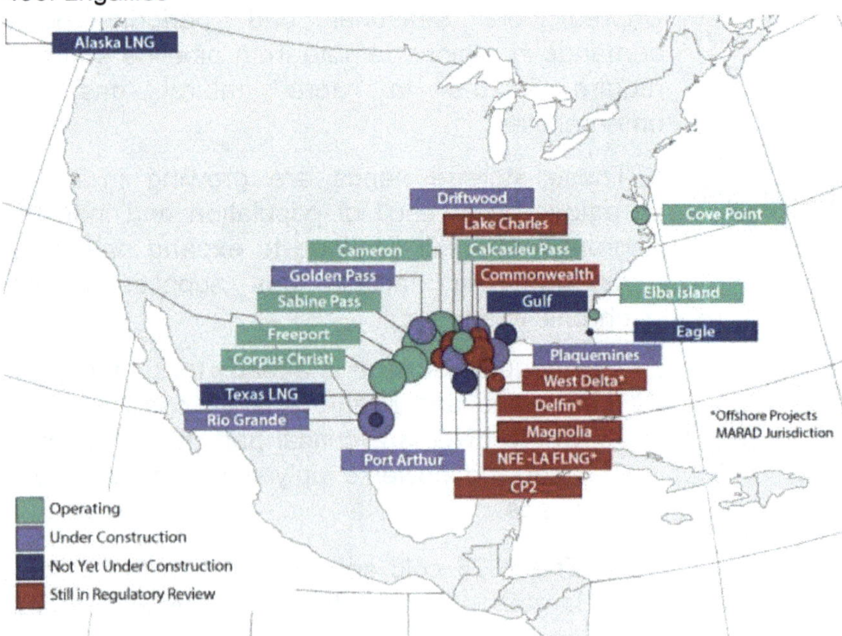

Active Liquid Natural Gas Facilities

Although the US currently obtains more than 99% of the natural gas it consumes from domestic and Canadian sources, the US DOE projects that liquid natural gas imports will account for about 25% of total US consumption by 2025.

- Most present LNG imports come from Trinidad and Tobago:
  - As North American gas fields are depleted at an increasing rate, the US has little choice but look to foreign supplies.
  - This is particularly urgent for some states, such as California which relies upon natural gas for more than half of its electricity, produces only about 10% of what it consumes, and is at the end of the US pipeline grid.

454. FERC

Increasing LNG imports will require development of new and expanded terminals to meet national and regional needs.

- The EIA projects that by 2025, new terminals will be required to collect 1,200 billion cubic feet annually to accommodate an estimated 70% increase, with 65% of that increase served by enlarging existing facilities:
  - Two deep-water LNG ports are proposed to be built eight miles off the coast of Gloucester, Massachusetts, and nearly a dozen land-based terminals are proposed from Rhode Island to eastern Canada.
  - Development of new terminals is facing strong opposition by local communities due to safety and terrorism concerns, and many of at least two dozen current proposals are experiencing legal challenges.

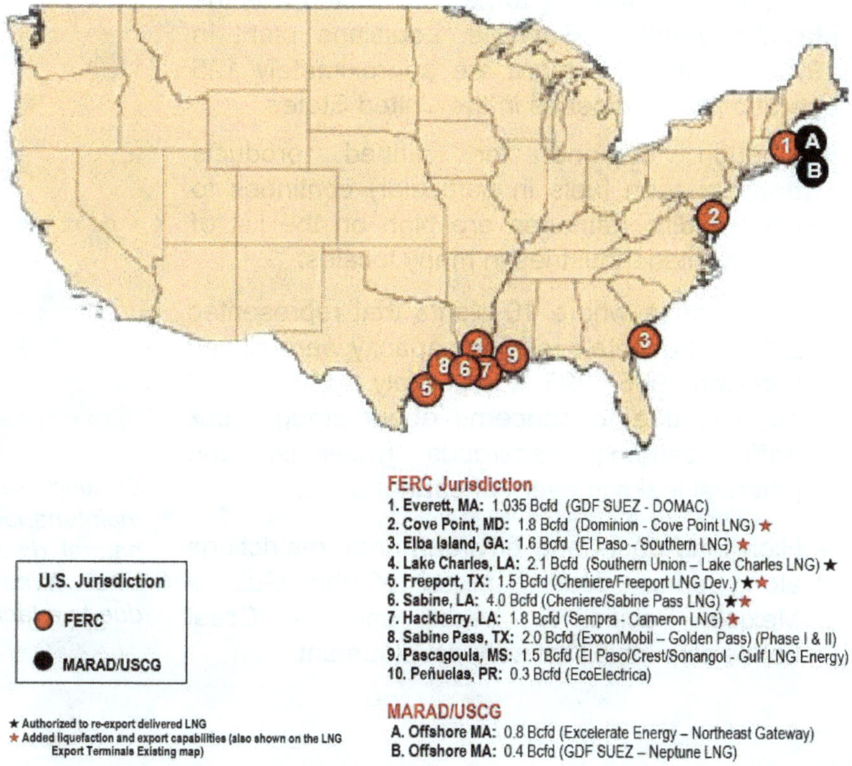

LNG Import Terminals

455. US DOE-EIA

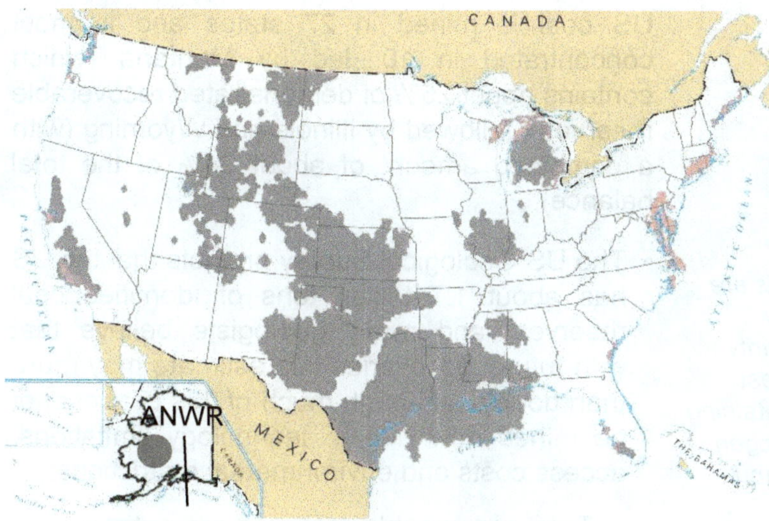

Intense controversy exists over whether or not a government moratorium should be lifted that prevents drilling for oil in Alaska's Arctic National Wildlife Refuge (ANWR) which is estimated by the US Department of Interior to contain between 9-16 billion barrels of recoverable oil.

Crude Oil Deposits

US crude oil reserves are located primarily in Texas, California, Louisiana, Oklahoma and Federal offshore locations, and oil infrastructures are most highly developed around these production centers.

- Canada and Mexico are key suppliers of crude oil to the US, where oil products flow back and forth between countries in trucks, pipelines and ships requiring large transportation and processing infrastructures:
  - The US transports oil over long distances from Alaska.
  - Canada's huge resources principally derive from oil sands that require major development and processing operations.
  - Mexico's heavy crude oils (Maya) also require significant development, transport and refinery adaptations to serve huge North American markets for lighter oils (gasoline, jet fuel and liquefied petroleum gases).

456. US EIA

The last US oil refinery to be constructed was the Marathon-Ashland Garyville, Louisiana plant in 1976. As of 2024, there are approximately 125 operational oil refineries in the United States.

- Although demand for refined products (transportation fuels in particular) continues to rise steadily, refineries are high on the list of least-wanted industries in many locales:
  - In California where 10 plants that represented 20% of the state's refining capacity were closed between 1985-1995, it is unlikely that more will be built due to concerns about smog, truck traffic carrying hazardous materials, and potential leaks in event of earthquakes.
  - High land costs and environmental restrictions along prime coastal areas of the Gulf of Mexico, Pacific, Northeast, and the Great Lakes also inhibit refinery development.

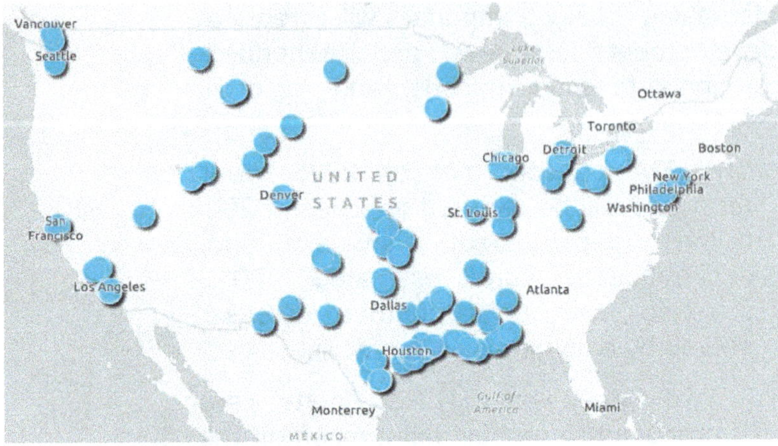

Existing refineries are vulnerable to disruptions due to maintenance and breakdowns caused by accidents and natural disasters that can create supply shortfalls and price hikes. West Coast and Midwest regions are particularly at risk due to a lack of easy supply accessibility.

Petroleum Refineries

457. The Kentucky Network

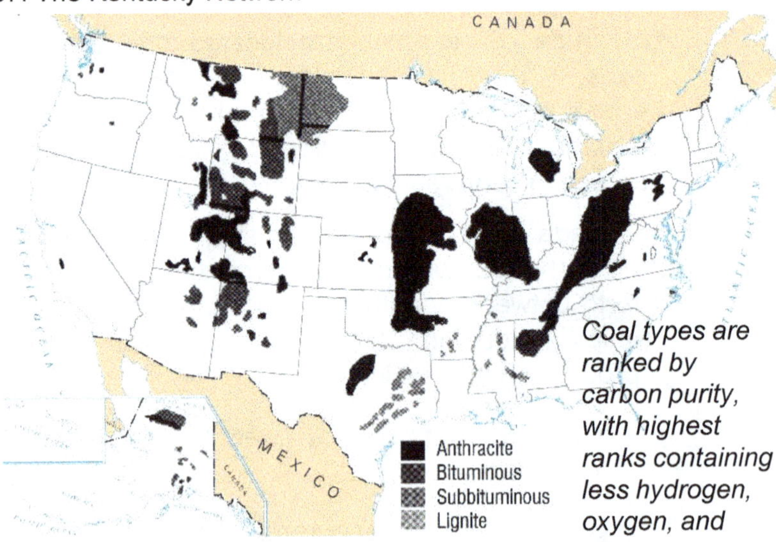

Coal types are ranked by carbon purity, with highest ranks containing less hydrogen, oxygen, and nitrogen:

- *Anthracite (95% purity and above) is primarily used for residential and commercial space heating.*
- *Bituminous (next in rank) is used for steam-electric generation,*
  combined heat and power, cement production and to make coke.
- *Sub bituminous and lignite (in rank order) are principally used for steam-electric generation.*

Coal Reserves

US coal is mined in 27 states and is most concentrated in 10, led by Montana (which contains about 25% of demonstrated recoverable reserves), followed by Illinois and Wyoming (with a combined amount of about 34% of the total balance).

- The US Geological Survey projects that the US has about 1.7 trillion tons of identified coal reserves, and many geologists believe that with future discoveries this estimate may more than double, although much of this coal cannot be mined now due to technology limitations, access costs and environmental restrictions:
  - Total recoverable reserves are estimated to be about 469 billion tons.
  - Since current mining techniques leave substantial amounts in place, near-term recoverable assets are estimated to be about 250 billion tons.

Coal production levels are impacted by influences of weather upon transportation (primarily rail and barge), mine accidents and safety conditions, and government permitting processes that can close or delay operations.

- Surface (or "open pit") mining which accounts for about 63% of total US recovery is used primarily in the Western Region where near-surface deposits can be up to 100 feet thick, presenting fewer safety problems than underground mining east of the Mississippi in Appalachian Mountain states:
  - The Western Region accounts for more than 56% of total production, with Wyoming (the largest US producer with 41% of the region's total) leading Montana.
  - The Interior Region is led by Illinois, followed by Indiana.
  - The Appalachian Region is led by West Virginia (largest producer in the region and second largest in the US) followed by Pennsylvania.

458. US DOE-EIA

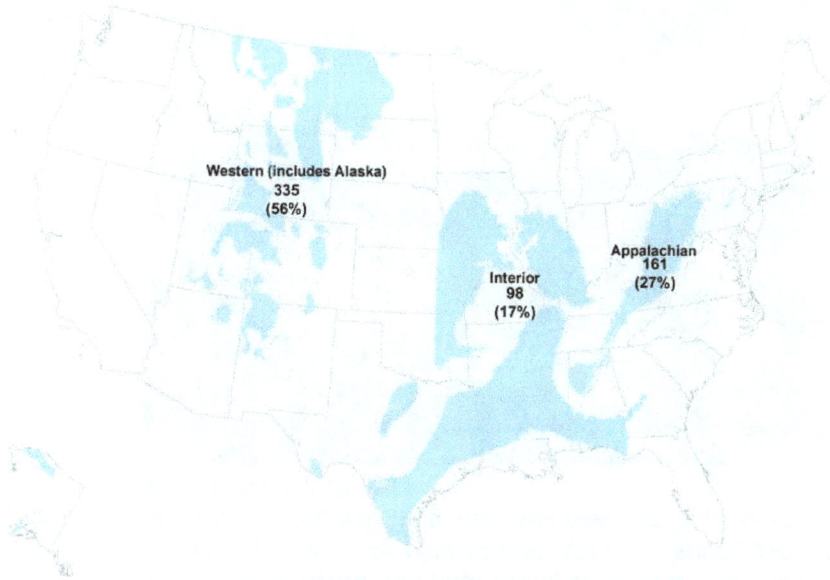

Production in the Appalachian Region continues to face challenges due to weather-related disruptions, legal issues, and safety concerns (roof collapse, high Methane gas levels). Flooding and severe weather events, such as those in 2022, have significantly impacted coal production and transportation.

Coal Production by Region, 2022

459. US DOE-EIA

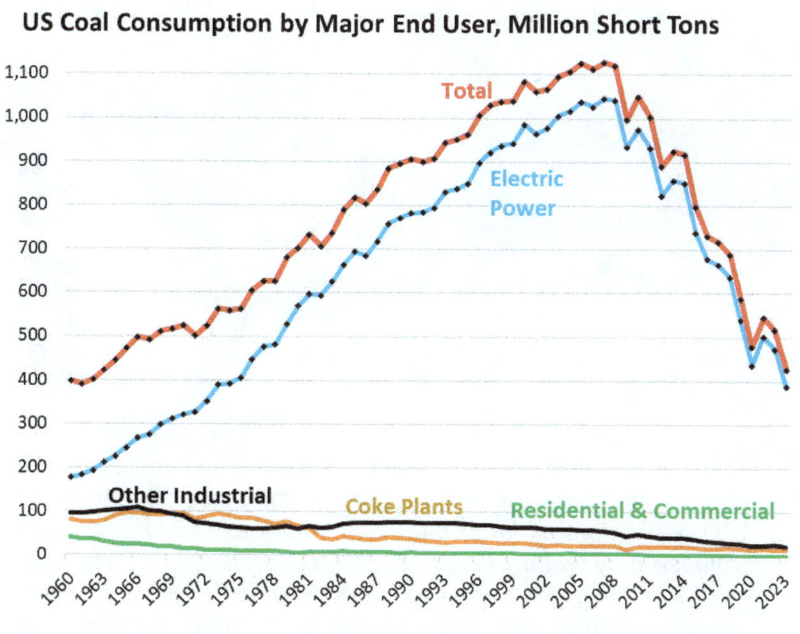

Coal Powered Electricity Consumption

Coal is the largest single source of fuel for electricity generation world-wide, and coke from coal processing is a vital component in the reduction of iron ore.

- An expanding US economy and warmer-than-normal summer weather in 2024 drove up coal demand for electrical power by 3.1% in 2024 from 2023, accounting for about 14% of all coal consumption.
  - While coal was a minor part (less than 20%) of the fuel mix for two census regions (New England and Pacific), it was a major source (more than 50%) in five others (East North Central, West North Central, South Atlantic, East South Central and Atlantic).
  - Coal was one of the two main electric fuel sources in the other two regions; the Middle Atlantic (competing with nuclear), and the West South Central (competing with natural gas).

460. State of South Dakota

Reclaimed Mine Area

Open-pit coal operations are now typically required to post bonds for each acre of land surface to be mined, and later restore soils as nearly as possible to original contours with native vegetation and trees replaced. Since 1977, more than 3 million acres of coal land have been reclaimed in this manner.

Despite enormous dependence upon coal for electricity, heating and important by-products, coal industries and utilities are broadly regarded as "dirty businesses."

- Public resistance to coal mining and burning relates to major environmental and safety concerns:
  - Coal-fired plants are the largest source of CO_2 emissions, and although capture has been proposed, it is not yet commercially used.
  - Coal contains low levels of uranium, thorium and other radioactive isotopes, and coal burning also produces such wastes as fly ash, boiler slag, sulfur, and a variety of heavy metals (including arsenic and lead).
  - Coal harvesting and abandoned mines release methane that can explode and put workers at risk.
  - Surface mining has had destructive impacts upon local land areas, polluted waters and destroyed natural habitats.

Coal Mining

Nuclear power development is gaining public advocacy in the US and abroad because of growing concern about CO_2 released by coal-fired plants, rising natural gas costs, and increased confidence in nuclear plant safety.

- Lithuania and France lead the world in percentages of nuclear power generated, and several other countries including Great Britain, Germany and China appear to be moving towards nuclear plant expansion:
  - Currently, 436 plants are in operation overseas, and the International Atomic Energy Agency (IAEA) says World nuclear capacity is now projected to increase by 2.5 times the current capacity by 2050.
  - Current US nuclear plants have demonstrated exceptional safety records that benefit from substantial technological advancements.

461. NEI

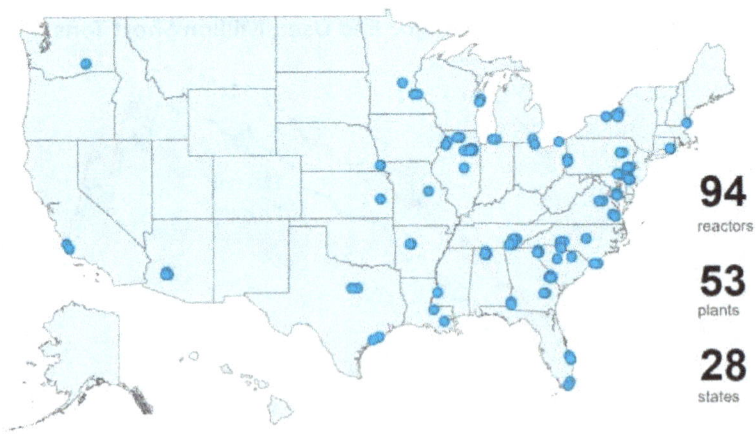

The US currently has 53 nuclear power plants located in 28 states that produce nearly 18% of the nation's electricity. Although more expensive to build than fossil fuel plants, they release only water emissions, use much less expensive fuels and are being leveraged to produce clean hydrogen, which could become a significant source of electricity for hydrogen production.

Power Plants

Coal and Nuclear

462. US DOE-EIA Grand Junction Project Office

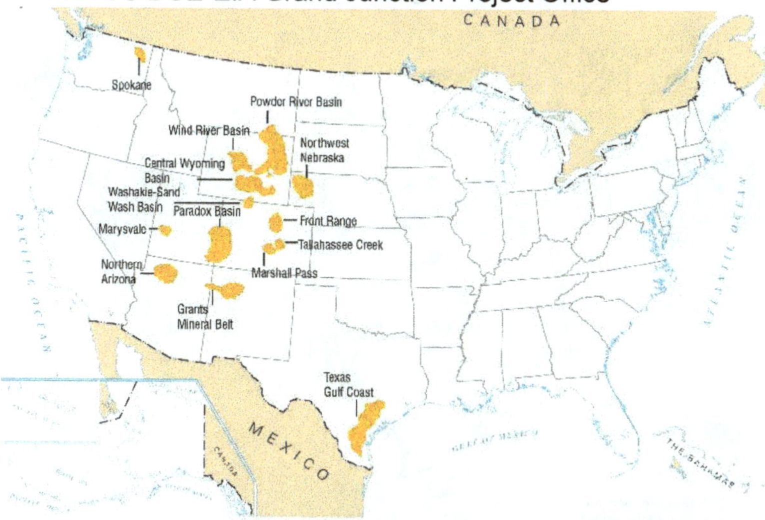

Most US uranium deposits are small and low-grade, but supply about 90% of the nation's production. Modern mining uses in-situ recovery (ISR) methods that pump water into sub-surface hydro chemical "cells" to dissolve uranium minerals, leaving the ore where it was naturally formed. Surface processing of the leached solution produces "yellow cake".

Major US Uranium Reserves

Following a period of very limited uranium exploration and mining between 1996-2002, global production and prices are rapidly rising due to a resurgence of interest in nuclear power.

- Primary production of 49,355 tons at the end of 2022 provided only about half of the world's reactor requirements, with the remainder coming from utility inventories, downgrading of military weapons and spent fuel reprocessing:

  - In the US, only seven in-situ leach (ISL) mines are operating in 2024.
  - US prices began to climb in January 2022 from $30/lb to $50/lb by the end of 2024 and are expected to go much higher.
  - New ISL mines are now being created or restarted, and about a dozen are now licensed in south Dakota, Wyoming, Nebraska and Texas.

While there have been no deaths resulting from US nuclear power plant accidents or incidents during the last 40 years or more and ISL mining is much more environment-friendly than past excavation methods, strong resistance to this industry continues to persist in many public quarters.

- Concerns about terrorism threats to plant security, safe containment of radioactive wastes, and lawsuits charging toxicological impacts, remain as development hurdles:

  - Contaminated sites near abandoned mines of past decades still require years of expensive cleanup operations, and lawsuits claiming related fatalities and illnesses have resulted.
  - Potential groundwater contamination from toxic chemicals released from ISL ore leaching presents an ongoing hazard.
  - Programs for transportation and storage of wastes face contentious environmental challenges in effected states.

463. DOE

Uranium Mining

Biomass power obtained from energy in plants and plant-derived materials is becoming a more important source of US energy and an integral part of the waste management infrastructure.

- It is predicted that future agriculture may greatly expand production of high-yield energy crops such as fast-growing trees and grasses, that coupled with high-efficiency conversion technologies may substantially supplement fossil fuels:
  - Wood is the oldest primary biomass source, and "black liquor" produced as a waste from wood pulp, paper and paperboard industries remain to be the largest.
  - Waste energy is the second largest biomass energy source, with municipal solid waste, manufacturing waste and landfill gas the main contributors.

464. The Global Change Program, University of Michigan

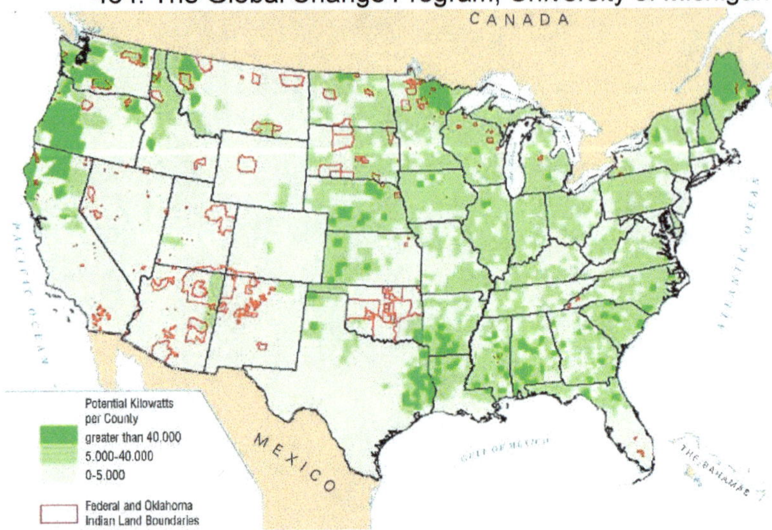

Biomass energy is considered to be nearly carbon-neutral in that it absorbs CO_2 from the atmosphere, during plant growth, approximately equal to the amount emitted in power conversion, producing fewer "new" pollutants than fossil fuels.

Potential Kilowatts of Energy per County

465. EIA

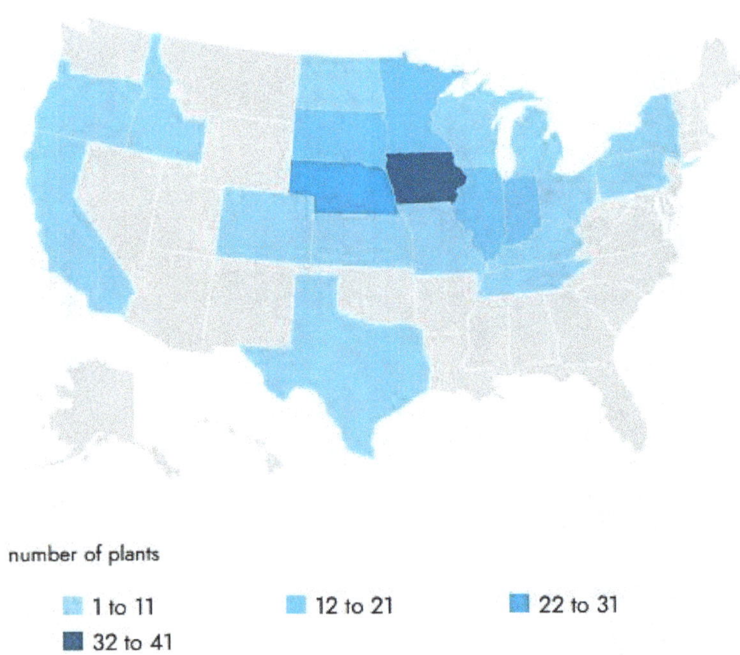

US Ethanol Plants

Ethanol and biodiesel production is often advocated as an important, cleaner-burning alternative to fossil fuels to reduce CO_2 emissions and dependence upon foreign oil.

- While at peak levels these alternatives might supply only 10%-14% of US transportation demands, many believe that they are a step in the right direction:
  - Corn-based ethanol can potentially pave the way for cellulosic ethanol constituted from switchgrass and various agricultural and forestry wastes that will be more substantial and sustainable.
  - Biodiesel made from processed vegetable oil, animal fat and recycled cooking grease can be useful either as a fuel additive (typically 20%) or in a pure form for diesel engines for agricultural fuel and local power generation, but is not generally envisioned as a primary national energy source.

Since cultivation of corn for ethanol is very energy-intensive, ethanol produced from cellulosic plants such as switchgrass is gaining popular interest.

- Ignoring crop rotation requirements, an acre of switchgrass harvested 1-2 years after planting might produce 5-10 tons of biomass annually, with a yield of 50-100 gallons (190-380 liters) of ethanol per ton:
  - According to these estimates, approximately 25 million acres of land (39,000 square miles) would be needed to displace 1 million barrels of crude oil daily.
  - At this rate, about 3% of all available US crop, range and pasture lands would need to be utilized just to reduce projected 2050 oil imports by about 10%.*

466. Texas State Energy Conservation Office

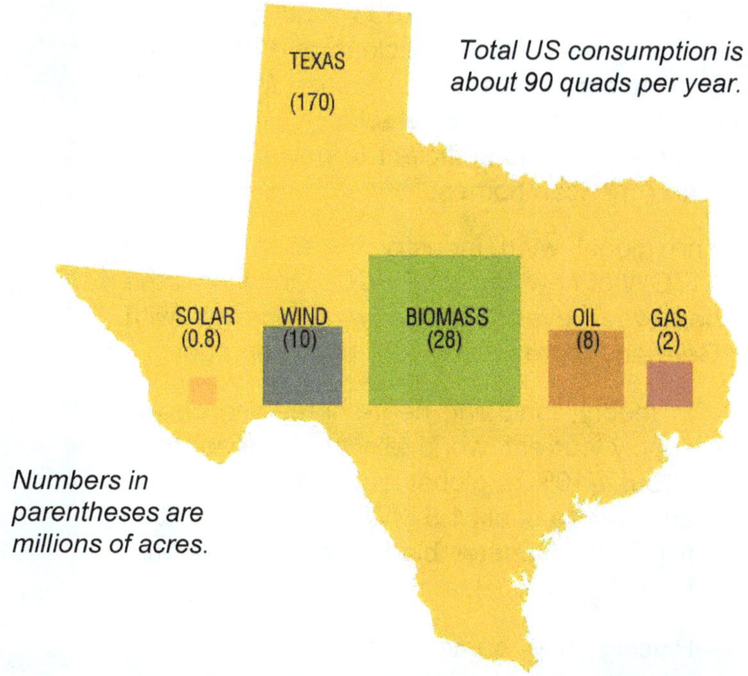

One quad (or quadrillion BTUs) equals about 1 TCF of natural gas.

Land Areas Needed to Produce One Quad of Electricity

467. Arizona Solar Center

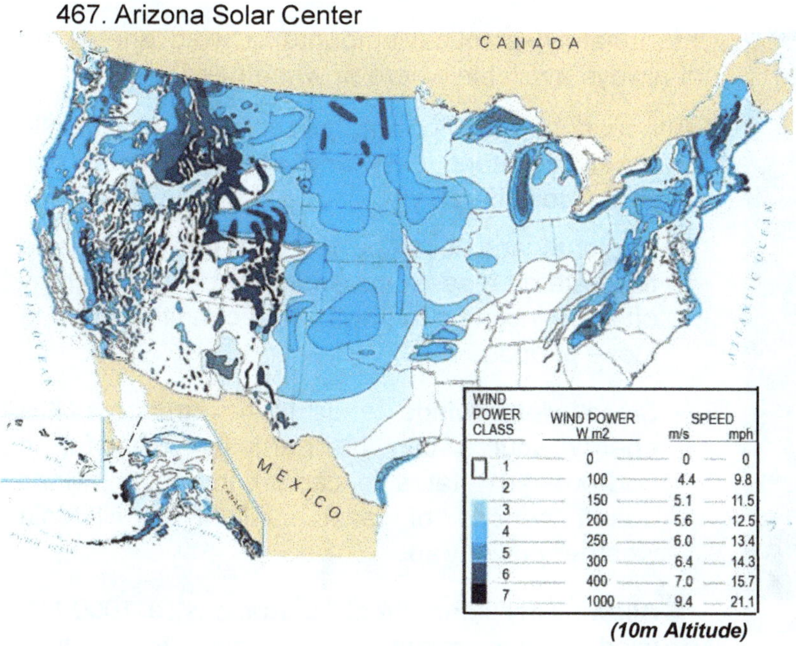

Annual Wind Power Resources

U.S. electricity generation from wind turbines decreased for the first time since the mid-1990s in 2023 despite the addition of 6.2 gigawatts (GW) of new wind capacity last year. Wind capacity increased steadily over the last several years, more than tripling from 47.0 GW in 2010 to 150.1 GW at the end of 2024.

- It is estimated that American wind farms currently save more than half a billion cubic feet of natural gas per day, and the rate of expansion is predicted to grow:
- More consistent wind resources are located in the Great Lakes Region, and from ocean breezes along the eastern, western and southern coasts.
- Important inland areas are in the Great Plains from northeastern Texas and eastern New Mexico northward to North Dakota and western Montana, in ridge crests and mountains of Appalachia, and throughout western states, particularly in mountain wind corridors.

* According to the US Deprtment fo Agriculture, the US has roughly 390 million acres of cropland (mostly used for animal feed and the meat industry), and another 655 million acres for grassland pasture and rangeland, about half of 2.26 billion acres of US land.

At present, more than 75,633 wind turbines are generating over 147.5GW clean and dependable electricity across the United States. Making it the fourth-largest source of electricity generation in the country, this is sufficient to power around 46 million American homes.

- The global wind industry installed a record 117GW of new capacity in 2023, making it the best year ever for new wind energy, with Germany representing 19% of this amount:
  - In theory, land and near-shore wind could supply current world electricity needs on about 4-10% of global land area (assuming an average of six 1.5 MW turbines with 253 ft [77 m] diameter blades for 0.38 square mile [1 square kilometer].
  - Placing the same number of turbines offshore could contribute about seven times more wind energy due to power multiplying effects of 90% higher wind speeds.

468. Iowa Energy Center

Wind Sites Can Share Other Uses

469. Bruce Center for Energy Research and Information

Wind Farms Tend to be Located Far from Power Demand Centers

Unfortunately, practical amounts of wind energy are not always available where or when needed.

- Although annual power capacities are usually relatively constant, daily and seasonal levels can fluctuate considerably:
  - Regional and local power demand levels also fluctuate on a seasonal and daily basis, and may often be incompatible with wind conditions during those periods.
  - Since winter winds tends to be strongest, this energy source can be particularly helpful in northwestern states to capture mountain winds during periods of least sunlight to fill cold weather power gaps.
  - Allowing for intermittent fluctuations, a 1000 kW wind turbine might be expected to produce roughly the same annual energy as a 500kW coal-fired generator.

Even though wind is characterized as a clean energy source, many new wind farm proposals face public resistance on environmental and aesthetic grounds.

- Numerous developments have been blocked by residents in prime coastal and hilly areas due to concerns about influences upon bird migration and nesting patterns and visual impacts upon picturesque vistas:
  - Seashores tend to be popular recreation and high-value real estate locations which also have good wind energy features caused by differential cooling of land and water over the course of day and night.
  - Offshore turbines (typically 6 miles [10 kilometers] or more from land) encounter less opposition, but are more expensive due in part to harsh, corrosive conditions with less maintenance accessibility, and greater water-to-shore transmission challenges.

470. Research Institute for Sustainable Energy
471. Greenpeace

Environmental and Visual Objections

472. Oak Ridge National Laboratory

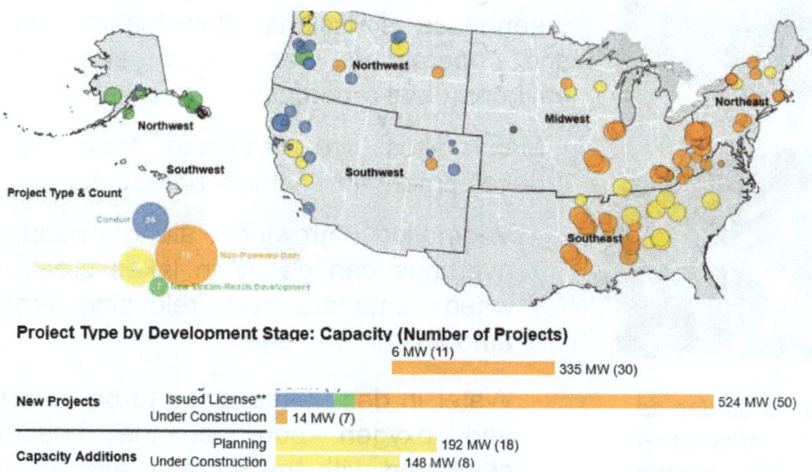

While most sites for large dams are already used or restricted, there appears to be significant opportunities to expand small installations, particularly in western states.

Current and Future Hydropower Capacity

With approximately 80,090 MW of generating capacity in 2023, hydropower is the largest US renewable electricity source.

- While a 2012 Department of Energy Study estimated that total potential in the contiguous 48 states might possibly double that amount using energy from rivers and steams, no feasibility studies were conducted to determine real practicalities:
  - “Low head” (less than 30 feet) and “low power” (less than 1 MW) resources constituted about 19,000 MW of this new potential.
  - Micro-hydro sites (less than 100 kW) make up an additional 9,500 MW possibility.

Hydropower generates about 50% of all electricity in the US Pacific Northwest Region, and the Idaho National Laboratory projects that there may be an underdeveloped capacity of 50% more on swift-flowing rivers and streams in western states.

- This potential might add small damless operations up to 30 MW, low-power installations and new powerhouses at existing dams, and expanded hydroelectric plant capacities, for a 30,000 MW increase:
  - Washington already has by far the highest generating capacity, and Alaska, California, Oregon and Montana have large new potentials.
  - It is also predicted that Hawaii can multiply its hydropower capacity many times

473. Oak Ridge National Laboratory

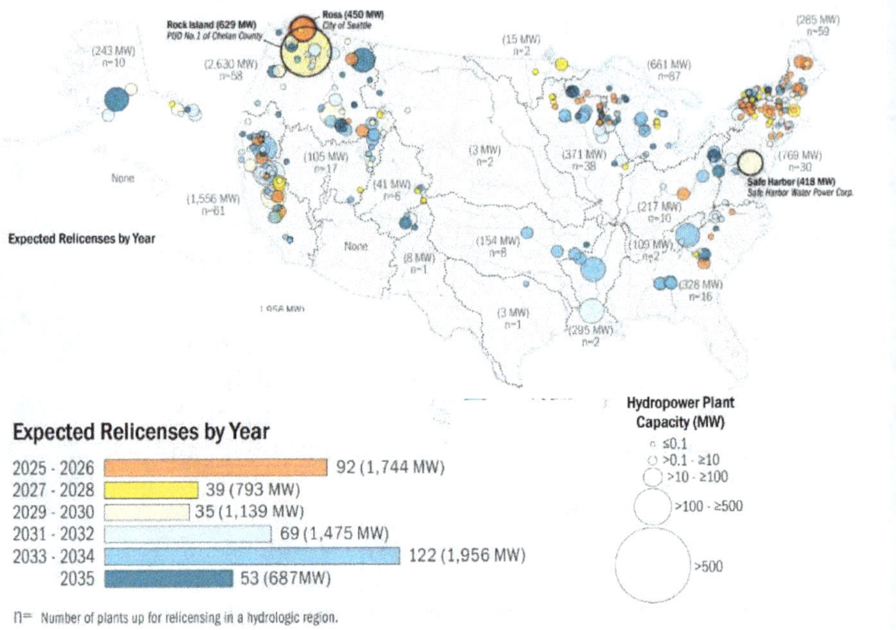

Hydropower Development Potentials

474. US Geological Service

475. The Atlas of North America

Many young salmon in the Northwest are killed by turbine blades as they swim downstream towards the ocean, and adult fish attempting to swim upstream to reproduce are blocked by dams. After salmon populations were dramatically reduced in the Northwest Columbia Basin, many fish ladders and side channels were built.

Although hydropower produces no greenhouse emissions, dam construction and operations can have significant environmental repercussions:

- Many sites are restricted from new development for a variety of reasons:
  - Vegetation growing along nearby riverbeds can decay in lakes created when dams are built, releasing large amounts of methane.
  - Water in dam lakes tends to be colder and oxygen poor at the bottom compared with the surface, and when rapidly released can kill fish and flood plant and wildlife habitats downstream.

Hydropower, Fish and Natural Habitat Issues

While solar power currently supplies about 7% of US energy, the potential for expansion promises to be large.

- A report issued by the Western Governors Association Solar Task Force in 2023 projects that 6,000 MW of PV, 5,000 MW of concentrating power and 3,000 MW of solar hot water can be easily developed:
  - An estimated 177 MW of domestic PV capacity in 2023 was expected to grow 26% by 2024.
  - Concentrating solar power has realized contracts to build about 4,000 MW of new installations in sunny southwestern states over the next decade.
  - Domestic solar water and pool heating is growing even more rapidly, and increased by about 23% from 2022-2023.

476. North Carolina Department of Administration

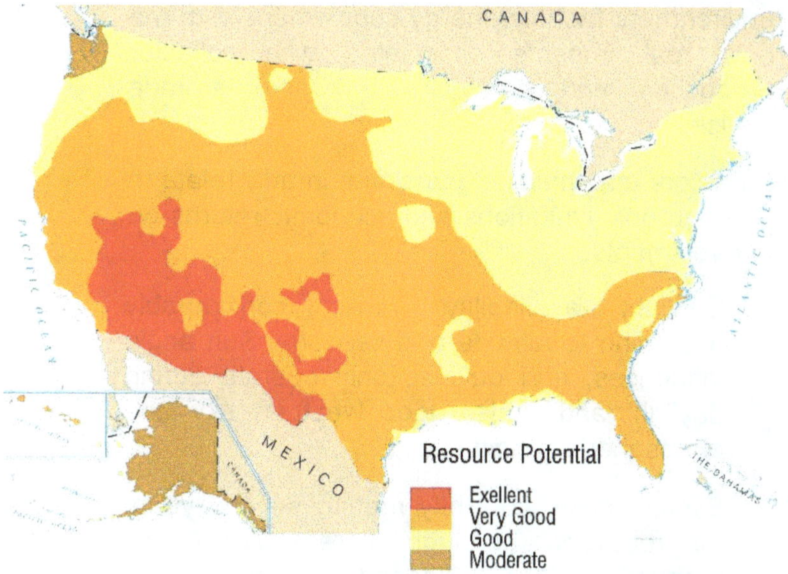

Solar Resource Potentials

477. WisconSUN

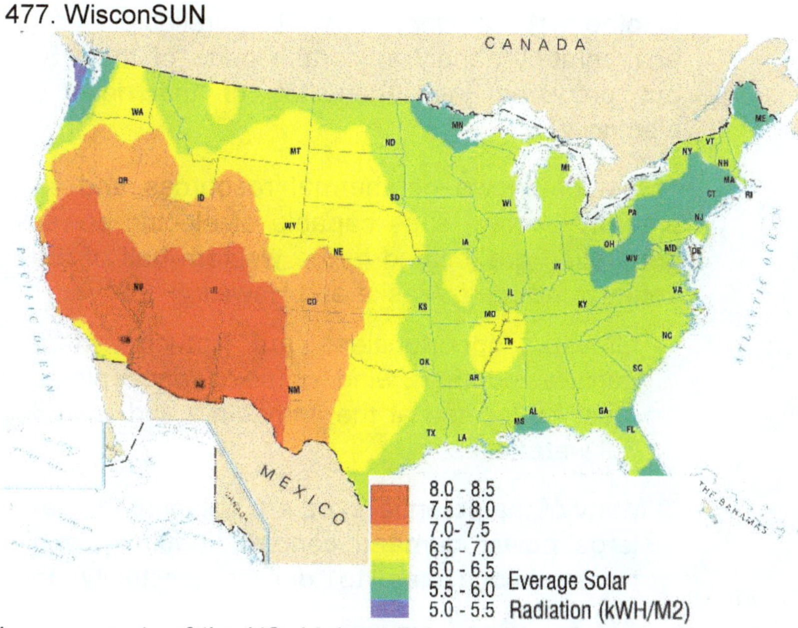

In many parts of the US, highest PV outputs and demands are on hot sunny summer days.

Solar Worst Case Month (June) Demands

Average Solar Radiation Received by a Horizontal Surface (kilowatt hours per square meter)

Much of the solar power industry growth can be attributed to rising energy prices along with federal and state government incentive programs.

- The US 2005 Energy Policy Act (EPAct) established tax credits for solar installations with apparent effectiveness, and 41 states enable customers to sell excess electricity back to grids, with 38 providing state-wide programs to do this:
  - In 2023, grid-tied PV surpassed off-grid installations, and this trend is likely to continue, the majority for homes and businesses.
  - California is dominant for grid-tied systems, with capacity of over 48GW.
  - Texas, second in the US for grid-tied PV, with capacity of roughly 32GW.

Solar power is broadly accepted as an environment-friendly energy supplement, with the only real impacts associated with collector production and disposal of battery storage devices

- Factors that inhibit expansion primarily relate to geographic limitations and economic returns on investments:
  - Sunlight availability varies considerably according to latitude and seasonal Sun angle influences, and devices only operate during daylight and under good weather, sky and atmospheric conditions.
  - System costs currently require several years for recovery (although increased PV cell production and planned expansion of polysilicon manufacturing combined with rising fossil fuel prices and solar tax incentives can greatly enhance economic benefits).

478. National Research Energy Laboratory

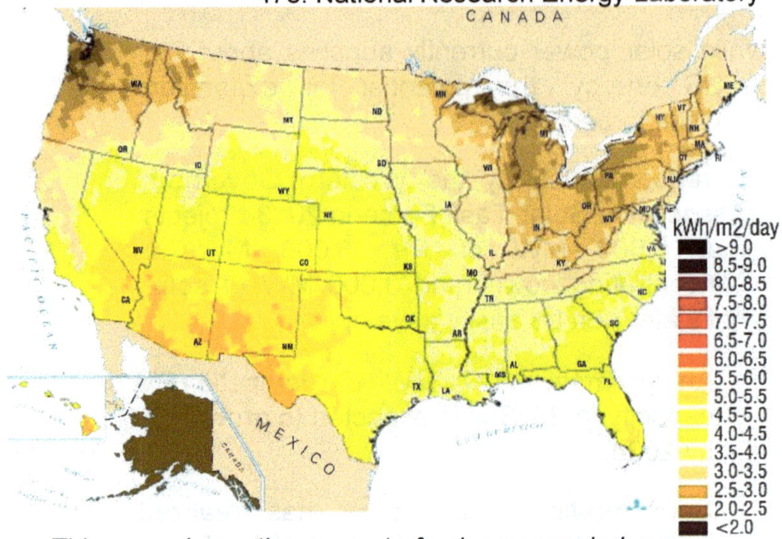

This map shows the amount of solar energy in hours received each day on an optimally-tilted surface.

Solar Worst Case Month (January) - Supply

PV Solar Radiation, Flat Plate Facing South Latitude Tilt

479. The Global Change Program, University of Michigan

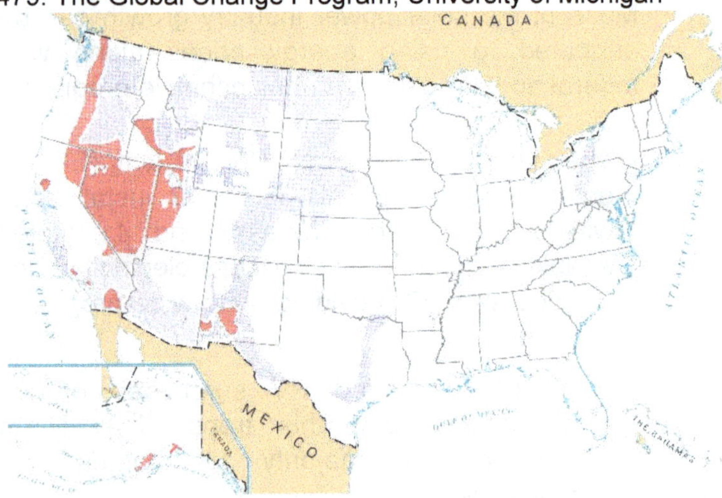

Economically-feasible geothermal resources are concentrated west of the Rocky Mountains, with California, Nevada, Utah, and Hawaii leading in geothermal power generation, and thermal springs are common in the West, including Montana, North Dakota, and Wyoming, while some low-temperature resources exist in Central Texas.

Geothermal

Geothermal power can be reliable and environment-friendly in certain parts of the US, but produces less than 1% of the nation's electricity.

- Most of known geothermal resources and all presently known sites capable of electric power generation are located in the western half of the country, including Alaska and Hawaii:
  - Growth opportunities are much more constrained than wind or solar since plants must be located at the immediate and limited hot water sites.
  - Many of the potential sites are not located near large power demand centers, requiring long transmission lines to deliver electricity to customers.
  - Diminishing water temperatures and pressures from plant operations can also impact production capacities and efficiencies over time.

In addition to hot water and steam geothermal resources, possibilities also exist to tap deep underground "hot rocks" for power generation.

- The heat collection process pumps high pressure water into dry wells, breaking up rock formations to collect heat and return it to the surface, driving turbines:
  - One barrier to commercialization is that wells must be much deeper than those required for conventional geothermal plants.
  - Another drawback is that since heat flow through dry rocks is slow, heat removal is also slow.
  - An additional limitation is that the majority of promising sites are in dry areas of the West where process water is scarce.

480. Los Alamos National Laboratory

The Fenton Hill Hot Dry Rock site plant in New Mexico uses a 11,500 ft deep well drilled into rocks in a 430° F environment. Water pumped into the well at 80° F returns to the surface at 360° F, producing up to 5MW of electrical power.

Geothermal Hot Dry Rock Plant

481. ORNL
483. MYAA Site
482. Western Libraries
484. The Livingroom

Meeting Demands

Safety and Security

Protecting Nature

Future Needs

Common Priorities

Energy alternatives and related needs, sources, environmental impacts and safety/security issues are subjects of hotly-contested disputes among groups with opposing viewpoints.

- Common aspects of these adversarial outlooks are that the passions expressed are justified, and the arguments are typically valid:
  - Public environmental and health safety concerns associated with all energy strategies are clearly legitimate, and are shared by most operating companies that are proud of progress and achievements in these areas.
  - Adequate, sustainable energy and conservation are critical needs recognized by the industries that must provide large investments to meet growing demands through efficient, reliable and economically-viable means.

485. Larry Bell

Advocates →		← Opponents
Is essential to reduce foreign dependence	Expansion of Oil & Natural Gas Drilling	Must protect limited existing supplies
New sources needed for economy / living		Limiting supplies promotes conservation
New development requires years		Improved conservation can solve the problem
Oil / natural gas are energy-efficient		Cleaner alternatives are available / essential
Essential to support growing import needs	New Ports, Terminals & Refineries	They all present major terrorist targets
Technologies are safer / cleaner than in the past		Offshore operations pollute marine life
Necessary to prevent regional power gaps		Don't put them in my region / backyard
Are vulnerable to disasters / breakdowns		Efficiency upgrades have kept pace with demands

Oil and Natural Gas Development

486. Larry Bell

Advocates →		← Opponents
Coal is our most abundant fuel source	Coal Mining & Power	Cleaner alternatives are available / essential
Clean coal technologies are reducing emissions		Gasification requires large CO_2 containment
Coal is a source of important derivatives		We can obtain them from biomass / biofuels
Coal mining is vital to regional economies / jobs		Mining operations harm people / ecosystems
Nuclear power has large expansion potential	New Nuclear Power Plants	Don't produce it in my region / backyard
Plants produce no CO_2 emissions, only water		Radioactive waste is an environmental hazard
Advanced technologies have proven very safe		Breakdowns / terrorism present large threats
Fuel reprocessing can greatly extend supplies		Breeder reactors can proliferate weapons

Coal and Nuclear Power Development

487. Larry Bell

Advocates →		← Opponents
Produces clean power	Hydrogen Fuel Cells	Production can pollute
Has good power density		Is a net energy user
Usable in different forms		The gas is explosive
A renewable alternative		Is difficult to transport
Are renewable sources	Corn / Cell. Ethanol	Can compete with food
Can reduce fossil use		Reduction may be limited
Releases only "new" CO_2		CO_2 is CO_2
Cellulose uses biowaste		Processing can be costly
A "free" energy source	Solar PV / Concentrat.	Devices are expensive
Solar Is broadly available		Not always optimum
Can reduce fossil use		Limited energy production
Offer versatile applications		Often requires storage

Hydrogen, Biofuel and Solar Power

488. Larry Bell

Advocates →		← Opponents
A clean power source	Wind Power	Not in my backyard
An abundant source		Depends upon location
A "free" energy source		Requires large investment
Can reduce fossil use		Intermittant availability
A clean power source	Hydro Power	Can pose ecohazards
An abundant source		Very limited locations
A "free" energy source		Plants can be costly
Can reduce fossil use		Limited growth potential
A clean power source	Geothermal Power	Sites are often restricted
An abundant source		Can deplete heat/pressure
A "free" energy source		Plants can be costly
Can reduce fossil use		Often not very efficient

Wind, Hydropower and Geothermal

Rather than seeking universal solutions to national energy challenges, it may be much more productive to concentrate more upon developing nationally-integrated special-case regional strategies.

- Strong evidence of opportunities and trends in this direction are already apparent:
  - Energy industry deregulation is opening markets for smaller power utility companies that take advantage of more efficient CHP production to serve local district needs.
  - Emergence of regional "boutique" fuel mixtures tailored for seasonal weather and state regulatory restrictions offers new advantages for smaller specialty refiners.
  - State governments and regional alliances are taking more active roles to promote and develop energy resources most appropriate to their unique resources, needs and economic opportunities.

489. Larry Bell

Priorities	Strategies
Develop appropriate energy supplies and technologies	Provide incentives for energy investments
Create and expand efficient distribution infrastructures	Coordinate state / regional programs and networks
Concentrate supplies / services where needed most	Optimize CHP and other shared-use opportunities
Anticipate demographic trends and future needs	Plan / implement long-term development initiatives
Transition to cleaner / safer energy solutions	Promote / facilitate green energy co-op programs
Encourage conservation in homes and businesses	Establish public information and education programs

Representative Regional Priorities and Strategies

490. Weber State University

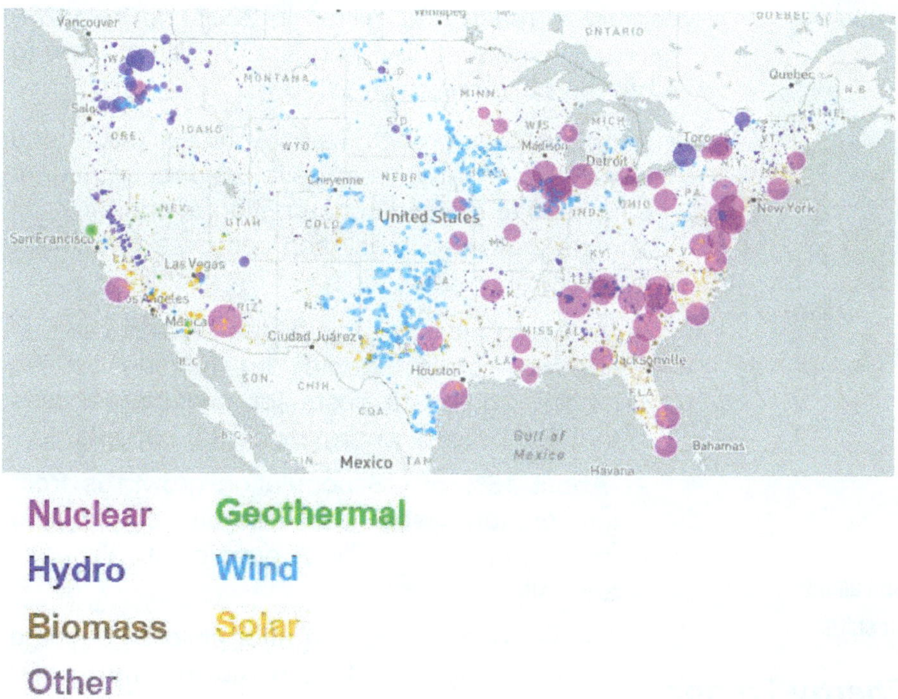

Renewable Energy Facility Locations

Essential progress is hampered when strong lobbies favoring one particular energy option as an ultimate solution work to block development of others that are also vital.

- In reality, widely different regional resources and needs in combination with changing demands posed by population and business shifts, global geopolitical developments, and technology advancements make reliance upon any single answer both impractical and dangerous:
  - True resourcefulness demands that all options be developed and optimized, including ways to make them more environmentally-friendly and efficient.
  - Efficiency includes means to consume less and put available resources to better use.

The US is largely an urban nation with more than 80% of its population living in and near cities.

- Households are becoming increasingly less densely concentrated in large metropolitan areas, and are shifting to more distant suburbs where real estate and services are less expensive, resulting in more energy use for travel and less efficient single-family dwellings:
  - Northeastern regions are by far the most densely populated, but also have the slowest growth (and in some instances are experiencing reductions) due to migrations to Sunbelt locations.
  - Southern and western states are realizing the highest influx of immigrants along with population gains from the Mideastern and Midwestern regions, often overtaking power and water supplies.

491. Wikimedia

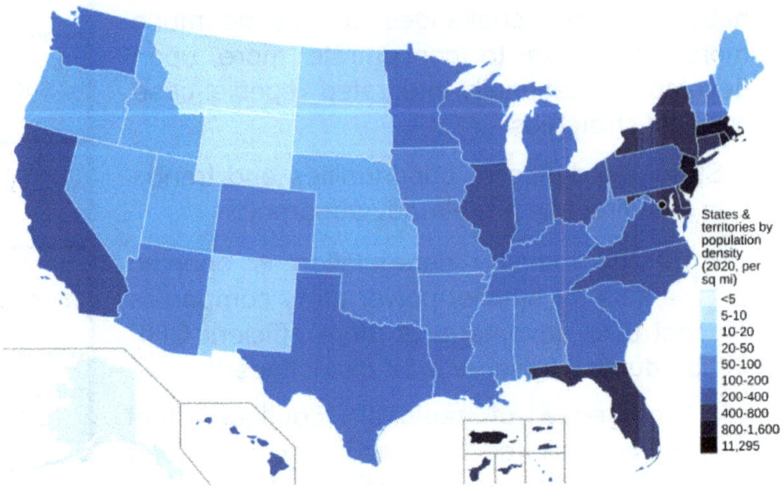

The US population is heavily concentrated around large cities in the Northeast, and slightly more than half is clustered in coastal counties. Rapid shifts are occurring from central cities to suburbs, and from the Northeast and Midwest to Southern and Western states.

Population Density by Counties, 2020

492. Larry Bell / US DOE - EERE

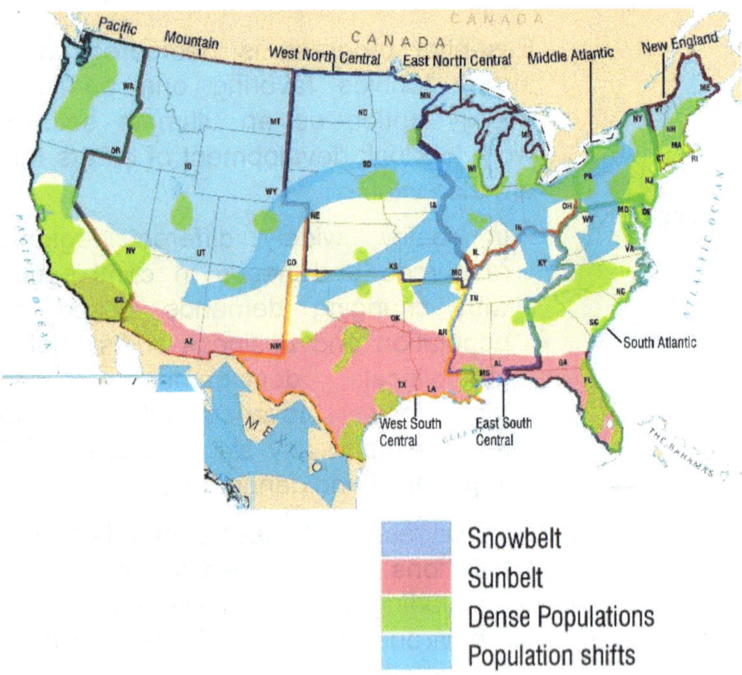

Population and Energy Trends

The US population more than tripled during the 20th century to reach 300 million in 2006, up from 200 million 39 years earlier, and is expected to grow to more than 370 million by 2050.

- Although this growth rate is nearly six times lower than many less developed countries, it continues to expand consumption of energy, food and natural resources, and removes land from agricultural production:
  - While population growth due to birth rates is relatively slow, people are tending to live longer, adding to costs for medical services that compete for energy and food budgets.
  - About 15% of US population growth is from immigration (legal and illegal), contributing more than 1.6 million people of all ages annually.
  - Rather than downsizing after children become adults, many parents are moving into larger homes, and retirees are purchasing vacation properties that consume fuel and space.

Population shifts from the Snowbelt to the Sunbelt create impacts upon regional energy demands, environments and economies.

- These changes present advantages and disadvantages for areas of gains and losses:
  - Shifts from energy use from fossil fuel heating in winter to more electricity for air conditioning makes better use of centralized power plants that are more efficient than stand-alone furnaces, and produce fewer CO_2 emissions.
  - On the other hand, relocations of people and industry from large metropolitan areas to rural locations create impacts upon aquifers and natural ecosystems, plus losses of prime farmlands and wetlands.
  - New households and businesses also contribute to greater energy consumption, while areas that lose them suffer job and tax base reductions.

493. US DOE-EERE

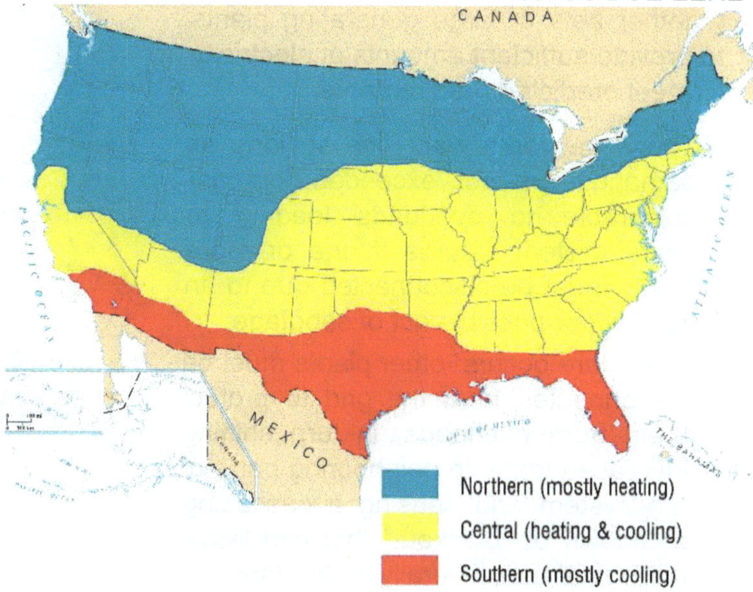

Population migrations from colder to warmer regions substitute fuel heating demands for increased electricity use.

Energy Needs

494. Larry Bell / ACF / US Census / US DOE-EERE

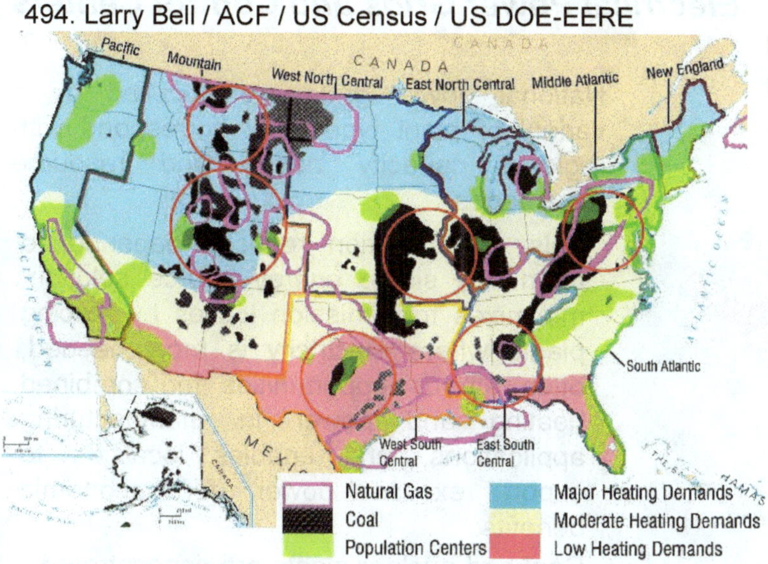

Heating Demands and Fuel Sources

Cold weather space heating must compete with year-round power generation for fossil fuel use.

- Various US regions present different opportunities posed by climate conditions and concentrations of people and industries:
  - The Pacific Region contains high population densities, with some coal in northeastern Washington, and natural gas primarily in southern California.
  - Mountain and West North Central states have abundant coal and natural gas, with modest population clusters in central areas.
  - West and East South Central states have large coal and natural gas deposits and moderate climates.
  - South and Middle Atlantic states have large coal and natural gas resources in northern areas.
  - New England, with cold weather and very limited fossil supplies, must depend upon other regions.

Power grids of different sizes connect together sets of large generating plants to provide sufficient amounts of electricity to meet predicted requirements.

- These networks work fine so long as demands are not exceeded, but are vulnerable to overloads leading to entire system failures if one or more plants must be disconnected due to an accident or possible act of sabotage:
  - If a failure occurs, other plants must be disconnected from the grid to protect them from overloads, in turn shifting increased loads to neighboring parts of the system and causing a cascading shutdown or "blackout" that can leave millions of users literally in the dark.
  - One solution is to provide significant amounts of excess capacity with more plants and transmission lines, an approach that requires major new investments as power demands grow.

495. US DOE / How Stuff Works

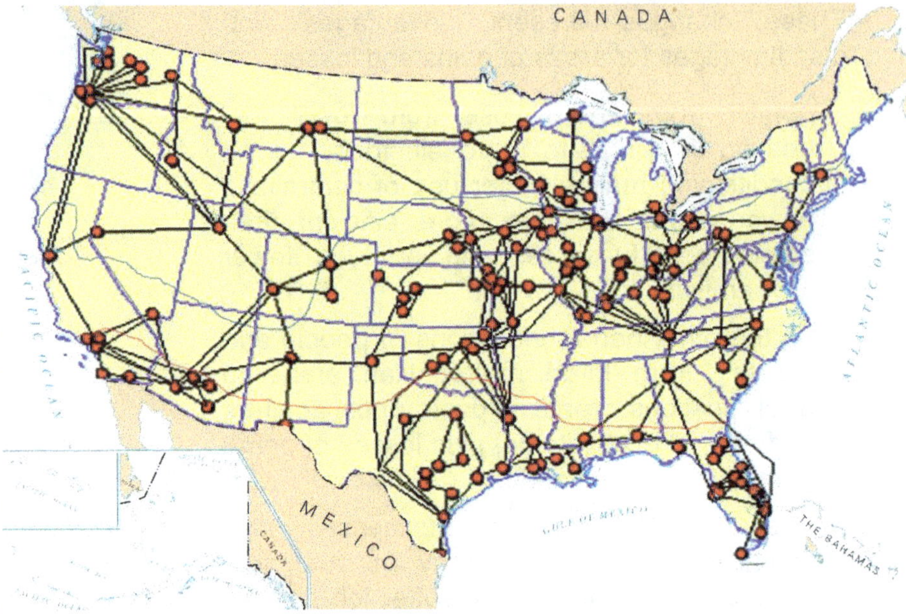

Computerized systems at each control area operations (CAO) center monitor power grid activity, balance supplies to meet demands, and prevent overloads. Population and industry shifts influence demand capacities that must be accommodated.

Electrical Power Grids and Control Centers

496. Weber State University

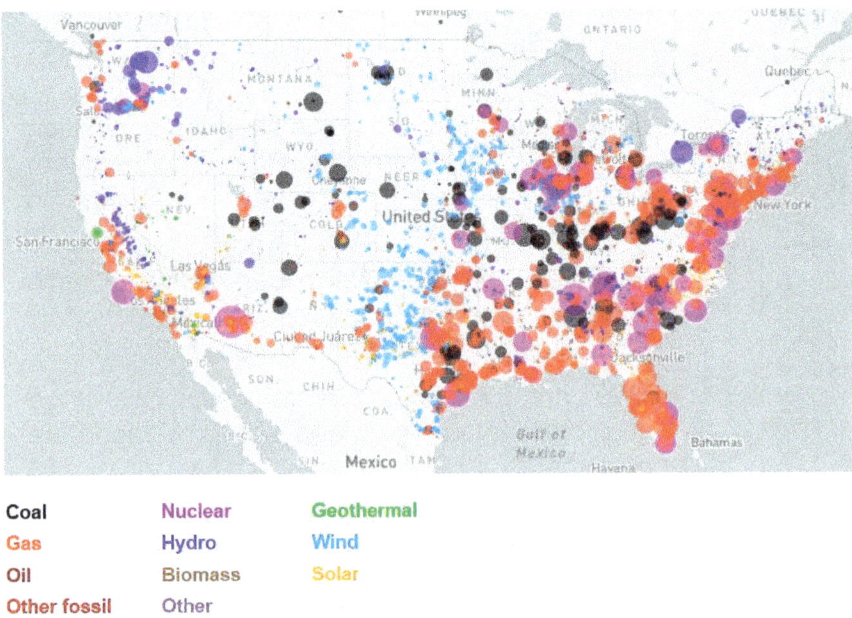

Diversified power sources draw upon regional advantages to support national needs most efficiently.

Major Power Plant Types

National power grids are supplied by a variety of plant types that correspond with regional capacity needs and resource advantages.

- This diversification reduces dependence upon any single energy source solution, minimizes transmission losses by placing plants where electricity is most needed, often affords opportunities for combined heating and power in manufacturing applications, and enables localities to export excess power for economic benefits:
  - Coal and nuclear plants are concentrated in the upper Midwest and North-eastern regions to support power-intensive industries and dense urban populations.
  - Hydropower, natural gas and oil-fueled plants are principally located in high-density coastal areas.

Large and small industrial plants place substantial demands upon fossil energy sources.

- Motors, steam, compressed air, pumps, process heating, combustion and combined heating and power account for about 80% of US industrial energy use:
  - The forestry industry is the largest user of energy from steam and CHP, and the petroleum industry is the largest user of fuel-fired heaters.
  - The metal industry is the third largest user of fired heaters, with major processing located in the West and East South Center, Middle Atlantic, East North Central and Mountain Regions.
  - Most primary aluminum producers are located in the Pacific Northwest, Ohio River Valley, Great Lakes, and Southern California.
  - Coal is a major source of industrial power, with leading consumer states located in the New England, West South Central, East North Central and Pacific Regions.

497. Larry Bell / US DOE - EERE

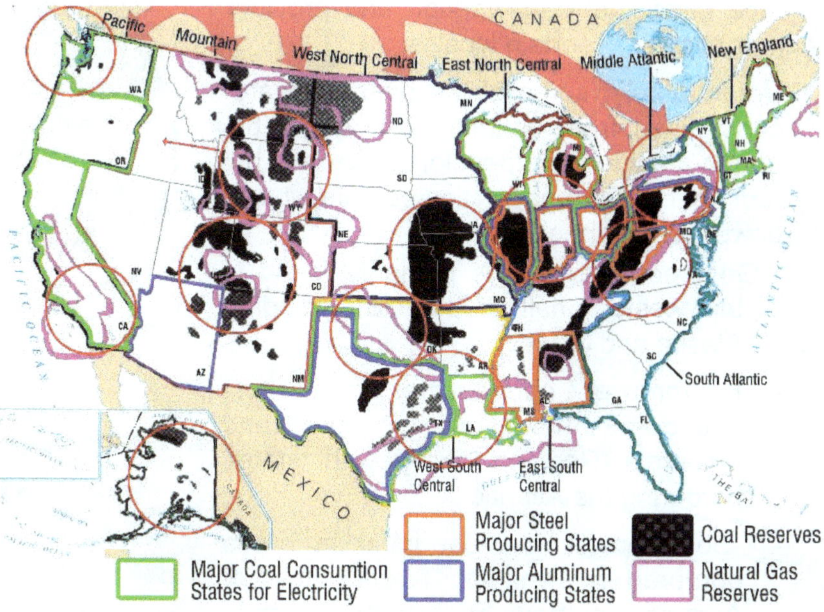

The three largest coal-producing states are Wyoming, West Virginia, and Kentucky, followed by Texas and Pennsylvania. Most primary aluminium producers are located in the Pacific Northwest, Ohio River Valley, Great Lakes Region and Southern California.

Fossil Fuel Energy Sources and Consumers

498. CNBC

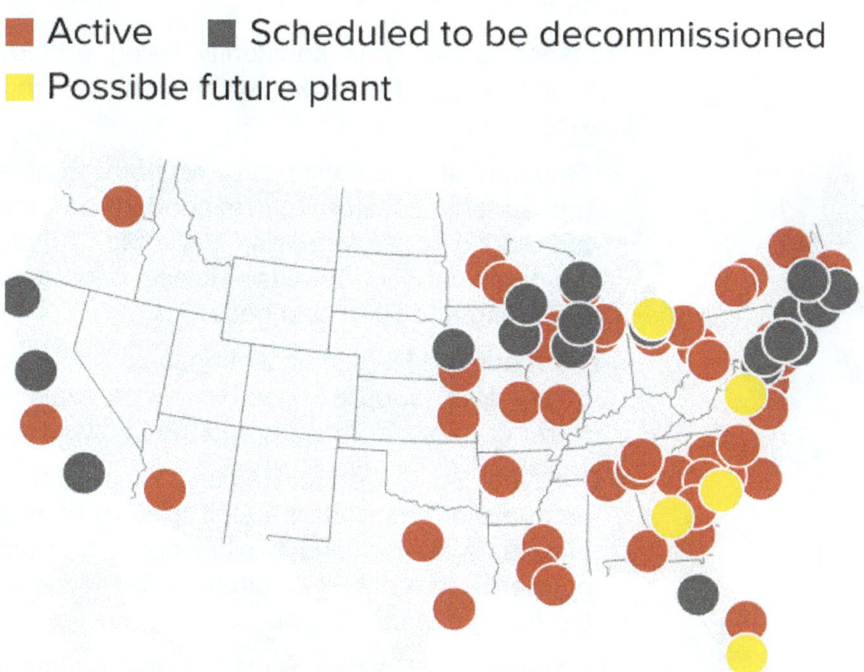

Nuclear Power

Followed by coal, nuclear plants are the second leading source of US electricity, providing nearly 20% of the nation's supply.

- Although broadly distributed throughout the country, nuclear energy is least prominent in the Mountain Region (only in Arizona), and most significant in seven Midwestern and Eastern states:
  - Illinois has 11 commercial reactors that produce nearly half of the state's electricity, and Pennsylvania ranks second with 4 plants.
  - South Carolina (about 50%), New Jersey (about 44%) and New Hampshire (about 40%); and New York (nearly 22%).

Renewable energy sources scattered throughout the US offer important means to reduce fossil fuel consumption and provide primary power in certain locales.

- Unfortunately, some of the best sites are constrained from optimum use either by heavy residential development or by remoteness from population centers where power is most needed:
  - Geothermal energy is primarily found in the western half of the US, including Alaska and Hawaii.
  - Hydropower opportunities are most abundant throughout Pacific and Mountain Regions along with states bordering the Atlantic.
  - Solar power is most ideal in areas bordered by western Texas, most of Arizona, southern borders of Colorado and Utah, and the southern half of Nevada westward to California.
  - Wind energy is prevalent around the Great Lakes, along ocean coasts, in Appalachia, and throughout the Great Plains.

499. Larry Bell / GEL / NREL/ ESRI

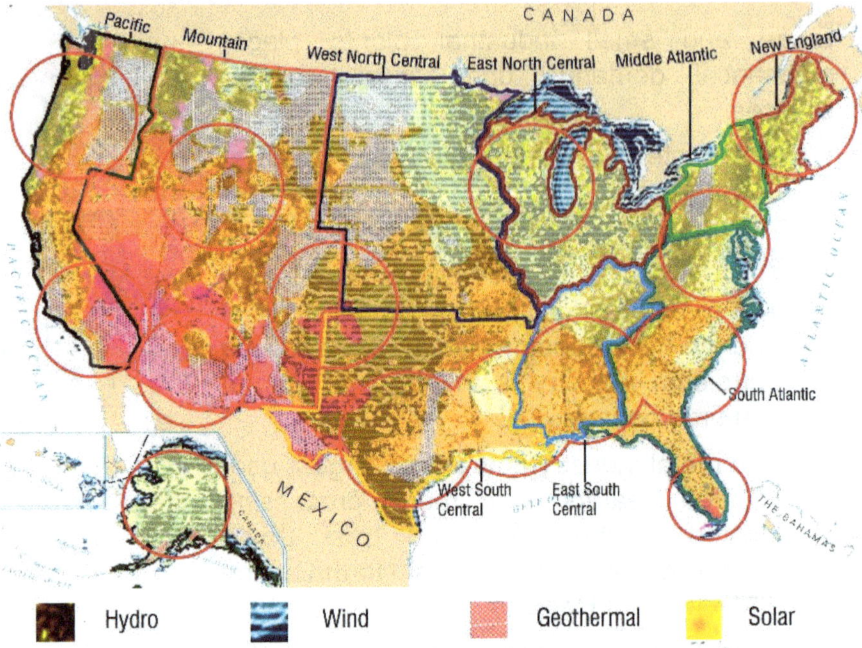

Clean renewable energy sources can support grid-connected and off-grid electrical needs. Circles show areas where multiple types converge to expand outputs and benefits.

Renewable Power

500. Larry Bell / US DOE - EIA

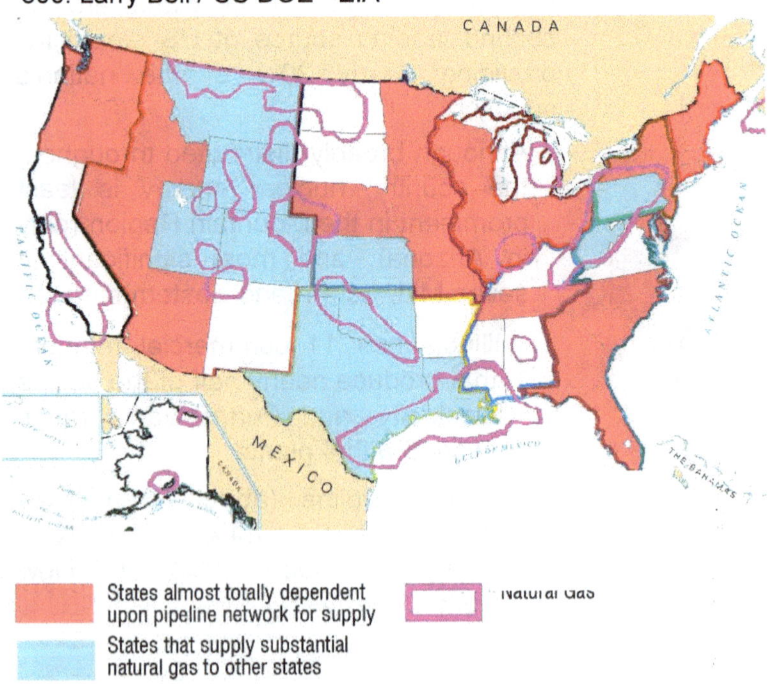

Natural Gas Infrastructure

The US has an extensive, well-developed natural gas infrastructure.

- Transport capacity is constantly being added, including expanded links with Canada and Mexico:
  - Substantial increases in western states (particularly California) responded to the 2000-2001 energy crisis, including supply connections with Canada, Mexico, Montana, Wyoming and Utah coal beds.
  - The Mountain Region, including Colorado coal beds, also supplies gas to meet rapidly-growing needs in Arizona and the Midwest.
  - The West South Central Region provides large amounts of gas from the Gulf area to central Florida, Alabama and Mississippi, and from northern Texas and southern Oklahoma to the North Central Region.
  - Northeastern, Southern Atlantic and adjacent Middle Atlantic Regions provide much of the gas needed to meet expanding northeastern and New England demands.

Liquid hydrocarbon fuels are likely to be dominant transportation energy sources for many future decades.

- Plant-derived ethanol and bio-oil produced throughout the US can extend diminishing fossil fuel supplies and benefit regional economies:
  - Corn ethanol opportunities are most prevalent throughout the West, and East North Central Regions, in north central Texas, and in Kentucky and Colorado.
  - Cellulosic ethanol possibilities are more broadly distributed in grassland and woodland areas found in most regions, and at all latitudes.
  - Crude oil reservoirs for petroleum refining are often associated with coal deposits in Mountain, and West South Central states, along the Pacific, Gulf Coast, and Alaska.
  - Additional opportunities for obtaining liquid fuels from coal gasification are present in Mountain states and Appalachian deposits.

501. Larry Bell / GEC / NREL/ US DOE

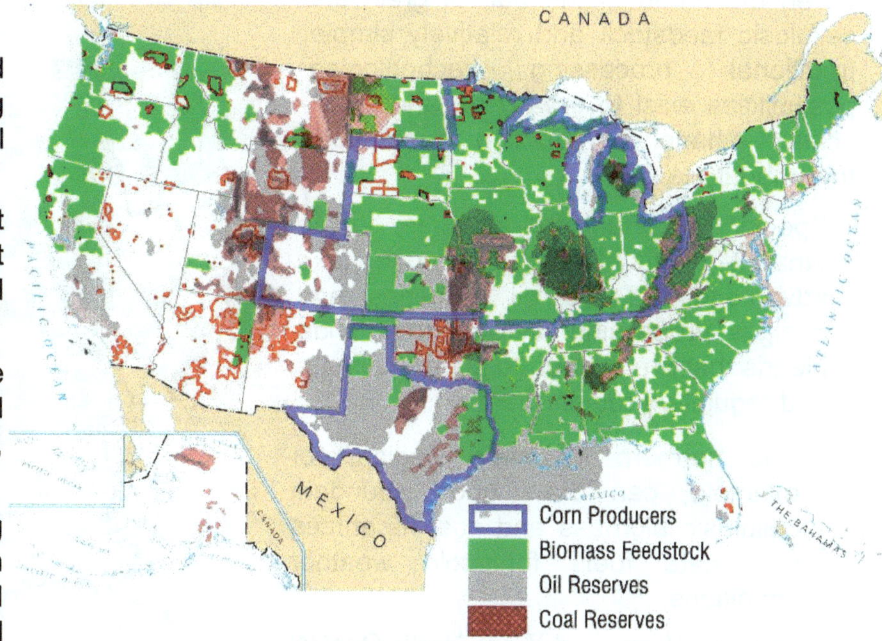

Liquid Fuel Sources

502. EIA

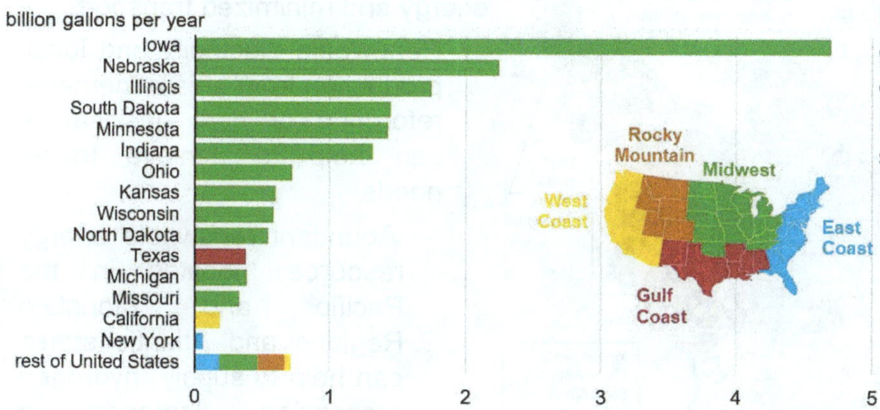

Ethanol Production

Ethanol production and use is rapidly growing, in part, as a gasoline octane enhancer to replace methyl butyl ether (MTBE) which has been banned in many states for environmental and health reasons.

- More than 200 ethanol production facilities currently exist in the US (about one-half of these farmer-owned), primarily in North Central and West North Central states:
  - About 10% of all US gasoline is now blended with ethanol, a practice that may become universal as trends continue.
  - Iowa and Nebraska are leading corn ethanol producers, followed by Illinois, South Dakota and Minnesota .

Given the broad distribution of grain and cellulosic feedstock and relatively simple, traditional processing technologies, possibilities exist to greatly expand small-scale ethanol production operations throughout the US.

- Special opportunities exist to locate ethanol plants where they can take advantage of area oil refineries and coalgas processing facilities to provide blends that are tailored to special climate and regulatory requirements:
  - The northern Mountain Region, for example, can draw upon abundant cellulosic biomass and coal resources to create fuels for cold weather conditions.
  - Northern West and West North Central areas can utilize corn crops, stover and other agricultural biowastes combined with coal gas resources in nearby southern East South Central and northern Middle Atlantic Regions.

503. Larry Bell / GEC / NREL/ US DOE

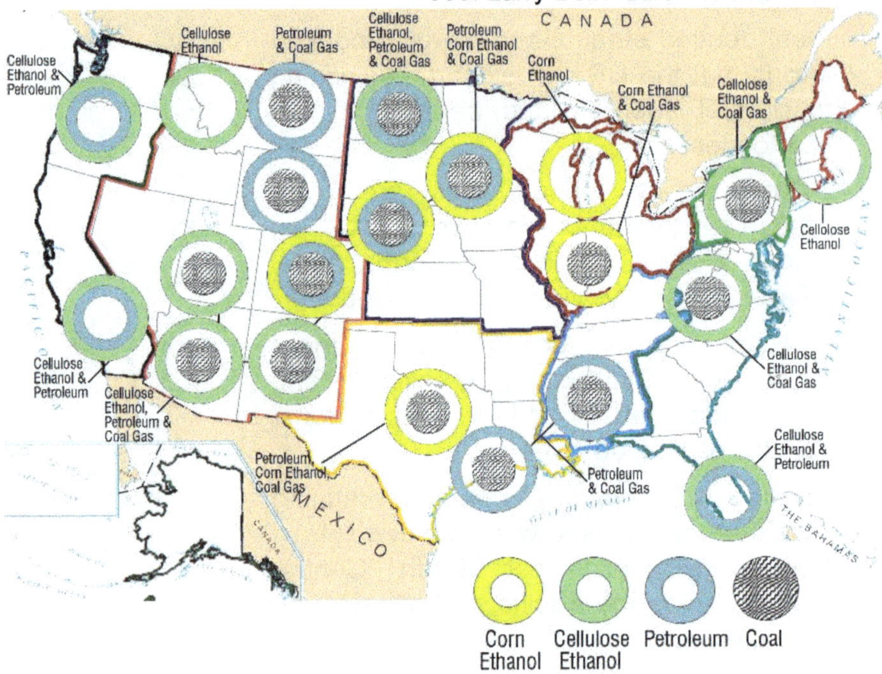

Combining Fuel Sources

504. Larry Bell / GEC / NREL / AFC

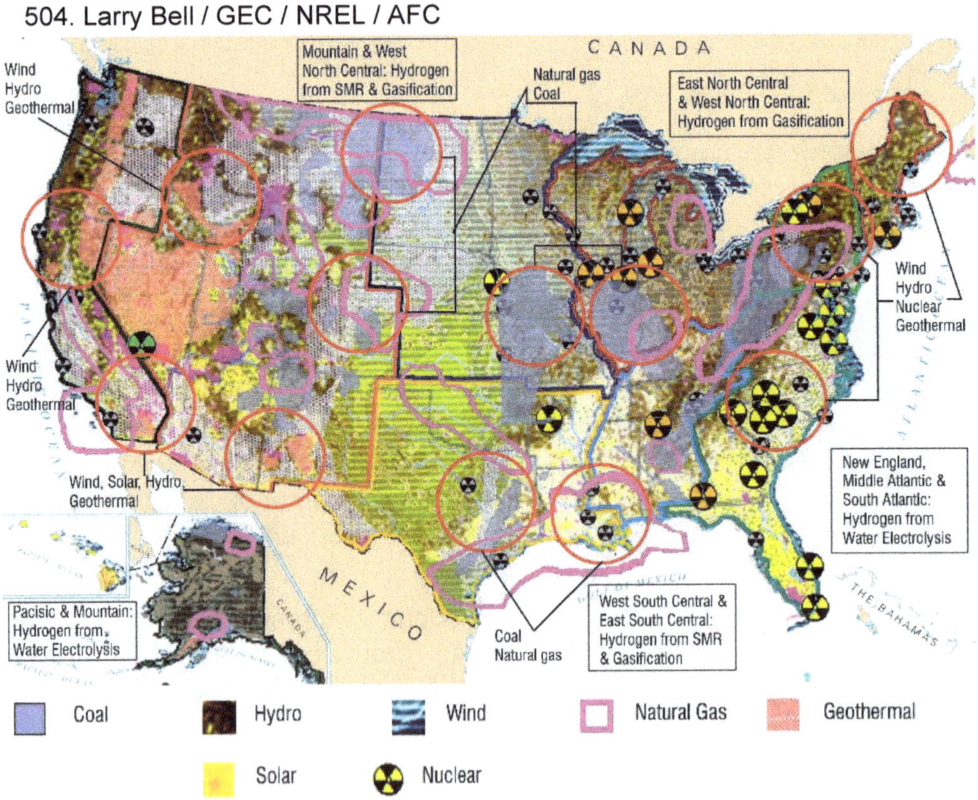

Hydrogen Production

Practical benefits of hydrogen depend upon affordable processing energy and minimized transport.

- Renewable electricity and local production from steam methane reforming or coal gasification can help to achieve these needs:
  - Abundant renewable energy resources located in the Pacific and Mountain Regions and Atlantic states can help to supply hydrogen processing demands in highly-populated western and eastern coastal areas.
  - Coal and natural gas in Mountain, West North Central, East North Central, West South Central and East South Central Regions can be used to produce hydrogen from SMR and gasification.

Hydrogen and/or methanol processed from fossil and biomass feedstock may eventually replace petroleum as predominant transportation fuels.

- Methanol offers advantages over hydrogen of easier and safer transport in liquid form, and can be used either in internal combustion engines or in direct methanol fuel cells (DMFCs):
  - Resources for hydrogen and methanol production are broadly distributed throughout the US, including green and nuclear power for hydrogen through electrolysis; and coal and natural gas for either or both.
  - Whereas hydrogen might be used near the production sources, methanol can be moved over much longer distances using existing fuel infrastructures.

505. Larry Bell / GEC / NREL/ US DOE

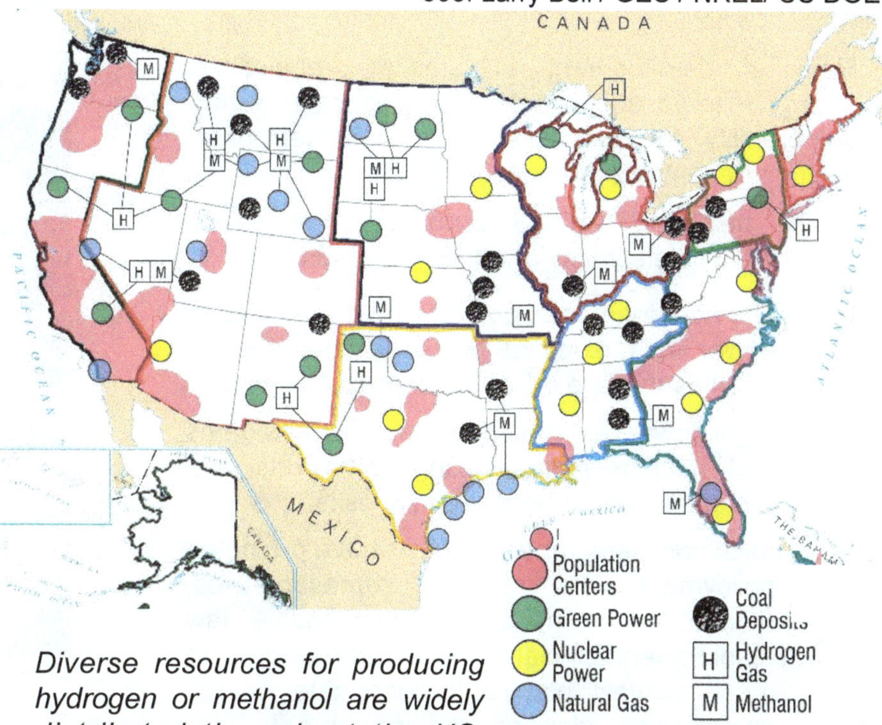

Diverse resources for producing hydrogen or methanol are widely distributed throughout the US.

Potential Hydrogen and Methanol Production Areas

506. Larry Bell / Olga Bannova

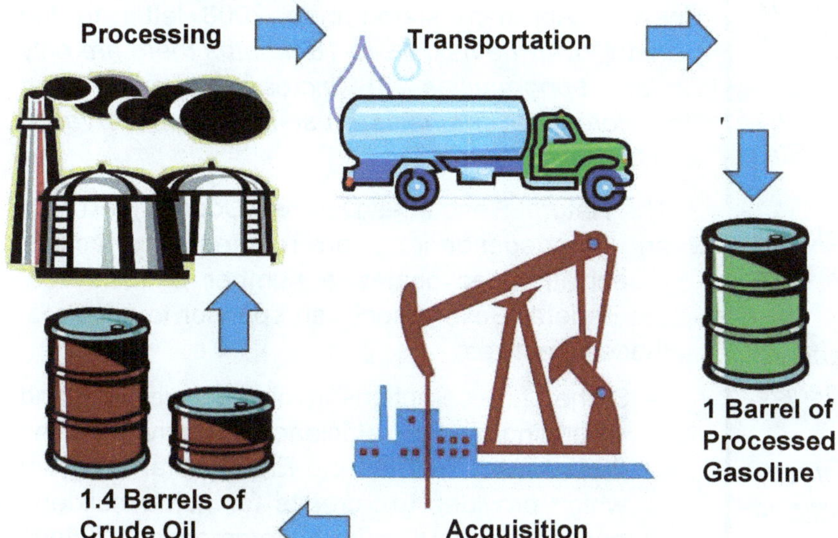

The EIA's Texas Energy Office estimates that every barrel of oil saved equals 1.4 barrels earned due to avoided costs of processing and transporting the energy.

Conservation

Although the US benefits greatly from diverse and broadly-distributed resources, these assets will be insufficient to meet escalating long-term demands.

- Notwithstanding continuing progress in developing new, more efficient and cleaner means of production (including renewable energy) it will also be essential to reduce consumption through more responsible personal, public and business conservation practices:
  - Some of these strategies will require substantial near-term technology and infrastructure investments to allow for transition periods that will avert avoidable crises later.
  - In the long run, every unit of energy saved pays substantial dividends through avoided costs of production, conversion and distribution.

507. Aachua County, Florida

Terrorist attacks on New York and Washington, DC, brought energy security and independence into the forefront of American consciousness.

Regional development of smaller, renewable energy resources can potentially reduce national vulnerability to incapacitating acts of terrorism through decentralization of supplies.

- The health of the US economy is ultimately linked to its electricity and fuel infrastructure:
  - The country relies upon thousands of miles of power lines, pipelines and associated support facilities that are difficult to protect.
  - Large power generating stations such as dams and nuclear plants, along with ports and refineries, also present special risks.
  - Attacks on September 11, 2001 have prompted policymakers, state representatives, emergency management and law enforcement officials and energy leaders to more actively plan and coordinate energy security and independence programs.

Energy Security

- *Expand information programs to promote energy efficiency in development and purchases of buildings and manufactured products.*
- *Provide additional performance-based incentives for existing home retrofits and Energy Star provisions of the EPAct.*
- *Fund the Low Income Home Energy Assistance Program (LIHEAP) and Westernization Assistance Program (WAP) already authorized.*
- *Enact a national renewable portfolio standard advocated by EIA requiring major utility companies to increase power from renewable sources.*
- *Extend production tax credits for electricity produced from wind and geothermal energy is extended at least till 2025.*
- *Issue minimum energy efficiency standards for certain appliances, equipment products and federal buildings through the Department of Energy.*

NRDC Recommendations

With regard to natural gas, US Secretary of Energy Spencer Abraham stated in a 2003 letter to the Senate "*over the next 12 to 18 months there are only limited opportunities to increase supply,*" and "*therefore, the emphasis must be on conservation and fuel switching.*"

- The Natural Resources Defense Council (NRDC), an independent environmental action organization, advocates a number of initiatives the Federal Government can sponsor to advance these objectives:
  - Some of these actions involve accelerating and extending energy efficiency incentives of the 2005 EPAct such as the Energy Star Program which provides tax credits for home builders, appliance manufacturers, commercial buildings and consumers of qualified products.
  - In 2023, state legislatures considered over 3,200 energy-related bills, with many focusing on clean energy development and energy efficiency

Federal Government Initiatives

Incentives and Developments

A growing number and variety of state conservation programs are being enacted with popular public and industry support.

- Many of these initiatives provide economic incentives:
  - Legislation establishing stricter environmental standards for equipment and building codes are realizing progress through tax credits and other benefits.
  - Many states are providing education and financial assistance for implementing efficient technologies and renewable energy solutions for farms, rural businesses, and agricultural and forest-thinning industries.
  - Forty-one states now allow green power homes and businesses to sell excess electricity back to grids, and 38 offer state-wide programs for this purpose.

508. SEIA

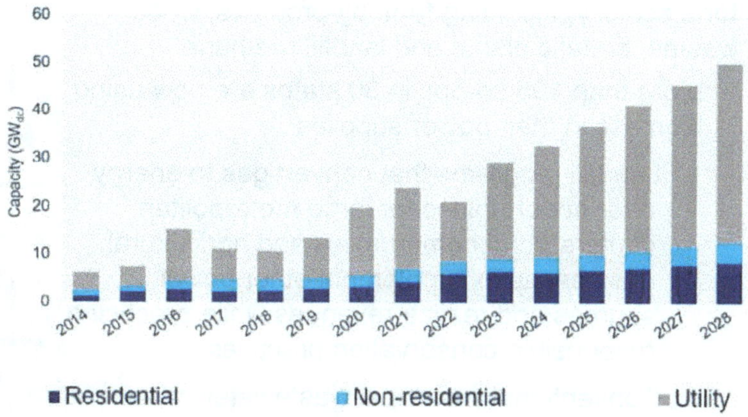

US Solar PV Installations and Forecast by segment, 2014-2028

Photovoltaic and other green power producers are supplying increasing amounts of excess electricity back to grids for energy credits.

State Government Initiatives

509. National Rural Electric Cooperative Association

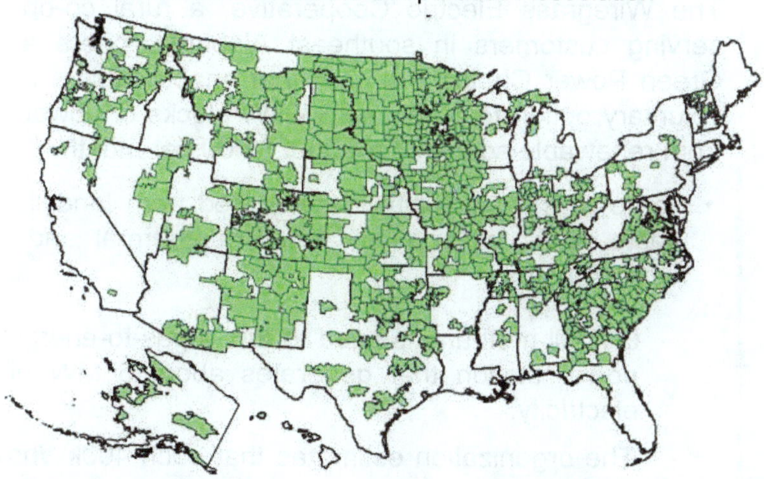

More than 900 rural electric systems offer green power through co-ops that supply energy to local grids. Member owners operating "backyard" systems receive revenue incentives.

Rural Electric Systems

"Green power co-ops" producing electricity from wind, photovoltaic and biomass are a popular trend in many parts of the country.

- Green tags are issued to the cooperatives by independent third-party entities to enable the co-ops to receive credits based upon the amount of electricity measured at the end of each year, and distribute checks to individual member producers that provide added benefits over energy they save in the process:
  - Typical home-sized solar power systems producing about 1kW and costing $12,700-$1700 might return approximately $300-$400 per year to help make installations more cost-effective.
  - Wind cooperative members often run larger 10kW turbines that produce more power, and co-ops sometimes help members finance wind surveys and provide low-interest loans to finance installation costs.

Green Power Co-Ops

Some energy co-ops focus upon biomass sources for electricity, including farm by-products, wood wastes, aquatic plants and landfill methane.

- More than 105 co-ops in 30 states are now using biomass in their power supplies:
  - Landfill programs that convert gas to energy offer opportunities for large metropolitan centers, rural communities and agricultural businesses to supplement other power sources and realize revenues while practicing responsible conservation practices.
  - Conventional municipal wastewater treatment processes can capture valuable methane for heating and methanol production, in combination with standard water purification for community use.

510. Larry Bell

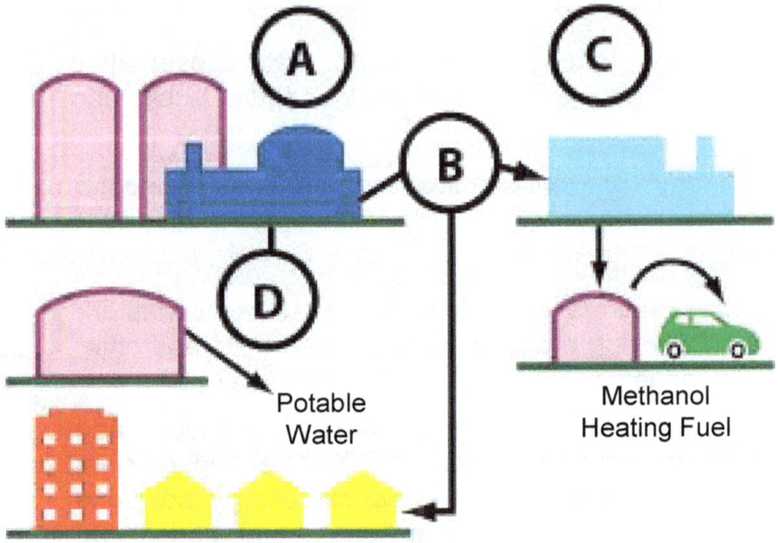

A - Wastewater Treatment
B - Methane Gas
C - Methane Plant
D - Water Purification

Municipal Waste Conversion

511. Wiregrass Electric Cooperative

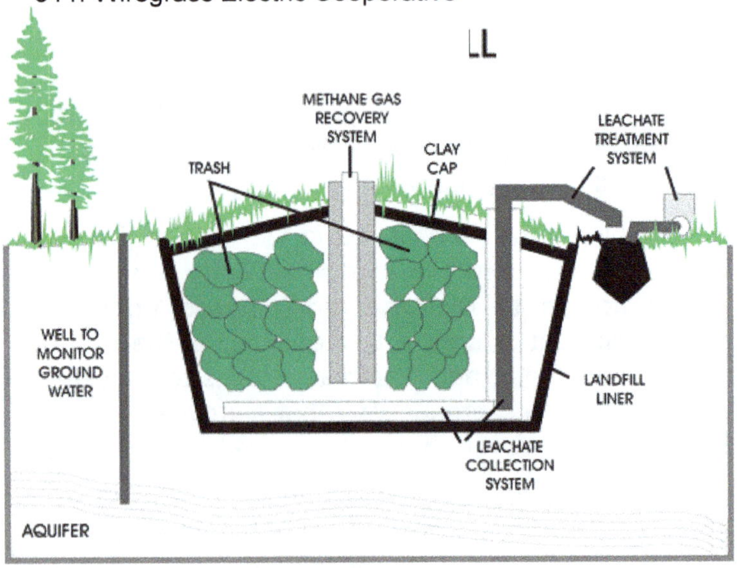

Methane

Landfill Methane

The Wiregrass Electric Cooperative, a rural co-op serving customers in southeast Alabama, offers a Green Power Choice Program that enables users a voluntary option to purchase 333kW blocks of power from renewable sources for about $4.00 per month.

- Some of the electricity is generated from landfill waste, along with solar, wind, geothermal and hydro sources:
  - Landfill methane is used to fuel a gas-to-energy power station that generates about 5 MW of electricity.
  - The organization estimates that each 400kWhs of electricity generated prevents 3.6 tons of CO_2 emissions each year, the equivalent of planting 1.2 acres (0.48 hectares) of trees, or not driving 9,600 miles (15,000 km) in a family car.

The Tokyo Bureau of Sewage (Japan) has developed plans to reduce greenhouse gases emitted during wastewater treatment and use sludge as a district power and heating fuel source.

- The biomass power production will capture and store methane from sludge biodigestion which is sent to a gas engine generator to create heat and electricity:
  - During nighttime hours when electricity demand is relatively low the power will be stored in sodium-sulfur (NaS) batteries for daytime use.
  - Heat produced from the power generating process will be fed back into sludge digestion tanks to accelerate methane production.
  - Sludge incineration at high temperatures will create additional power with about 70% lower nitrous oxide (N_2O) emissions due to carbonization of the material.

512. Bureau of Sewage, Tokyo Metropolitan Government

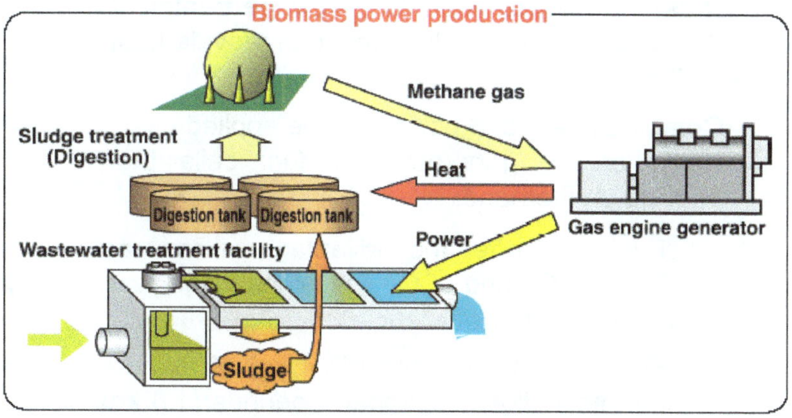

Biodigestion of wastewater sludge can produce methane fuel for district power and heat. The heat can also be used to carbonize sludge as a low-ash emission fuel source for power through incineration.

Municipal Waste Conversion

513. Practical Action

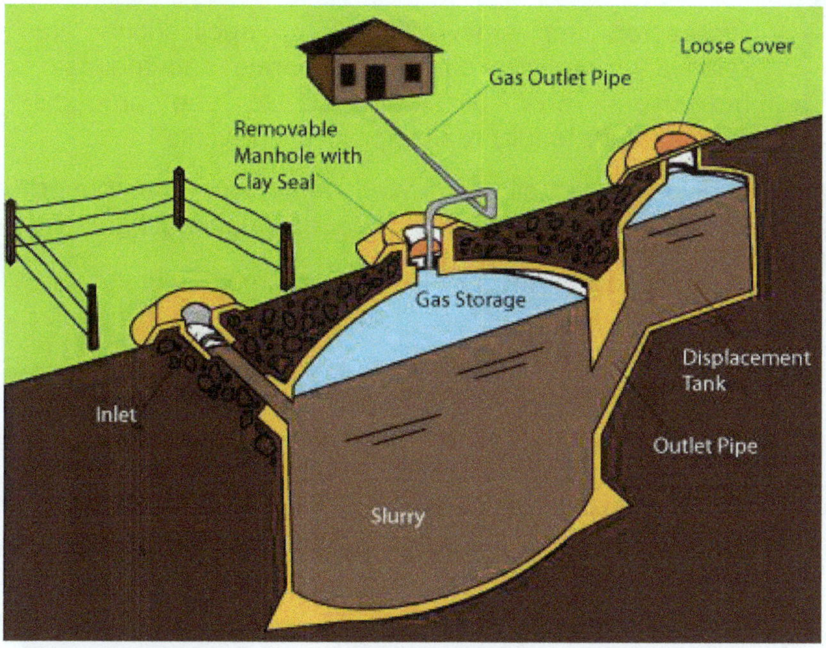

Manure bioreactors can provide biogas and fertilizer for on-site farm use.

Manure Bioreactors

Bioreactors are also used to maximize benefits of agricultural resources in rural areas by producing biogas and fertilizer that recycles animal and plant waste.

- Livestock manure is a good source of bioreactor methane and nitrogen:
  - Product gas is relatively free of pathogens and odors, and can have a methane content of 67%-71%.
  - In rural locations the biogas can be directly used for domestic heating fuel, and also for electricity generation to power water pumps and other equipment.
  - Bioreactors can greatly reduce total fecal coli forms, and all salmonella organisms, enabling safe recycling of water content for irrigation.
  - Integration of energy, fertilizer and water recycling process inputs/outputs can optimize economic and practical benefits.

Just as in municipal water and sewage treatment systems, farm-scale biodigesters can provide heat as well as power.

- Combined heat and power can be applied using either engine generators or fuel cells that operate on methane:
  - Fuel cells offer the advantage of reduced CO_2, N_2O and CO emissions, and can provide higher efficiencies.
  - For each 100 pounds (45.4 kg) of manure fed into a digester, about 4 pounds (1.8 kg) is converted into methane and the rest leaves as effluent, a process that requires about 20-25 days, with fresh manure added and effluent removed daily.
  - Process heat can be used for space and water heating, and electricity can be used for milk coolers, vacuum pumps, lighting and other purposes.

514. Larry Bell

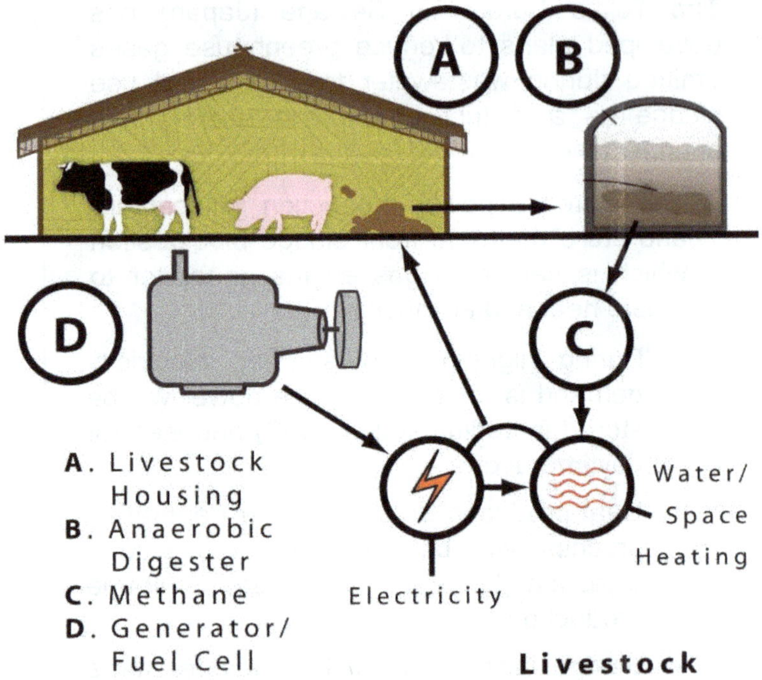

Bioreactor Power and Heat

515. Larry Bell

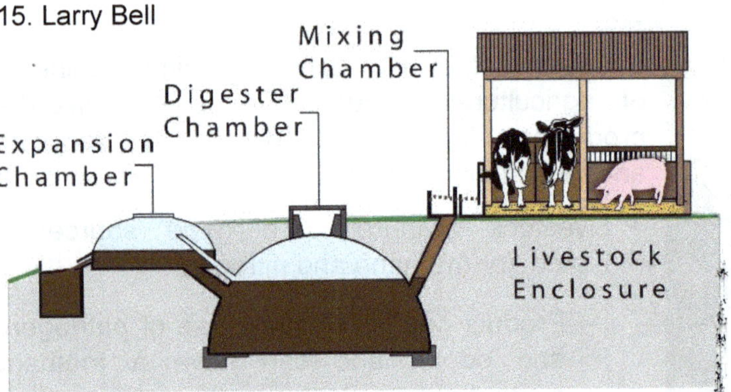

- *Must have a minimum of three cows or ten breeding pigs to sustain operations*
- *Livestock enclosure must be within 65 ft (20 meters) of the bioreactor*
- *Animals must be in the enclosure all night (a minimum of 12 hours)*
- *Requires year-round access to ground water within 65 ft (20 meters).*
- *Biogas should be used within 330 ft (100 meters) of the reactor*

Manure Bioreactors

Bioreactors for small-scale farm applications are relatively simple to construct, operate and maintain, but users must be motivated to use products resourcefully to fully realize practical benefits.

- Typically, small-scale fixed dome types are buried underground, and have three main connecting parts:
  - **Mixing chamber**: Animal excrement is first mixed with water before it is poured into a digester chamber.
  - **Digester chamber**: As the wet excrement ferments, methane and other gases push the manure and slurry that settle to the bottom into an expansion chamber.
  - **Expansion chamber**: excess manure and slurry are collected and flow back into the digester chamber to push biogas up for use. (When excess manure exceeds the chamber volume the manure is drained out.)

Although relatively few places have suitable geological characteristics for major geothermal power production, abundant opportunities exist to use this resource type for small-scale building heating, cooling and hot water applications.

- Hot springs and wells can provide heat for greenhouses, fish farms and other facilities, and relatively constant year-round near-surface ground sinks can afford economical space heating and air conditioning benefits:
  - Some cities such as Klamath Falls, Oregon and Boise, Idaho have used geothermal water to heat homes and other buildings for more than a century, and new housing developments in Reno, Nevada are using geothermal wells.
  - Subsurface land heat approaches can be applied in many areas of the US with high energy-saving efficiencies.

516. Lewis County Rural Electric Cooperative

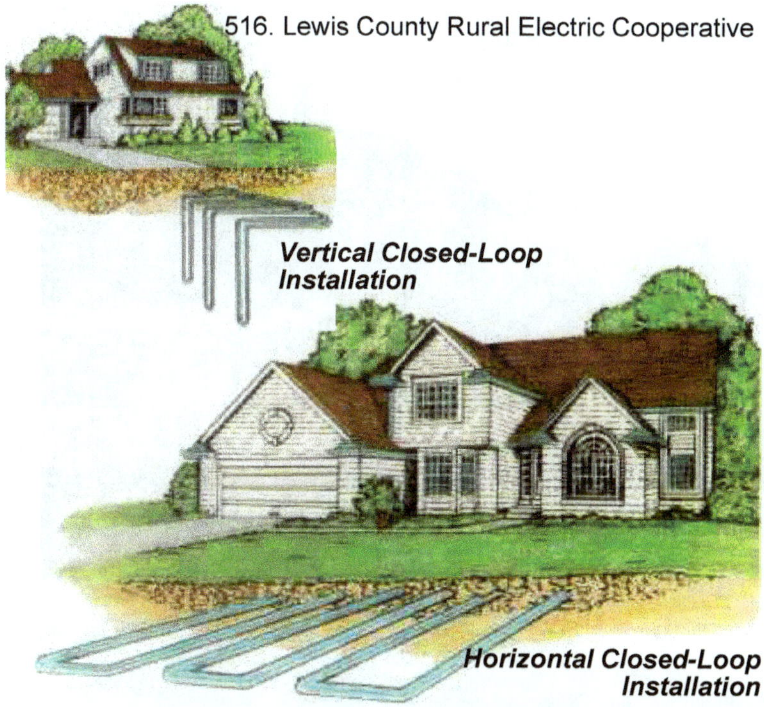

Geothermal heating and cooling can be more efficient than electric resistance, gas, or oil-fired systems, and can save homeowners 30% - 70% of heating and 20% - 50% of cooling costs according to EIA estimates.

Small-Scale Applications

517. Virginia Tech Department of Mines, Minerals and Energy

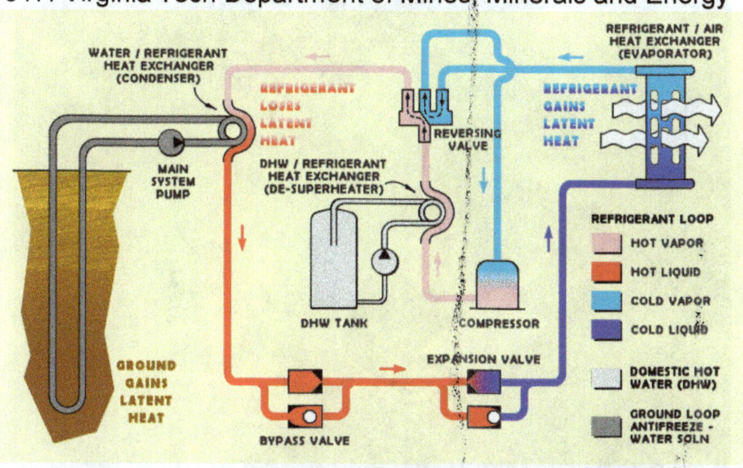

Heat pipes move heat from a building into the ground or a well in summer, and reverse the process to provide heating in winter. Key system elements typically include a heat exchanger, variable-speed blowers and multiple-speed compressors to circulate air and fluids.

Typical Heat Pump System

The most conventional way to tap geothermal energy for small-scale applications uses heat pumps that take advantage of stable temperatures a few feet below the surface (often about 50°F [10°C]).

- Heat pumps are relatively recent technology devices that are used in two general types of system applications:
  - Open-loop systems draw well water as the heat source or heat sink, and after use, the water is sent to another well.
  - Closed-loop (or "earth-coupled") systems use a water or antifreeze solution circulated through vertical or horizontal subsurface ground pipe loops to extract heat from the earth.

518. Water Wheel Factory

Waterwheel

Small-scale hydro systems apply very old and inexpensive conventional technologies that can be implemented by people with very limited means.

- Two principal types of technologies are waterwheels and turbines:
  - Waterwheels are familiar variations of massive antique wooden wheels that move slowly, but with great torque as streams pour over them.
  - Turbine approaches collect water from an upstream intake pipe which is forced down through a turbine by gravity pressure.
  - All available micro-hydro generators use a small turbine connected to an electrical generator or alternator

While the efficiency of a Banki turbine is lower than other designs, it enjoys a niche market for some low-cost and home-made installations.

- Construction is only slightly more complex than creating a simple waterwheel using means available in small workshops:
  - A turbine runner can be made from pipe sections that are slit to form blades.
  - The Venturi and associated plumbing can be fabricated with sheet metal tools.
  - Generators can also be homebuilt permanent magnet alternators (very similar to automotive brake disc alternators).
  - A small installation can be fed by a 4 inch (9.8 cm) diameter pipe in a low head of water with a 20 gallon/minute (76 liter/minute) flow.

519. Other Power

Small Home-Made Banki Turbine

520. ABS Energy

Small-Scale Applications

Wind has been an important source of energy for small-scale agriculture, industry and domestic applications for hundreds of years, and remains a valuable resource for electricity in many rural and remote sites.

- Wind can be particularly useful for off-grid circumstances, especially in combination with solar and other renewable sources:
  - Wind pumps for irrigation are viable options in some locations where breezes are constant and adequate for demands.
  - Wind can also be an excellent power source for telecommunication sites where height and exposure of antennas optimize conditions.
  - Since wind is not always present when needed, some means of electricity storage is often required, and use to electrolyze water into hydrogen for fuel cells represents one option.

In selecting appropriate sites, hills, valleys, river bluffs, and lakes can produce highly variable wind regimes, and trees and buildings can add to this complexity on a smaller scale.

- Each geographical feature influences wind in certain ways:
  - Hills, plateaus, and bluffs provide high grounds which raise wind speeds.
  - Valleys which are sheltered often have lower wind speeds, but when oriented parallel to wind flow can channel and increase wind velocity.
  - Constrictions in valleys can further increase wind flow by funneling air into a smaller area.

521. Canadian Renewable Energy Network

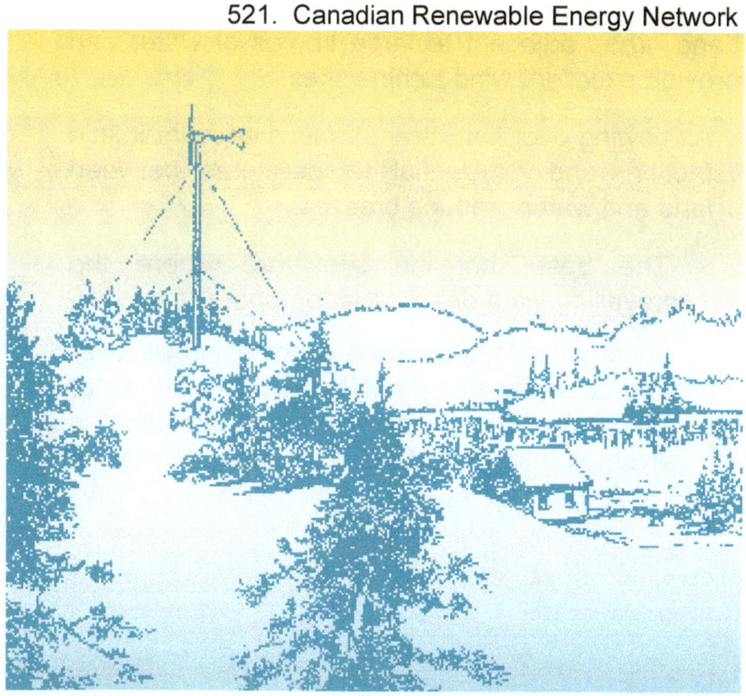

Surface-Driven Effects

Valleys often experience calm conditions at night, even when adjacent hilltops are windy.

- Cool, heavy air moves down from hillsides and collects below:
  - The resulting layer of cool air is removed from the general air flow above to produce calm lowland conditions.
  - Wind energy benefits can be most substantial when high-terrain features are at least several hundred feet above the surrounding land.
  - Wind turbines on a hill may produce power all night, even while those at lower elevations are idle.

522. Britannica

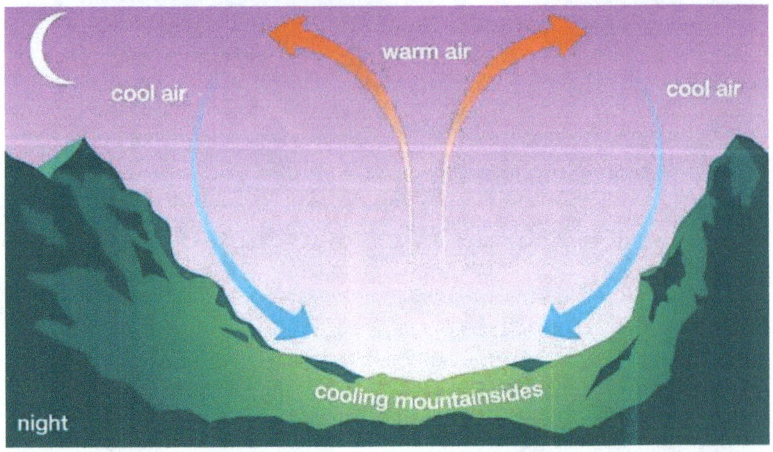

Evening Wind Effects

Land areas adjacent to large bodies of water can provide excellent wind turbine sites.

- Air flowing over flat water surfaces encounter little friction, and differential temperatures between land and water produce breezes:
  - The best shoreline sites are where the prevailing wind direction is “on-shore”.
  - Local winds occur when warm rising air over land is replaced by cooler air over the water during the day, and when this condition is reversed at night as land cools more quickly.

523. Britannica

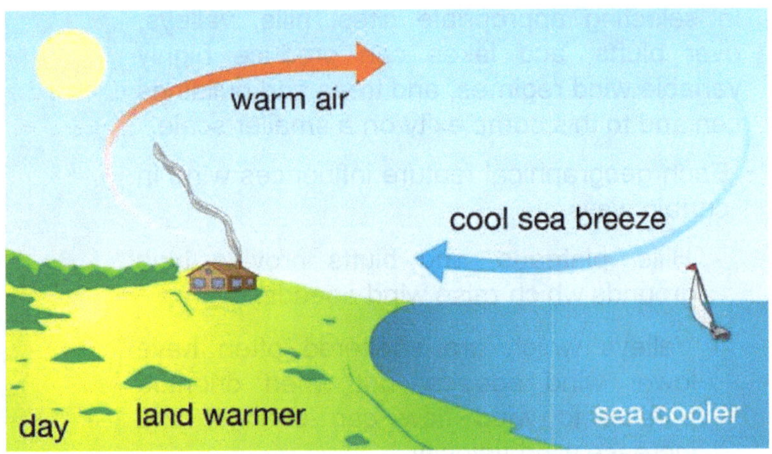

Sites Near Water Bodies

High terrain features can accelerate the flow of wind.

- An approaching air mass is often squeezed into a thinner layer so that it speeds up as it crosses over a mountain summit.
  - Over a ridge, maximum acceleration occurs when wind is perpendicular to the ridge line.
  - Isolated hills and mountains may accelerate wind less than ridges because more air tends to flow around the sides.
  - The downward ("lee") side of high terrain features can produce high turbulences that are not conducive for wind turbines.

524. Iowa Energy Center

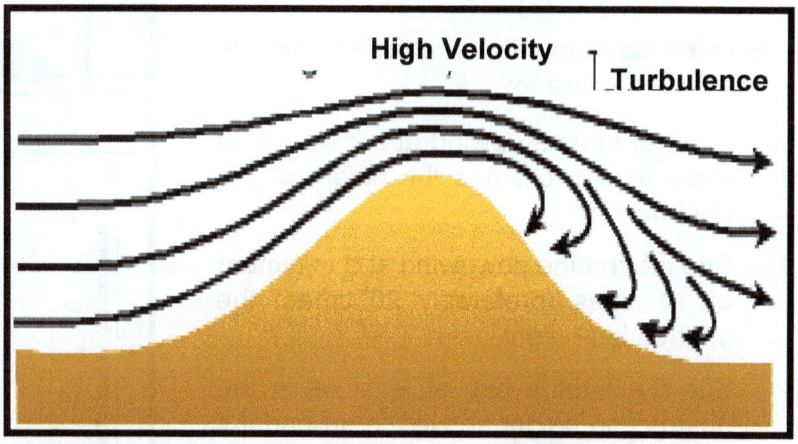

Air flow accelerates when squeezed as it passes over hill and mountain ridges, producing turbulent areas on the lee side.

Terrain Effects

Rough surfaces created by trees, buildings, and other features produce more friction and turbulence than smooth surfaces such as lakes or open croplands.

- The approximate increase of wind speed with height for different surfaces can be calculated by the equation:

$v_2 = v_1 \times (h_2/h_1)^n$, where:

v_1 = known (reference) wind speed at height h_1 above the ground

v_2 = speed at second height h_2

n = the exponent of determining wind change based upon different types of ground cover.

Values for n:

Smooth ocean sand: n = 0.1 — Low grass area: n = 0.16

High grass/low crops: n = 0.18 — Tall crops/low woods: n = 0.2

High woods/forest: n = 0.3 — Small towns: n = 0.3

Example wind speed calculations:

- Assume an average 10 mph wind speed measured at 30 ft above the surface in an area of low woods (n = 0.2) and you wish to estimate wind speed of a 100-foot high wind turbine:

$v_2 = 10 \text{ mph} \times (100/30)^{0.2}$

$v_2 = 10 \times (3.33)^{0.2}$ $v_2 = 10 \times 1.27$

$v_2 = 12.7$ mph (average at wind turbine)

Surface Effects of Wind Speeds

Trees and buildings are common wind obstacles at a wind turbine site. They disturb air both upwind and downwind of the obstruction, reducing wind speed and increasing turbulence.

- Some rules of thumb for reducing obstacle interferences are to:
  - Site the turbine upwind at a distance of more than two times the height of the obstruction
  - Site the turbine downwind at a minimum of 10 times (preferably 20 times) the obstruction height.
  - Site the turbine hub at least twice the obstruction height above the ground if it is immediately downwind from the obstruction.

525. Iowa Energy Center

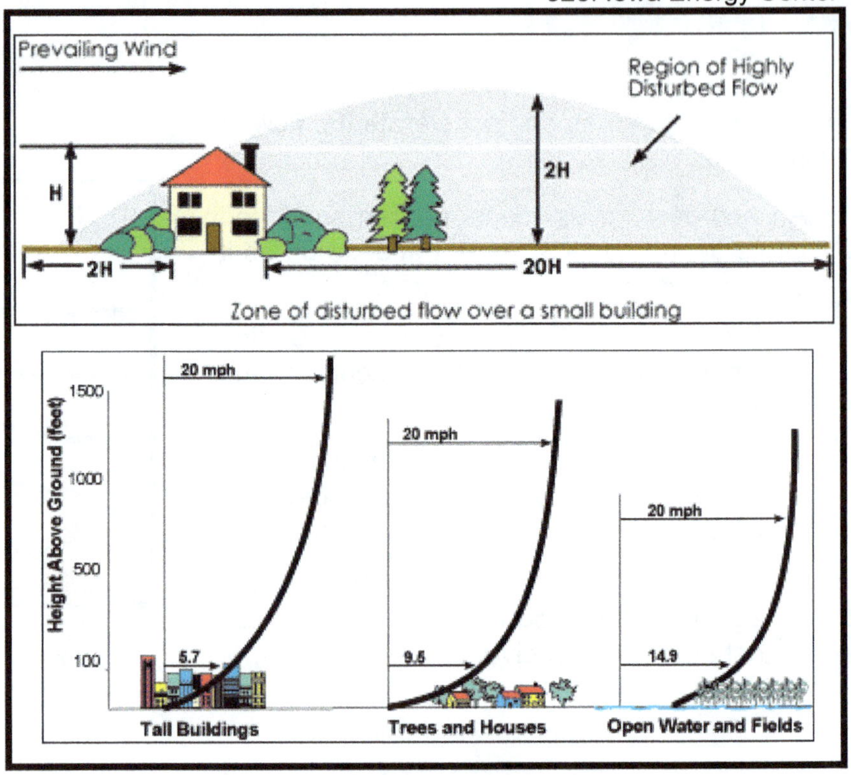

Minimizing Obstruction Influences

526. Iowa Energy Center

Trees as Wind Indicators

Tree deformation can be used as a rough guide for determining good wind turbine sites:

- Trees growing in high wind areas reveal wind effects when trunks bend or branches extend away from prevailing wind directions:
  - "Brushing" or "flagging" typically occurs when trees are exposed to average wind speeds of 15 – 18 mph (24 – 28 km/hr), but can often be seen at average wind speeds as low as 8 – 10 mph (12 – 16 km/hr) when dominant wind direction is constant.
  - Brushing is common in deciduous trees (maple, oak, elm) where branches and twigs bend downward like a brushed fur pelt.
  - Flagging is common in coniferous trees (pine and spruce) with short/missing upwind branches and long ones downwind.

Millions of PV systems have been installed world-wide, ranging in size from a fraction of a watt to multi-megawatts:

- Mass production has driven cost down, and for many applications they offer the most economical and practical power option:
  - PV systems are ideally suited for many remote locations ranging from recreational cottages to orbital and planetary space environments.
  - Recreational vehicles, boats and scientific expeditions use PVs for direct power and to recharge batteries.
  - PVs are used in agriculture to pump water for livestock and to power electric fences.
  - PVs can be adapted for very small and large equipment, including calculators and watches, telecommunications, highway construction signs and navigational warning lights.

527. NASA
529. Europ. Solar Ener. School

528. IEA PV
530. Energy Supermarket

Diverse Applications

531. Fraunhofer ISE
532. Fraunhofer ISE

Roof-Mounted PV Installation

Two-Array Ground-Mounted Installation

Off-Grid Applications

Large populations throughout the world depend upon off-grid autonomous and hybrid "island" PV systems to power their homes, farms and other businesses.

- Residents of southern Europe make extensive use of autonomous PV-battery systems due to a relatively warm climate and frequent access to sunshine.
- Residents of central Europe often use hybrid systems which pair PV with wind or other generators to accommodate the "solar gap" in winter. (In Germany, the amount of solar energy is six times lower in December than in June.)
- Off-grid stand-alone and hybrid systems are also used world-wide for remote site applications to provide electricity for summer cottages, livestock watering and irrigation, public and private areas such as parks and campgrounds, emergency facilities, and operations, and communications.

533. Provision Technologies, Inc.
534. Sky Built Power

Mobile Power Station developed by SkyBuilt Power, Arlington, Virginia.

Solar Powerhouse developed by Provision Technologies, Hilo, Hawaii.

Some transportable off-grid PV systems have been developed for temporary power applications where utility services do not exist or are unreliable.

- These units are typically designed to be transported by a variety of means, including truck, train, or even parachuted by airplanes to remote locations.
  - A Mobile Power Station (MPS) developed by SkyBuilt Power of Arlington, Virginia, is a solar-wind hybrid unit providing air-conditioned space for telecommunications, medical, emergency/command operations and other functions housed in a standard shipping container.
  - The Solar Powerhouse developed by Provision Technologies generates AC electricity for diverse purposes, including construction sites, water pumping, and emergency and village sites.

Off-Grid Applications

Grid-connected PV installations are used to reduce power consumption from public utilities, while ensuring continuous energy access.

- An important advantage is elimination of a need for storage batteries during periods when PV is ineffective, along with benefits of clean, renewable power:
  - These systems are most appropriate for applications requiring high winter or year-round energy demands, and where power must be constantly available.
  - Although installation costs can be relatively high, these up-front expenses can often be offset by energy tax credits and opportunities to sell excess electricity back to public providers.

535. National Photovoltaics

Typical Home Installation

Grid-Connected Applications

536. Lawrence Berkeley National Lab

Building Integrated Photovoltaics

For large building applications, PV systems can be integrated directly into structures in a number of ways:

- They can be incorporated into weather membranes, or as external architectural cladding features.
  - The cells can be integrated into a laminate made from two sheets of glass, or comprised of a sheet of glass and a sheet of foil.
  - Glass or foil behind the cells can be colored, tinted or screen-printed to create a variety of visual effects.
  - Solar glazing can be used in double-glazed units, curtain walls and glazing systems mounted onto a rooftop frame or shade device.

537. University of Cambridge UK

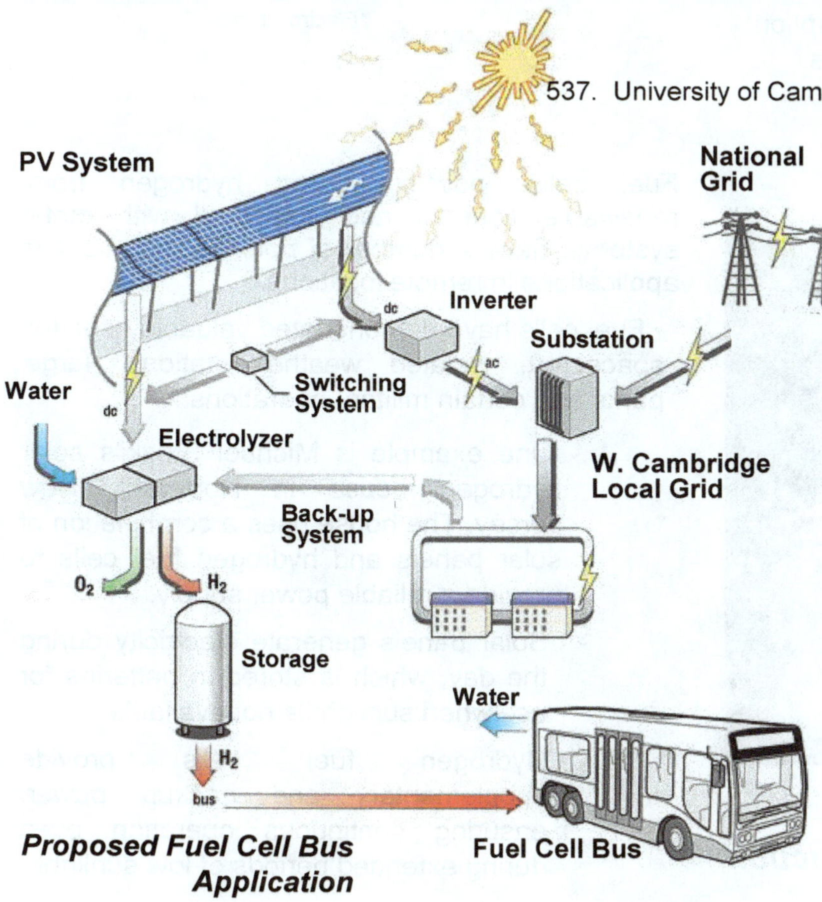

Solar Hydrogen Production

Solar energy can provide an inexpensive source of power to produce hydrogen through electrolysis.

- An example is a proposal to supply hydrogen for fuel cell buses in Cambridge, England:
  - A PV system integrated into a colonnade roof at the site would connect with national and local power grids.
  - DC electricity would power both the hydrogen electrolyzer processing and cryogenic storage equipment, with DC-AC conversion of excess power to the grids.
  - Backup grid power can also be sent directly to the electrolyzer and storage system.

Since electrolyzer systems don't store fuel themselves, they can be supplied with electricity from renewable sources which also compress or liquefy the hydrogen gas.

- Although electricity delivered directly from one or more renewable sources to an end use would be much more efficient than using the power to electrolyze water for hydrogen fuel cell conversion back to electricity, this approach can sometimes be very beneficial:
  - Unlike battery storage of electricity when power is not available or needed, fuel cells do not run down or require recharging, offer much higher energy densities plus CHP cogeneration opportunities, and can respond to changing loads by adding more hydrogen fuel.
  - Use of multiple renewable sources for hydrogen production can provide larger and often more continuous product outputs for local consumption or commercial export than wind or solar alone.

538. Watt Pyle, et al., Home Power

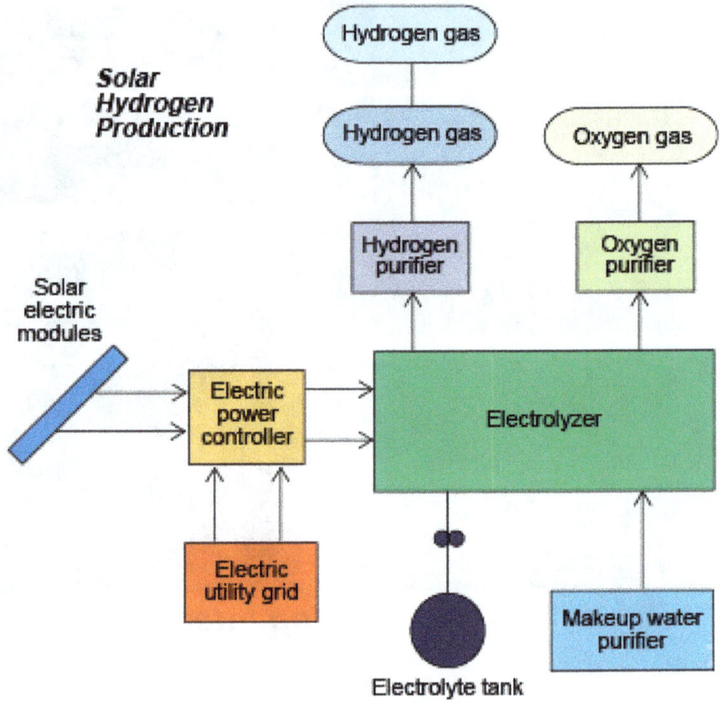

539. NJ Tech Weekly

Solar Hydrogen Production

Fuel cells, possibly using hydrogen from renewable sources and combined with other systems, have a number of potential small-scale applications in remote locations.

- Fuel cells have demonstrated valuable uses for spacecraft, isolated weather stations, large parks and certain military operations:
  - One example is Michael Strizki's solar hydrogen house in Hopewell, New Jersey. The house uses a combination of solar panels and hydrogen fuel cells to provide a reliable power supply.
  - Solar panels generate electricity during the day, which is stored in batteries for use when sunlight is not available.
  - Hydrogen fuel cells provide supplementary and backup power, ensuring continuous operation even during extended periods of low sunlight.

A recent fuel cell application is for family home, office and factory combined heating and power use which might obtain hydrogen from PV or other renewable sources.

- Matsushita Electric (Panasonic) in Osaka, Japan has developed a compact CHP system that can generate enough electricity and heat to supply an average household, with backup power provided from a public utility grid:
  - The system's membrane-electrode assembly (MEA) structures offer an industry-level power-generating efficiency of about 35%, along with substantial thermal benefits.
  - Operating noise levels are comparable to a typical air conditioner, acceptable for use in densely populated areas and at night.

540. Panasonic

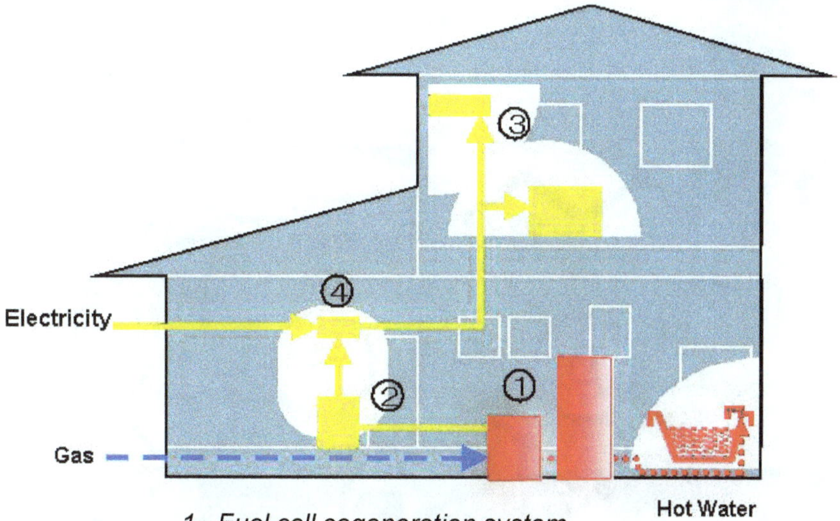

1. *Fuel cell cogeneration system*
2. *Commercial power splitter (for connection to a system-linked inverter.*
3. *Electrical appliances (such as air conditioners)*
4. *Electrical distribution panel*

Matsushita Home CHP Fuel Cell System

541. Larry Bell

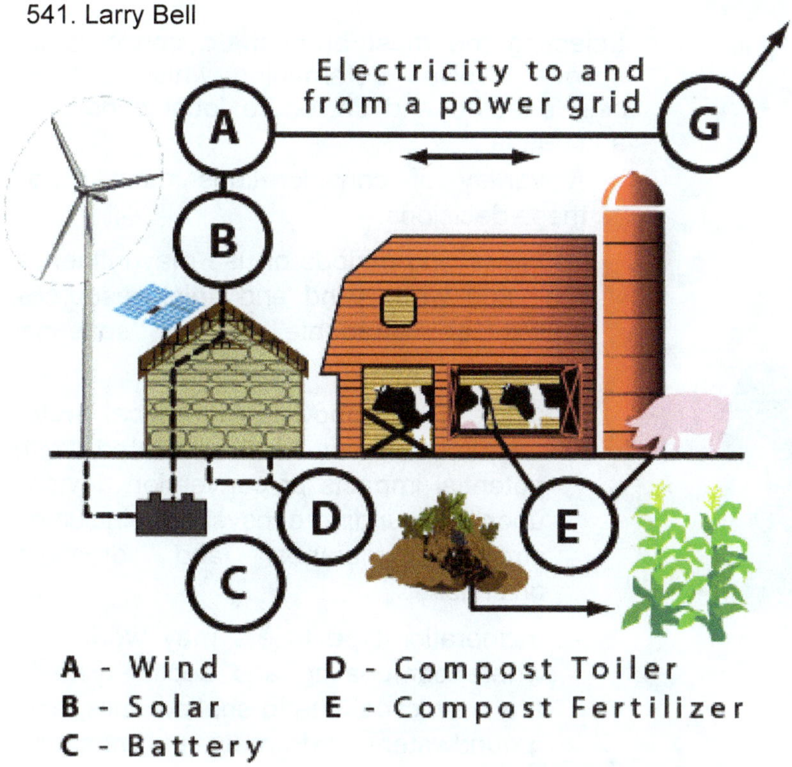

Private Farm

Optimal selection and use of renewable resources depends upon a variety of circumstance and needs that are specific to each site and application, including geographic location, topography, and the nature and scale of end use.

- A small family farm in an area with favorable wind and Sun conditions, for example, may combine turbines and solar PV panels with battery storage for electricity, along with composting for waste reduction and fertilizer:
  - Wind and solar energy may reduce or avoid utility power use for domestic services, water pumping, milk cooling and other purposes, potentially releasing excess electricity back to a grid for credit.
  - Composting can convert waste biomass into safe plant nutrient supplements.

542. Larry Bell

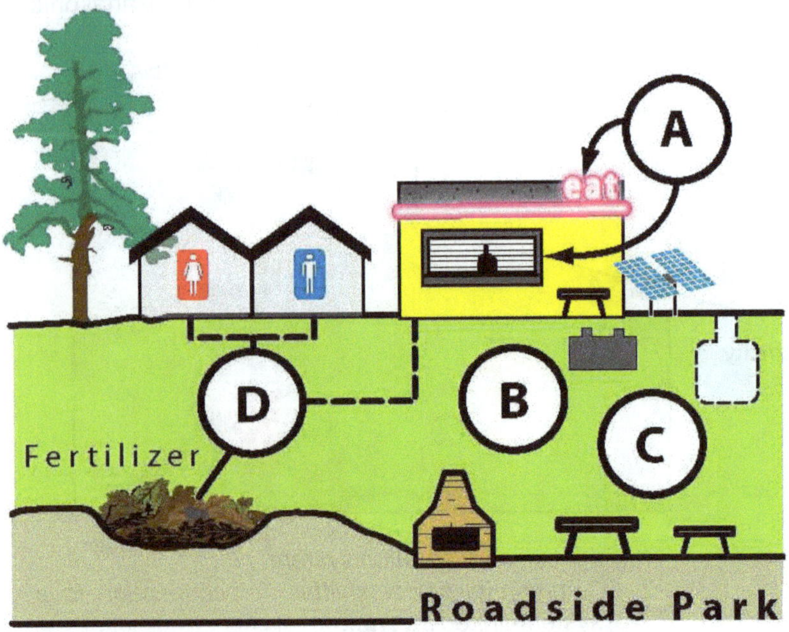

A. Solar Signage / Service
B. Battery
C. Water Pump
D. Compost Toilet / Waste

Public parks, rest areas and other facilities in locations unserved by power and water lines can combine renewable resources for electrical and sanitary purposes.

- Such green applications make practical sense and demonstrate good examples of environmental consciousness:
  - -Photovoltaic systems, potentially supplemented by fuel cells or gas generators, can provide clean power for signage, lighting, food vending machines, water well pumps and other uses.
  - Odorless compost toilets can convert human and food wastes into fertilizer that can be used for horticulture to beautify the sites as an alternative to conventional types that consume water and require septic leaching fields and sludge removal.

Public Facilities

543. Larry Bell

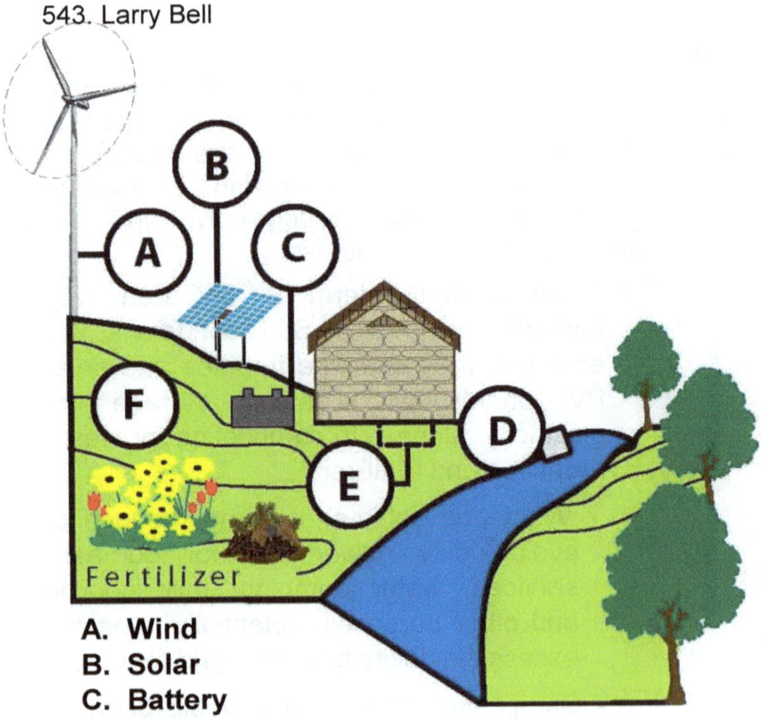

A. Wind
B. Solar
C. Battery
D. Hydro
E. Compost Toilet
F. Compost Garden

Remote Cottage

Selecting the most appropriate conservation options involves determining which can be best adapted and scaled to local conditions and needs.

- A variety of considerations must guide these decisions:
  - Times and periods of use may influence if and when wind and solar resources will be adequate, taking seasonal climate variations into account.
  - Use of hydropower from local water resources must be correlated with potential impacts of conversion devices upon surrounding ecosystems, including streams, marshes and drainage diversions.
  - Incineration-type toilets may work best where composting and septic systems are impractical due to shallow soils, high groundwater or extremely cold weather.

Vacation Residences

544. The Apache Trail Archives

Anasazi Indians of Colorado applied effective solar design principles to cliff dwellings thousands of years ago.

Vitally important resource conservation benefits are realized through simple methods that reduce energy required to heat and cool buildings and water, often with little or no added up-front expenses.

- As energy costs continue to rise, it is only logical to expect that the desirability and market value of energy-efficient buildings will be dramatically enhanced:
  - Prudent planning to effectively utilize and control direct solar gain can provide major results for diverse building types, large and small, in all climate regions.
  - While some approaches apply "active" technology devices, basic natural principles proven over many centuries by world-wide civilizations universally apply.

Applying Natural Principles

545. National Renewable Energy Lab

Solar thermal radiation is broadly used in small-scale active and passive applications to reduce water and space heating costs.

Active Solar Collectors

Active systems tend to be somewhat technologically-dependent and use-specific, often using pumps and fans to circulate heat transfer fluids and air.

546. Powered Living

Active vs. passive uses are generally characterized according to the degree to which they depend upon technology to function:

Passive Solar Design

Passive solar systems rely primarily or entirely upon direct solar gain through natural conduction, convection and radiation.

The major economic application of solar energy is to heat buildings and to provide hot water, often using rooftop solar collectors along with backup systems for supplementary power.

- Used directly as a thermal source or combined with PV cells for electricity, solar energy can significantly reduce heating bills and conserve costly resources:
  - During summer, most locations between northern latitudes of 20°-58° can use passive solar panels to meet some electrical needs, and most hot water requirements.
  - Advanced technology and production economies are continually reducing system costs, with residential swimming pool heating applications constituting an important market.

547. European Solar Thermal Industry Federation

Solar systems and effective passive planning can significantly reduce energy costs for home water and space heating, air-conditioning, and electricity consumption for power.

Reducing Power Costs

548. Larry Bell/Olga Bannova

Active Residential System

Active solar systems are the most efficient way to heat water and air, producing low-grade thermal energy of a low temperature (less than boiling) direct from the Sun.

- Typical basic components are:
  1. **Water supply**: potable water from a utility source or well.
  2. **Flat plate collectors**: use radiant energy to heat water.
  3. **Heat transfer and storage**: transfers hot water from collectors to a storage tank.
  4. **Domestic hot water**: for kitchen, bathroom, washing machines, etc.
  5. **Space heating**: hydronic heating systems (radiators).

549. Green Energy

Active solar heating systems are used as a sustainable alternative to typical household water heaters.

- Utilizing a minimal amount of components, the system layout remains relatively simple:
  - A controller constantly compares the temperature of the solar collectors with the temperature of the water in a tank cylinder.
  - A circulating pump is switched on by the controller whenever the temperature of the collectors rises above that of the tank.
  - A water/ antifreeze mixture is then circulated through the collectors and heat exchanger to heat the tank.

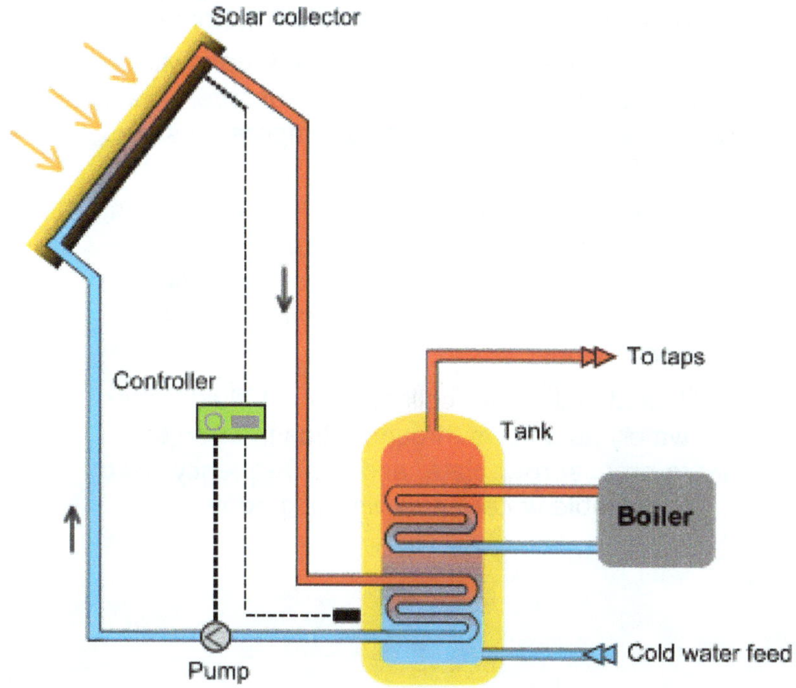

Active Hot Water Circulation

550. Larry Bell / Olga Bannova

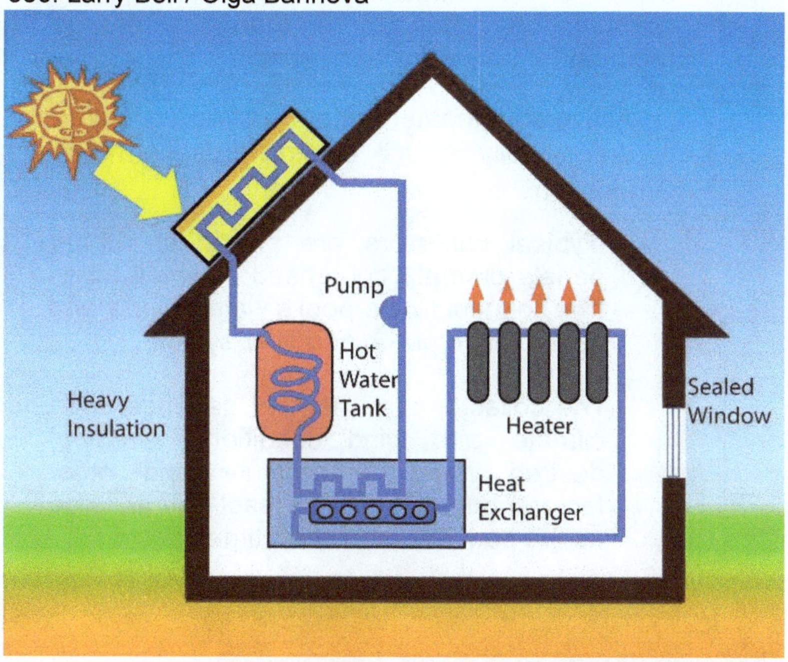

Active Air-Based Heating System

Space heating can also be accomplished using solar-heated water.

- This approach is similar to a conventional domestic hot water system, but uses a large flat plate collector array:
  - Such a system requires a large hot water storage capacity, and usually employs an air-to-water heat exchanger to supply the hot water during sunny periods.
  - Heat from the storage tank is typically transferred to a radiant floor heating system, baseboard radiation heating system or through a fan coil to a forced-air system.
  - Solar systems are often combined with supplementary gas or electrical water heaters to accommodate times of highest demands.

551. US DOE-EERE

Flat plate collectors are the most commonly used types for solar water and space heating in homes.

- A typical collector is an insulated box with a glass cover and dark-colored absorber plate which can heat air or liquid to temperatures less than 180°F (82°C):
  - Air flat plate collectors primarily used for solar space heating are typically made of metal sheets, layers of screen or non-metallic materials that absorb heat by natural convection.
  - Liquid flat plate collectors heat fluid (often water) as it flows through tubes in or adjacent to an absorber plate, and are frequently used for potable water and swimming pools.

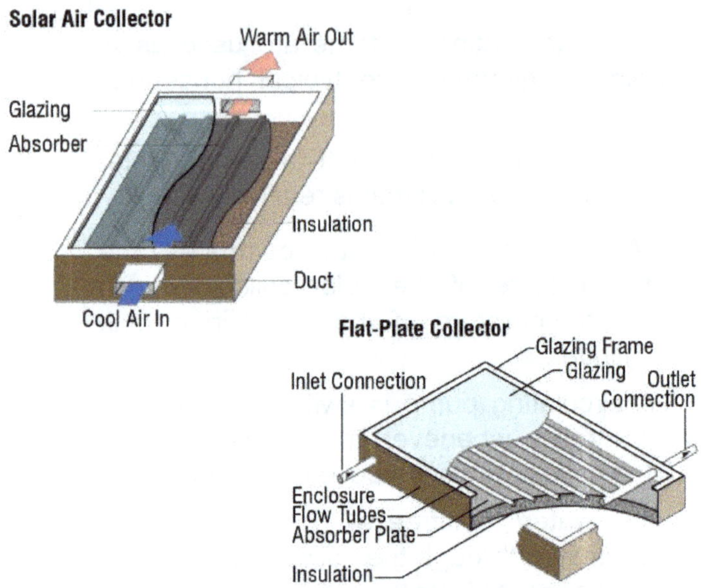

Active Space Heating Collectors

552. Norbert Lechner, Heating, Cooling, Lighting

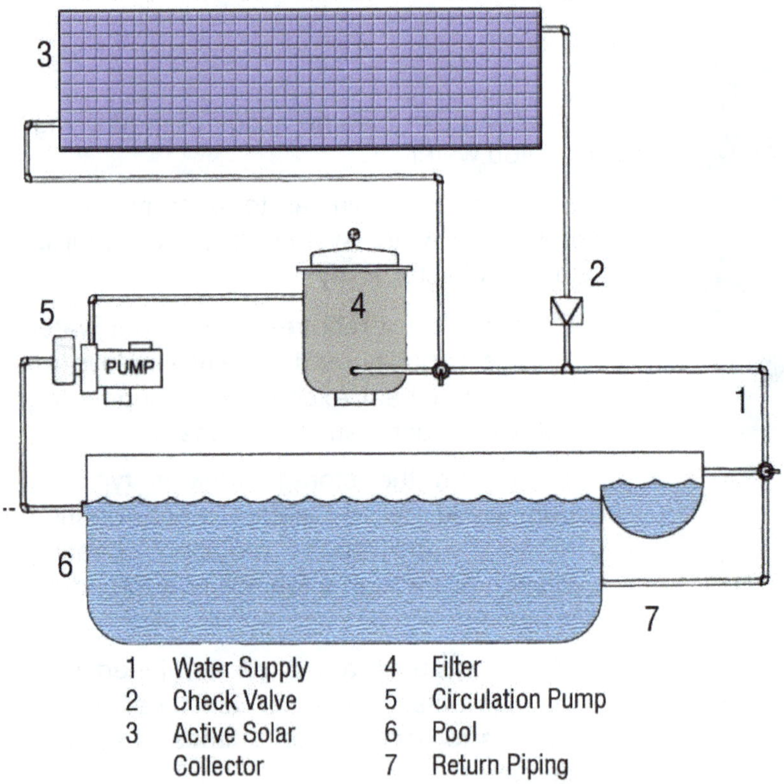

Swimming Pool Heating

Swimming pool heating can account for 50% or more of a household's summer power consumption, often more than the energy required for home winter heating.

- Active solar heaters offer an effective way to dramatically reduce electricity use in many locations:
  - Typical collectors are plastic or rubber panels or mats comprised of small tubes that connect to the pool's water supply and recirculation pump and filter system.
  - The collector area needed depends upon climate, and wind conditions, shading, desired water temperature and other factors, but is usually at least half as large as the pool size and sometimes equal.

553. US DOE-EERE

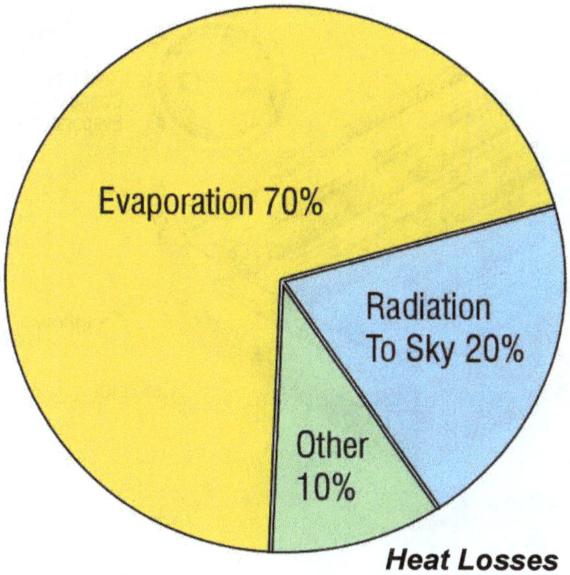

Heat Losses

Swimming Pool Heat Retention

554. US DOE-EERE

An inexpensive translucent pool cover is another way to dramatically increase energy conservation by reducing evaporative cooling while permitting radiant heating.

Swimming Pool Covers

555. Norbert Lechner, Heating, Cooling, Lighting

Cold water from the supply circulates through the collector to a heat exchanger / storage tank.

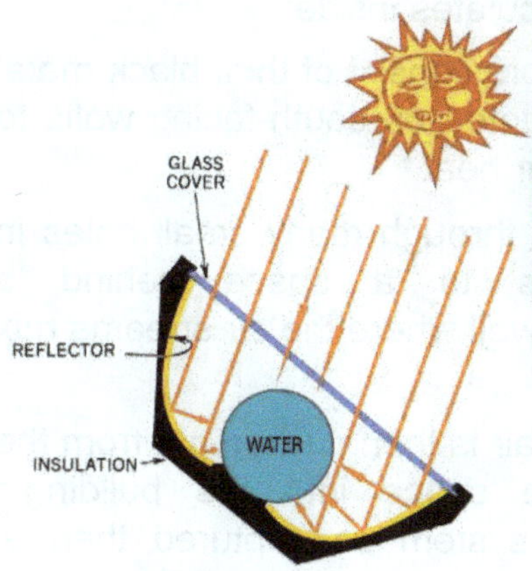

Reflected sunlight heats the internal tank.

Integral Collector Storage

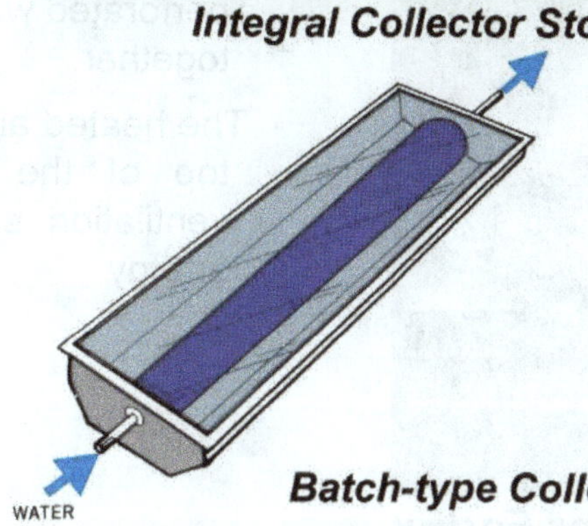

Batch-type Collector

Active Stand-Alone Water Heaters

Solar hot water systems are also available as stand-alone units that can be used for external building and outdoor applications.

- Some systems use conventional flat plate collectors, while others concentrate heat using reflectors:
  - The integral collector storage (ICS) is a flat plate type that incorporates a hot water storage tank for outdoor use such as an external sauna.
  - A batch-type solar concentrator system provides a collector water tank and reflector in the same glass-covered and insulated unit, and multiple units can be connected in series to increase throughput capacity.

Some active solar heating systems are designed to provide large quantities of hot water for commercial and industrial uses.

- Evacuated-tube collectors are efficient at high temperatures (170°-350°F or 77°-177°C), but can be about twice as expensive as flat plate types:
  - Collectors are shallow boxes full of many glass, double-walled tubes and reflectors that heat a fluid inside.
  - A vacuum between outer tube walls with absorbing coatings and inner fluid tubes covered with a copper sheet provides insulation that holds in the heat.

556.US DOE-EERE

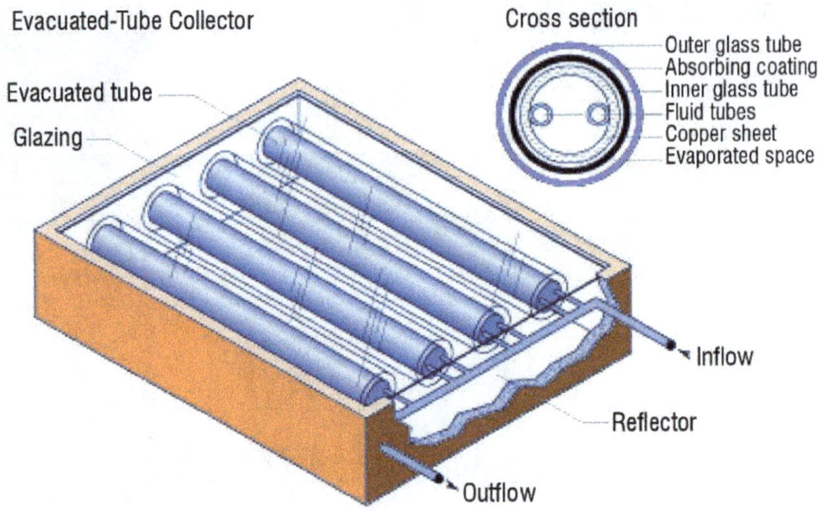

Evacuated-Tube Collector for Commercial/Industrial Use

557. US DOE-NREL

Transpired Collector at the Denver FedEx Building for Space Heating

Large buildings can consume vast amounts of energy to heat ventilated air, particularly in cold climates.

- A simple transpired collector system can save energy and money by preheating the air before it circulates inside:
  - The collectors consist of thin, black metal panels mounted on south-facing walls to absorb solar heat.
  - Air passes through many small holes in the panels to a space behind a perforated wall where the air streams mix together.
  - The heated air is then sucked out from the top of the space into the building's ventilation system as captured thermal energy.

558. Norbert Lechner, Heating, Cooling, Lighting

Since heat is fundamentally an energy source, solar thermal energy can also be used to cool building spaces much like a familiar home air conditioner uses electricity to accomplish this purpose.

- Solar thermal air-conditioning techniques can be very basic, using direct evaporative processes to cool water or air, or can be more complicated, applying mechanical refrigeration systems with special heat transfer fluids:
  - An example of a simple approach is to pump water from a cistern that serves as a cooling medium for a home during hot summer months, and a source of heat from solar gain during colder seasons.

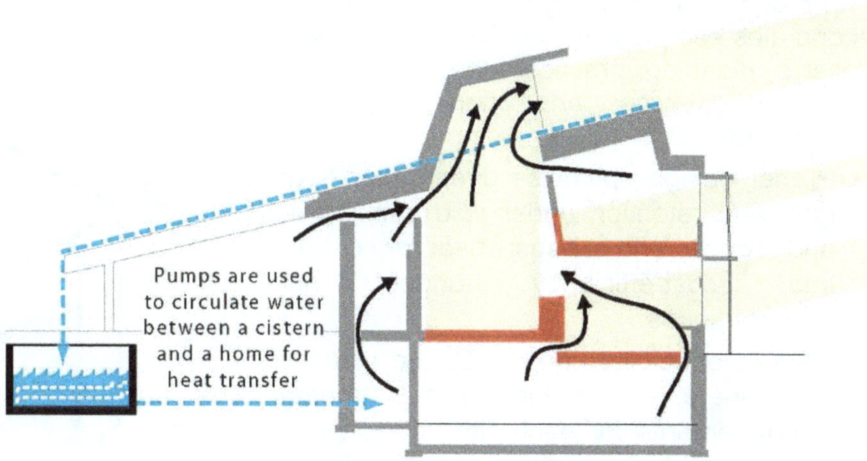

with roof evaporation for summer cooling, and direct solar gain from winter Sun for heating. Natural convection distributes the conditioned air.

Chilled water from a roof evaporator reservoir can supplement cooling effects to further reduce summer air-conditioning energy loads.

Active / Passive Hybrid Approach with Water Transport from a Cistern

Space cooling can also be accomplished using a solar-driven thermally-activated cooling system (TACS) which functions much like electrically-powered air conditioners.

- TACS apply thermal absorption or evaporative cooler devices (often called "swamp coolers"):
  - The systems are typically sized to provide 30%-60% of a building's requirements using solar, supplemented with natural gas energy for the balance.
  - TACS have not yet found broad use due to high current initial costs, the need for large solar fields to drive them, and limited ideal locations where they can be most effective.

559. US DOE-EERE

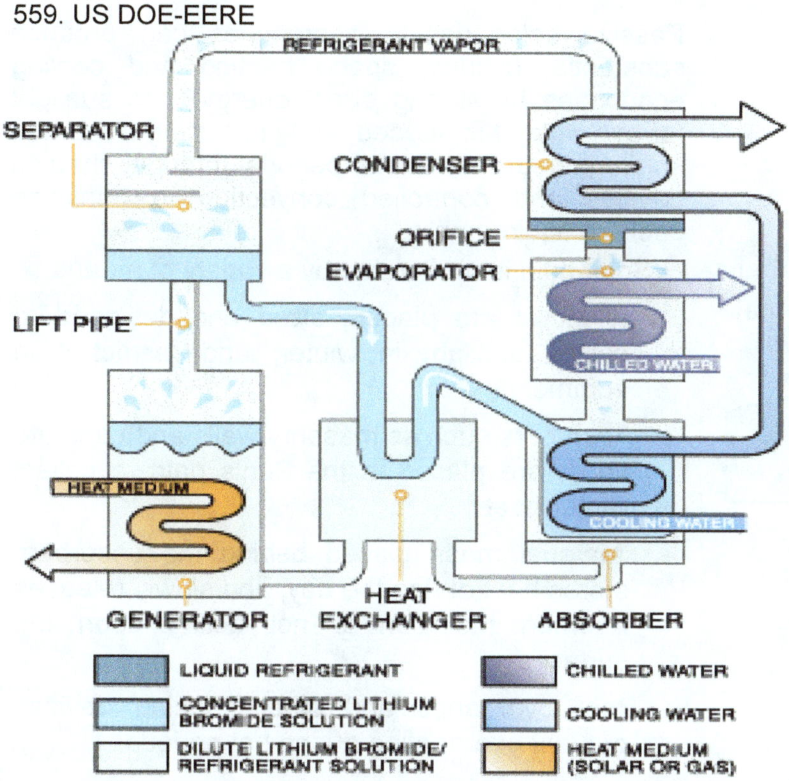

Active Thermally-Activated Cooling System (TACS)

Enormous space heating and cooling energy economies can be achieved through sensible passive planning practices that require no mechanical devices and often impose no added construction costs.

- Proper design optimizes direct solar heat gain and retention under cold conditions, and reduces and exhausts heat for comfort and cost-efficiency under hot circumstances:
  - Basic principles can be applied virtually anywhere, balancing energy requirements for each site location and building type with special climate conditions and user needs.
  - Appropriate design and placement of windows and shading elements, provisions for natural convection and ventilation, and appropriate thermal insulation, are key influencing factors.

560. Larry Bell

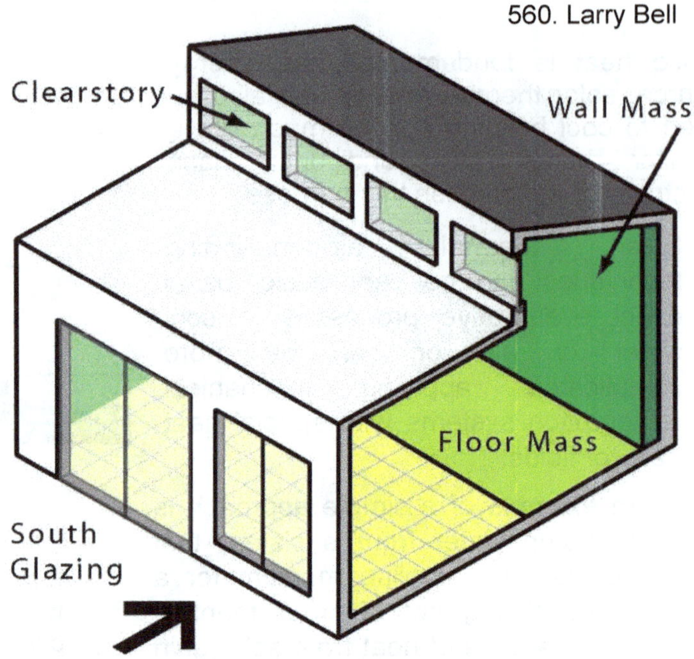

Direct solar gain utilizes sunlight that shines directly into a building to heat interior materials which release energy at night.

Direct Solar Gain

561. Norbert Lechner, Heating, Cooling, Lighting

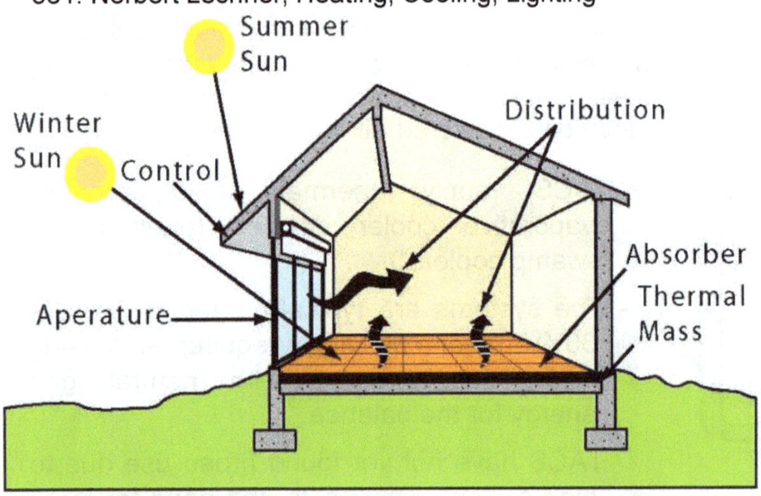

Distribution of solar heat throughout a building uses three different transfer modes: radiation, conduction, and convection.

- *Solar radiation is admitted by windows and is also stored in thermal mass.*
- *Thermal mass conducts heat to absorbers, which radiate it when needed.*
- *Convection circulates the warm air.*

Moderating Solar Gain

Passive solar design approaches can produce substantial building space heating and cooling economies by storing direct energy from sunlight for use when it is needed, and preventing unwanted solar gain during warm periods through shading devices and controlled convection to minimize cooling requirements.

- Solar gain is moderated by a variety of means:
  - Windows are placed, sized and designed to collect sunlight in winter, and restrict it in summer.
  - Absorbers such as masonry walls and concrete floors are placed in the Sun's path to collect winter heat.
  - Material mass placed behind the absorbers stores heat during the day, and slowly releases it when the Sun is not acting upon the surfaces.
  - Roof overhangs and other shading devices prevent overheating during hot periods.

562. Larry Bell

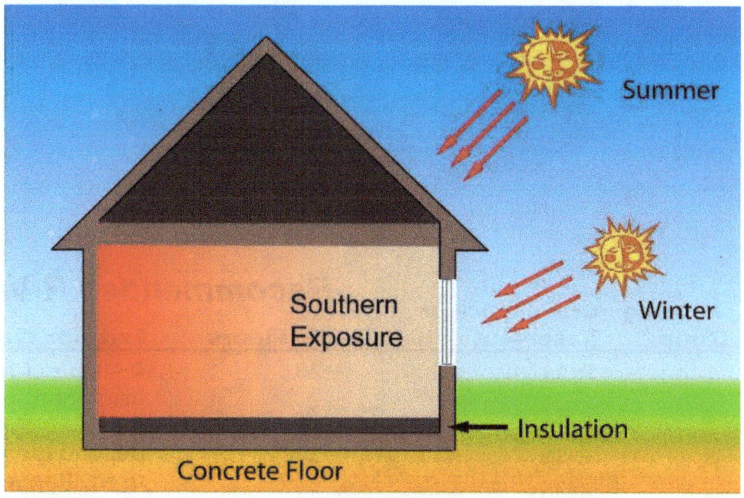

Large window areas oriented to the south will allow low winter solar angles to heat interior floors and walls. Good insulation and well-sealed building envelopes hold heat inside.

In cold latitude locations where winter heating creates high energy demands it is important to take advantage of direct solar gain through windows.

- The most efficient heat orientation in the Northern Hemisphere places windows due south, although any orientation within 30º of due south is generally considered acceptable:
  - To admit sunlight, a ratio of about 8% window to floor area is usually recommended for south walls to enable heat to be absorbed by floors and walls.
  - Latent heat produced by lighting, appliances and people can also provide some benefits.
  - Well-insulated, air-tight building envelopes can substantially reduce heat loss through exchanges with colder outside air to conserve energy.

Space Heating

563. Norbert Lechner, Heating, Cooling, Lighting

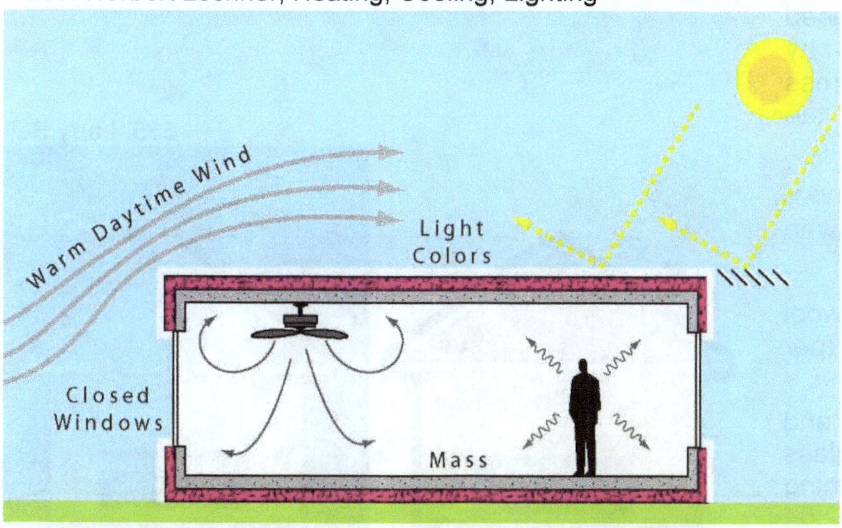

Good insulation, shading, closed windows, and light reflective colors can minimize heat gain. Fans can enhance comfort and reduce condensation, but do not save energy except through reduced air-conditioning use by residents.

In hot regions where space cooling is most essential, the building envelope and floor mass provide a useful cooling strategy only for dry climates because mass does not "insulate" in the same sense that thermal insulation materials do.

- Here it is most important to provide high heat insulating materials in roof and wall structures to delay heat entry until cooler evening hours, and delay the impacts of cold nights until hot afternoons:
  - A time lag between heat gain and heat release affords combined warming and cooling benefits.
  - Light colors, shading and closed windows also minimize heat gain.

Space Cooling

564. Accurate Building Inspectors

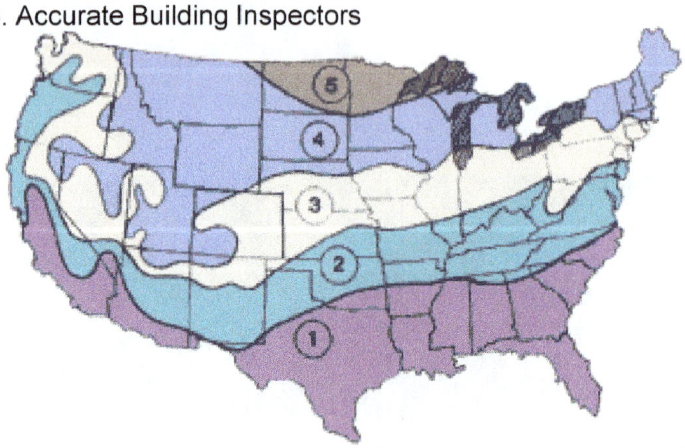

Recommended R-Values

Heating Zone	Crawlspace or Basement Ceiling	Attic Floors	Exterior Walls
1	R-11	R-26	R-Value of full wall insulation, which is 3 1/2" thick, will depend on material used. Range is R-11 to R-13
2	R-13	R-26	
3	R-19	R-30	
4	R-22	R-33	
5	R-22	R-38	

R-Values rate the minimum recommended total thermal resistance for insulating materials (higher numbers have better resistance).

R-Values

Good insulation with radiant barriers and vapor barriers is very important for building energy conservation:

- Different heating zones present special requirements for each of these components:
  - Minimum insulation thermal resistance R-Values are influenced primarily by temperature characteristics of various heating zones.
  - Radiant barriers (often reflecting foil) placed under attic insulation in hot climates can reduce heat transfer through a roof by 40%.
  - Vapor barriers placed on the inside of insulation in cold climates and on the exterior side in hot regions can help seal buildings from moisture and air, and prevent condensation.

R-Values for various insulating materials are based upon their resistance to thermal transfer by conduction through solids, thermal radiation across attic and other spaces, and convection by the motion of air across surfaces.

- Common insulation types are flexible sheets, non-structural rigid boards and spray - or blow-in products:
  - Flexible sheets include glass fiber and rock wool batts ranging from about 3.2-3.6 R per inch (per 2.54 cm) thickness.
  - Rigid boards include expanded bead and extruded foam polystyrene, isocyanate and glass fiber board, and fiberglass wall sheathing ranging from about 3.5-5 R per inch (per 2.54 cm).
  - Spray and loose-fill insulation includes polyurethane foam (5.5-6 R per inch [2.54 cm]), and cellulose and mineral fiber (about 3.1-3.7 R per inch [2.54 cm]).

565. Larry Bell

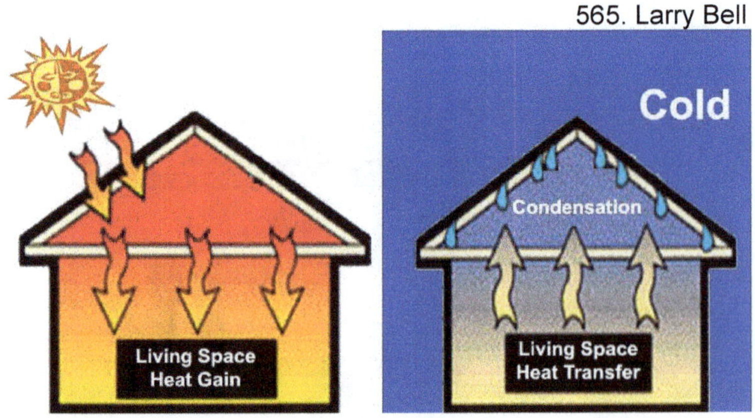

Hot weather conditions demand thermal barriers to prevent attic heat transfer to lower levels, and cold weather barriers must prevent condensation from rising heat.

Heat Transfer Barriers

566. US Department of Housing and Urban Development

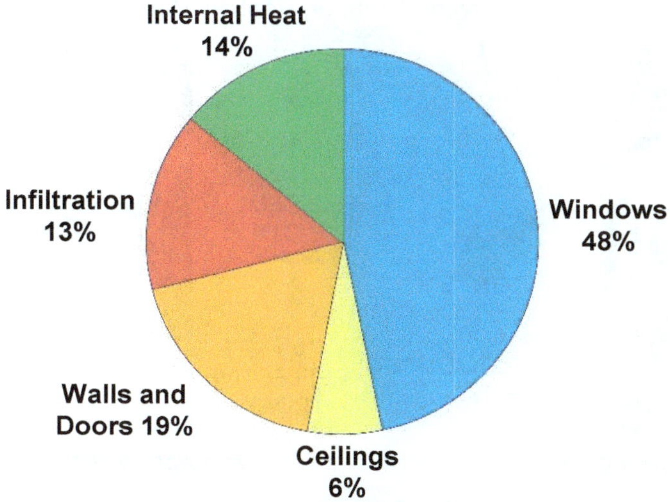

Sources of Summer Heat Gain in Homes

Window design, orientation to summer and winter solar conditions, and proper shading are important energy conservation considerations. All orientations except south receive maximum summer radiation, and east and west window orientations are least desirable from heating and cooling standpoints.

Windows

A fundamental measure to avoid indoor overheating and to control energy costs is to prevent solar radiation from entering a building during summer, and allowing it to enter during the winter.

- Solar radiation control devices associated with window treatments and shading can go a long way to bring about large economies:
  - Generally speaking, summer heat gains through windows account for nearly half of the total thermal load.
  - Internal and external shading devices can block additional solar radiation using fixed and movable approaches.
  - Special window selection and treatments can dramatically reduce solar radiation penetration into interior spaces, while also allowing good daylighting conditions.

Heat transfer through windows can be minimized through the use of coverings such as outside shutters or transparent sheets, or using multiple-pane installations or units with insulating air spaces between glass panels.

- In addition to energy efficiency, insulated windows offer excellent means to avoid outside noise and dust intrusion, and are comfortably warm to touch in cold weather:
  - Since conventional single-pane windows have thermal resistance values that are typically much lower than surrounding insulated walls, adding a second pane can make a large difference.
  - Triple glazing is generally used only in severe climates where higher costs are justified.
  - Some insulated window units contain gases such as argon or krypton within spaces to further reduce heat transfer through conduction between panes.

567. The Worlds of David Darling

Multiple-pane window units have 1/4 in – ½ in (63.5 mm – 127 mm) insulating air spaces between glass panels. These spaces are sometimes filled with special gases to further reduce heat transfer through conductance between panes and convection within the spaces.

Insulating Windows

Much solar radiation can be blocked to reduce heat gain by using tinted glazing and reflective films.

- Ordinary window glass is transparent for shortwave radiation emitted by the Sun, but is opaque to longwave radiation emitted by objects in a room so that once it passes through and enters the space, it becomes trapped:
  - Tinted glazing (green, grey, brown, blue, and browns tones) can reject 66% - 75% of solar heat, and reflective film applied directly to the inside of windows can be as effective for blocking heat as solar shade screens.
  - Tinted glass and reflective films work best on east and west windows, but such treatments are not recommended for south-facing windows because they prevent passive solar gain in winter.

568. Norbert Lechner, Heating, Cooling, Lighting

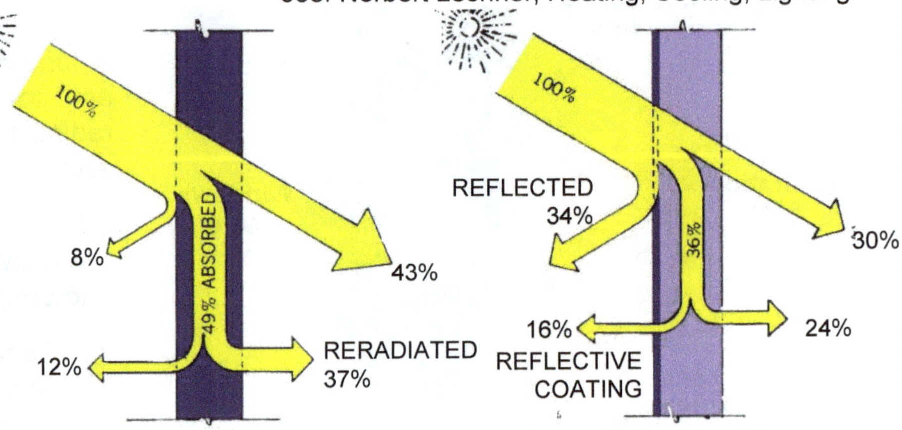

Tinted Glass

Tinted glazing reduces light transmission, but doesn't significantly decrease heat gain because much solar radiation absorbed is re-radiated indoors. Tinted curtain wall glass became popular in the 1960s for efficiency and appearance.

Reflective Glass

A glazing surface covered with a thin metallic coating can significantly increase the amount of solar radiation that is reflected while allowing outside viewing. This can be combined with tinted glazing to improve efficiency.

Window Glazing Treatments

569. Kby research

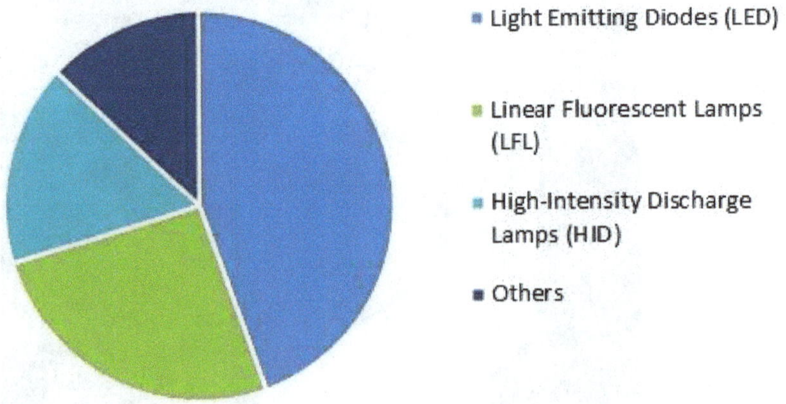

Energy Efficiencies of Various Lighting Technologies

Lighting systems operate on the same resistance principles that electrical heating units do and consequently release a lot of thermal energy along with illumination.

- Most often this heating poses a considerable conservation detriment because it uses costly power to inefficiently benefit space cooling objectives and imposes large AC power and heat loads during hot periods:
  - Fluorescent lights use about 20% of the power they consume for lighting, with the rest emitted as heat. LED lights are highly efficient, using at least 75% less energy than traditional lighting technologies.
  - Linear Fluorescent Lights (LFL) are more efficient than older lighting technologies. However, they still consume more energy compared to LEDs.

Lighting Systems

A variety of design strategies can be incorporated into buildings to substantially improve natural and artificial lighting efficacy and efficiency.

- Many of these approaches apply simple and inexpensive methods:
  - Industrial facilities often incorporate vertical or angled "saw tooth" windows to supply natural lighting and enhance ventilation, and can be provided with operable louvers to avoid unwanted glare and harsh shadows.
  - Skylights can be designed to redirect light onto light-colored ceilings and walls to diffuse illumination into spaces.
  - Automated controls can turn lights and other equipment off when people aren't present and also adjust thermostats accordingly.

570. Utah Efficiency Guide

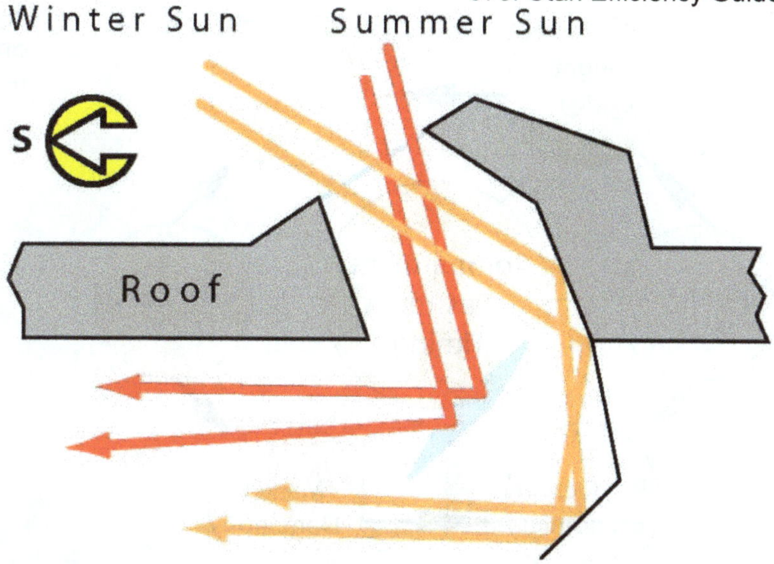

Rooftop windows can be designed to enhance natural interior illumination and avoid glare by redirecting sunlight. High-angle summer light can be dispersed using specular reflectors, and low-angle winter light can be redirected by solid reflectors for seasonal adaptation.

Conserving Lighting Energy

571. Oak Ridge National Laboratory

Hybrid solar lighting distributes natural sunlight through fiberoptic bundles to reduce electrical power consumption and heat loads. The full-spectrum light energy exactly matches sunlight red shifts that occur during the day.

Hybrid Solar Lighting

Hybrid solar lighting (HSL) which combines direct fiberoptic distribution of sunlight within a fixture also containing fluorescent tubes can reduce power costs as well as heat loads.

- A recent system developed at the Oak Ridge National Laboratory primarily consists of a rooftop-mounted collector and secondary mirror that tracks the Sun throughout the day, focusing optical light frequencies onto bundles of optical fibers equipped with diffusion rods:
  - Originally developed for commercial buildings, HSL may also have residential applications if high current costs can be adequately reduced.
  - Use is limited to spaces immediately below a roof and to sunny periods, with fluorescent lighting offering a backup.

572. Larry Bell

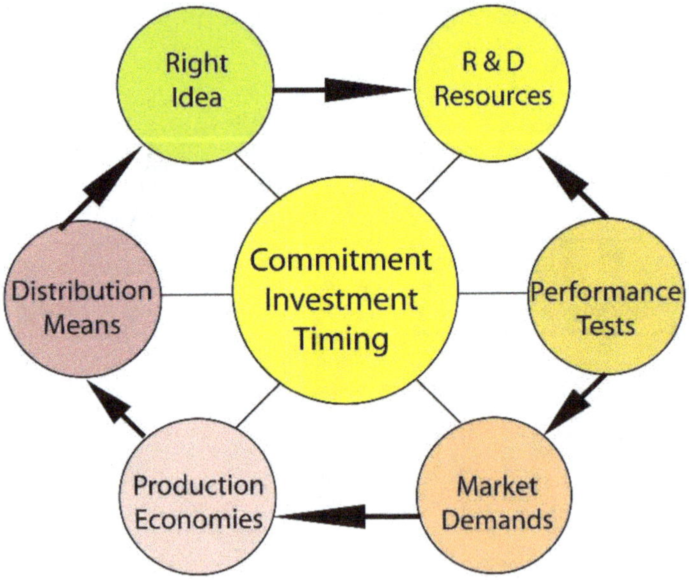

New ideas must compete with existing products that have established manufacturing capabilities, market demands, and distribution and service networks which offer advantages.

Innovation Requirements

Hybrid solar lighting is but one example among a great variety of new technologies that are emerging in response to conservation needs.

- Many of these innovations require substantial development and implementation investments before they can demonstrate performance and be economically competitive with more conventional approaches:
  - Equipment systems often represent only a small part of these investments since essential mass production, distribution and support infrastructures can be even more costly.
  - Inevitably, public interest determines which alternatives will realize sufficient market demands to support economical scales of production and profitability essential for success.

Computer and telecommunication advancements have afforded much more resource-efficient ways to obtain and exchange information.

- In addition to providing real-time access to global data sources, they also conserve resources:
  - Electronic media and web-based programs have significantly replaced energy-intensive newspaper print publications that consume pulp and impose fuel-consuming freight requirements.
  - An increasing number of distance-learning programs enable students and professionals to take courses from homes and businesses with no travel or residence relocation necessities.

573. Georgetown County School District

Electronic Media and Web-Based Learning Conservation Influences

574. U of Illinois College of Veterinary Medicine

Technological developments in certain fields can have influences upon others that may appear unrelated.

Computer and Internet Impacts on Travel Habits and Business

The recent events of COVID in 2020 exploded popularity of personal computers and internet-based telecommunications and it may eventually have large impacts upon American commuting and long-range transportation habits.

- Telecommuting and teleconferencing offer potentials to often make much travel unnecessary altogether:
  - Telecommuting has already become a viable alternative to driving or public transit for some jobs, hundred of thousands of US and European employees have already been replaced by Indian workers that telecommute over thousands of miles.
  - Teleconferencing with electronic data transfer holds large potential to minimize national and global business travel, saving time, money and energy in the process.

For many, however, life without personal automobiles is unthinkable, and though a variety of compact, fuel-efficient cars are gaining markets, larger conventional vehicles continue to dominate US streets and highways.

- The new Corporate Average Fuel Economy (CAFE) standards require an industry-wide fleet average of approximately 49 mpg by 2026. Despite these improvements, SUVs, minivans, and pickup trucks continue to dominate U.S. streets and highways, making up 75% of total U.S. car sales in 2024:
  - Relatively low petroleum prices and high costs of hybrid alternatives have caused some consumers to question whether the miles per gallon savings warrant the substantial up-front investments.
  - Such assessments may not take greater future resale value into account, particularly if fuel prices rise as predicted.

575. Direct Vehicle Leasing
576. Phronk

Economy of Motion

The hybrid Toyota Prius mid-sized hatchback can achieve up to 55 mpg using either an internal combustion engine, an electric motor or both.

The "Smart For Two" microcar produced by Daimler Chrysler can achieve more than 70 mpg using a turbocharged 3-cylinder internal combustion engine.

Economy of Scale

While solely battery-powered electric vehicles have not gained broad market interest due to limited ranges between recharges, plug-in hybrids appear to be showing real promise.

- Plug-in hybrids that can be recharged from 110-volt, or even 220-volt outlets may be able to cut fuel consumption by more than 50%, particularly for drivers who travel relatively short distances each day:
  - Many of these experimental vehicles have onboard computers that determine when power can be most efficiently obtained from electricity or a fuel.
  - They typically have larger battery packs than standard hybrids, sometimes offering 30-35 mile (47-55km) ranges on electricity alone.
  - Future successes will be influenced by needed battery technology advancements to increase range with reduced weight and costs.

577. US DOE - NREL

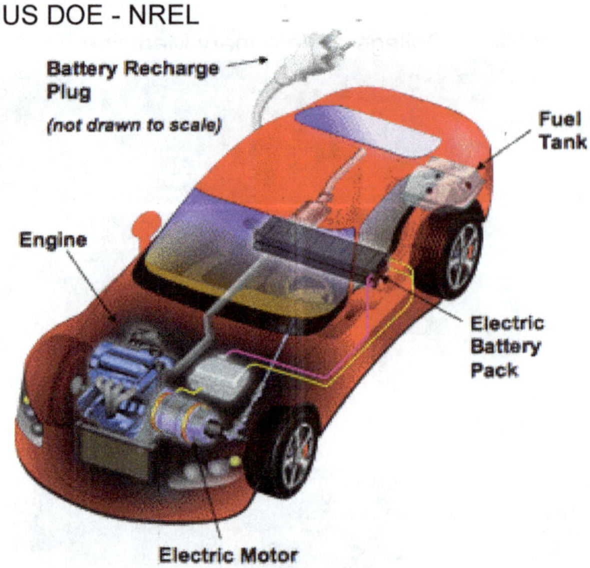

Plug-in hybrids are applicable to vehicles with gasoline or diesel engines using either fossil or biofuels. Large batteries enable them to be powered totally by electricity over limited ranges. Improved battery technologies are key to expand utility and market interest.

Plug-In Hybrids

578. All 4 Engineers

HCCI engines which combine features of diesel and gasoline systems may eventually be perfected to increase fuel efficiency and significantly reduce combustion pollutants.

Homogeneous Charge Compression Ignition Combustion Engine Systems

Diesel engines afford efficiency advantages over gasoline for cars, trucks and other uses, but emit higher levels of pollutant gases and particulates.

- Several automobile manufacturers, including General Motors, Ford, Nissan, Toyota and Volkswagen are pursuing development of homogeneous charge compression ignition (HCCI) compression engines that may be substantially more efficient than current diesel and gasoline engines, and eliminate most NOx and other pollutants:
  - This relatively recent technology uses a lean, diluted mixture of fuel and air that auto-ignites without spark plugs similar to a diesel, but using a pre-heated fuel mixture or warm combustion products reused in the cylinders.
  - Several obstacles remain to be overcome, such as difficulty in controlling the moment of auto-ignition and energy release rates, mechanical and chemical complexities, and costs.

US interests in progressing towards greater independence from oil imports are driving legislation to promote biofuels through rapid transition to flex-fuel vehicles.

- The Inflation Reduction Act of 2022 extended several fuel tax credits through December 31, 2024, including credits for biodiesel and renewable diesel. Additionally, the U.S. Department of Energy (DOE) has been making significant investments in clean energy and alternative fuel infrastructure:
  - The DOE's Advanced Technology Vehicles Manufacturing Loan Program offers direct loans to eligible manufacturers for up to 30% of the cost of re-equipping, expanding, or establishing manufacturing facilities in the United States used to produce qualified alternative fuel vehicles and infrastructure.
  - Ford and GM have been actively involved in developing vehicles that run on ethanol and other alternative fuels.

579. National Alternative Fuels Training Coalition

Flex-fuel vehicles and supplies provide a means to transition to renewables and offers the public opportunities to take advantage of the least expensive options.

Flex-Fuel Vehicles

580. Fair-PR
581. Mitsui – Bussan
582. APTA
583. Direct News

Fuel cell buses and electric trains can offer high passenger mile efficiencies with low pollution. Commercial airlines are often passenger mile-efficient, but release pollutants into the stratosphere.

Use of public transportation can be an effective way to reduce energy consumption and vehicle maintenance expenses:

- Buses, commuter trains and commercial airlines can offer good efficiencies provided that they operate at high average capacities:
  - While diesel-powered metropolitan buses produce substantial pollution levels, they replace large numbers of low-occupancy cars which are cumulatively dirtier and consume more fuel, particularly under city driving conditions.
  - Electric surface, elevated and subway trains offer cleaner, more efficient and often more expeditious travel options than other urban systems.
  - Passenger airplanes can realize average capacities of 50 passenger miles per gallon or more with 80% occupancy levels, but unfortunately spew NOx pollution directly into the stratosphere where ozone layer destruction is most problematic.

Public Systems

584. Nat. Sustain. Agri. Info. Svc.
585. GE Dolbear & Assoc.

Plant biowaste ethanol can offer a more land- and energy-efficient alternative to grain sources.

Coal-derived liquid fuels can be substituted for petroleum vehicles and aviation fuels.

As petroleum supplies continue to be depleted, it will be essential to transition to liquid transportation fuel substitutes derived from biomass and coal.

- Advancements in biowaste ethanol and coal-to-liquid fuel processing are becoming increasingly vital:
  - Cellulosic ethanol avoids large areas of cropland and amounts of energy required for cultivation of grain ethanol, but is more difficult to process, a problem that is being addressed through genetic engineering and other approaches.
  - Conversion of coal into diesel fuel, liquid petroleum and jet fuel is currently possible, but expensive, and many researchers are pursing methods to reduce these costs for commercial production.

Liquid Fuels

While natural enzymes can readily break down food crop biomass into simple sugars and ferment the sugars into ethanol, cellulosic biomass materials are not easily digested.

- Many research organizations are exploring biotech approaches that may reduce the harsh pre-treatment required to dissolve solid cellulosic feedstock, increase the amount of ethanol produced, and efficiently separate the ethanol from the liquor:
  - A key challenge is to genetically engineer microbes that produce enzymes capable of digesting cellulosic feedstock such as corn stalks and switchgrass into liquid biofuel in an efficient manner.
  - Possible breakthroughs may ultimately replace foodcrop-derived ethanol, conserving enormous amounts of land and energy for other purposes.

586. US DOE - EERE

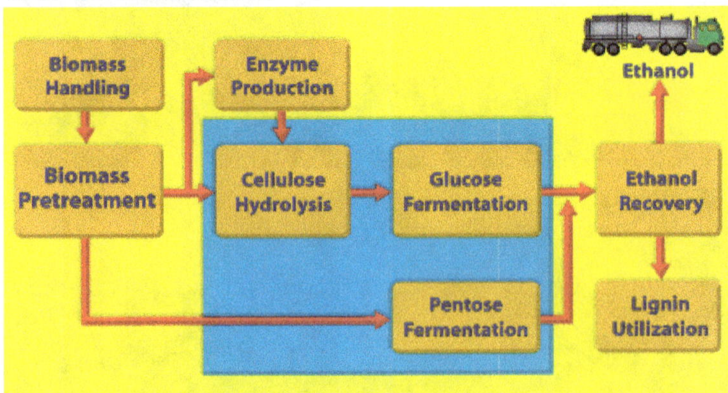

Biotech processing uses enzymes to break down difficult-to-digest feedstocks into simple sugars that can be fermented into alcohol and other biofuel products. The undigested lignin can be utilized as a combustion fuel for process heat to conserve energy.

Cellulosic Biofuel

587. Intertek Petroleum Industry Applications

Coal-derived fuels, including hydrogen, can potentially be processed in conventional petroleum refineries.

Processing of coal to create liquid synthetic fuels may significantly extend and eventually replace crude oil-derived petroleum products.

- Although this gas-to-liquid capability can be traced back to the Fischer-Tropsch process development of the 1920s, continuing research is revealing new possibilities:
  - Penn State University experiments that are yielding coal-derived jet fuel, a high-grade kerosene that remains stable at high temperatures, may prove essential to the future of commercial and military aviation.
  - Some coal-derived fuel processes can be accomplished in conventional petroleum refineries for production of gasoline, diesel, fuel oil, and even hydrogen for fuel cells.

Coal-Derived Fuels

588. US DOE

The US has large reserves of kerogen-rich oil shale that can be used as a source of liquid and gas petroleum products. A major economic obstacle for capitalizing upon this resource is the large amount of electricity required for thermal extraction (approximately equivalent one barrel of oil used for every three obtained). Use of an alternative renewable power source can potentially reduce this disadvantage.

Petroleum products can also be obtained from oil shale, rocks rich in an organic material called kerogen that yield oil when heated.

- Although oil shale can be mined through traditional surface or underground operations, more environmentally-friendly approaches avoid excavation:
  - The Shell Oil Company was conducting in-situ electrical heating extraction method in Colorado that lowers a heating element into a well for about four years to convert kerogen into oils and gases that are pumped to the surface.
  - Another "staged" approach gradually lowers an element that increases heat with depth to double as a refinery that causes oil components to separate through distillation, ranging from ethane and methane (higher up) and oils (at lower levels).

Oil Shale Fuel

589. Larry Bell / Candy Feuer

The Road to Renewables

Growing interest in renewable energy alternatives is spurring technological advancements.

- Improved performance can be expected to continually drive costs down and expand markets for many of these systems:
  - Although initial up-front expenses are often quite high, rising electricity costs and a variety of government incentives are making investments more attractive on a life cycle basis.
  - Photovoltaic systems for both utility and small-scale domestic applications have particularly promising potentials due to active development in this field.

Spectrolab, Inc., a Boeing subsidiary company, has demonstrated the development of a new photovoltaic system with 40.7% solar energy conversion efficiency for space and terrestrial purposes.

- The device uses a special class of metamorphic semiconductor materials that are much more efficient than conventional silicon cells which are mounted in concentrator units:
  - Called the "Pyron Solar Generator," a platform 23 feet (6.9 meters) in diameter and 12 inches (30.5 cm) high can generate 6.6 kW of peak power (a record for its size).
  - Higher cell efficiency enables fewer cells to be used to achieve equivalent power, resulting in lower system and electricity costs.

590. Pyron Solar

The Pyron Solar Generator is a new high-efficiency photovoltaic system that combines advanced semiconductor cells and short-focal-length lenses that concentrate sunlight. The system can produce 800 times more electricity than conventional non-concentrating cells of the same size.

Pyron Solar Generator

Nuclear Reactors

Alarm about contributions of coal- and oil-fired plants to planet warming, and soaring electrical demands, are influencing many world governments to turn again to nuclear power which has been shunned for two decades.

- Globally, 59 nuclear plants are being built, and more are planned:
  - India and China are rushing to construct dozens (India has 7 reactors under construction).
  - Japan is currently building two reactors, and plans to create more during the next decade.
  - The US 2024 ADVANCE Act provides streamlined regulatory processes to encourage utilities to move forward with new nuclear plant development.

591. ESI Africa

China currently has the only operational pebble bed reactor, the High Temperature Gas-Cooled Reactor - Pebble-bed Module (HTR-PM), which entered commercial operation in December 2023 in Shidao Bay, Shandong Province. China plans to build several more of these reactors as part of its broader nuclear expansion strategy.

Pebble Bed Nuclear Reactors

592. Wikipedia Commons

Graphite pebbles containing nuclear fuel in ceramic form are self-controlling because as the reactor gets hotter, the energy spectrum of neutrons released by fission are randomized by the Doppler effect. As the fuel heats, the reaction rate lowers the fuel temperature, placing a natural limit on power produced without mechanical aids. As a result, a containment failure would release at most only a 0.02 in (0.5 mm) sphere of radioactive material.

The pebble bed reactor (PBR) is an advanced technology that offers significantly higher safety and efficiency levels than conventional light water nuclear reactors.

- Instead of water, PBR uses pyrolytic graphite as the neutron moderator, and an inert or semi-inert gas such as helium, nitrogen or carbon dioxide as the coolant that directly drives a turbine:
  - Uranium, thorium or plutonium fuels are fixed in ceramic form within spherical graphite "pebbles" that function like complete mini-reactors that are combined together in quantities necessary to achieve desired power outputs.
  - The design is relatively simple and self-controlling, avoiding complex water cooling systems with piping that can become brittle over time and expensive backup safeguards.

593. Energy

Nuclear power reactors use heat from atomic fission to boil water and produce pressurized steam to spin large turbines that generate electricity.

- Fission occurs when a neutron slams into a larger atom, splitting it into two smaller atoms, releasing neutrons that initiate a chain reaction releasing a tremendous amount of energy:
  - The reactor core contains ceramic Uranium pellets, each producing about the energy equivalence of 150 gallons of oil stacked end-to-end in 12-foot fuel rods.
  - As of 2023, 93 total fission reactors were operating at 54 U.S. nuclear plants in 28 states.
  - Nuclear energy has supplied about 20% of total annual U.S. electricity since 1990.

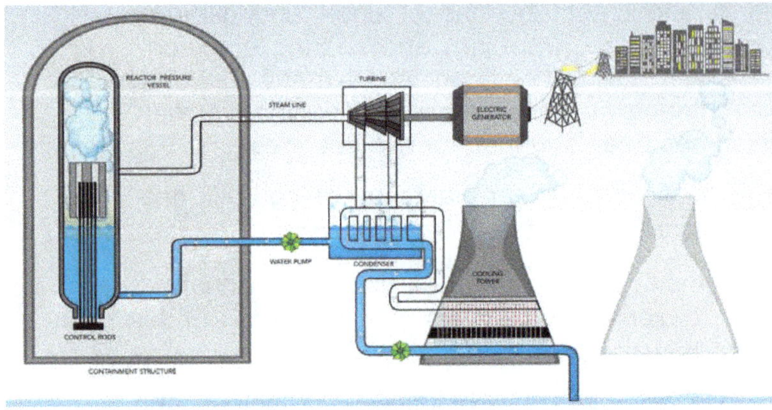

The largest nuclear fission plant currently in operation is the Kashiwazaki-Kariwa Nuclear Power Plant in Japan. It has a total capacity of 7,965 megawatts (MW) from its seven reactors, making it the world's largest nuclear power station by net electrical power rating

Nuclear Fission Plant

594. IAEA

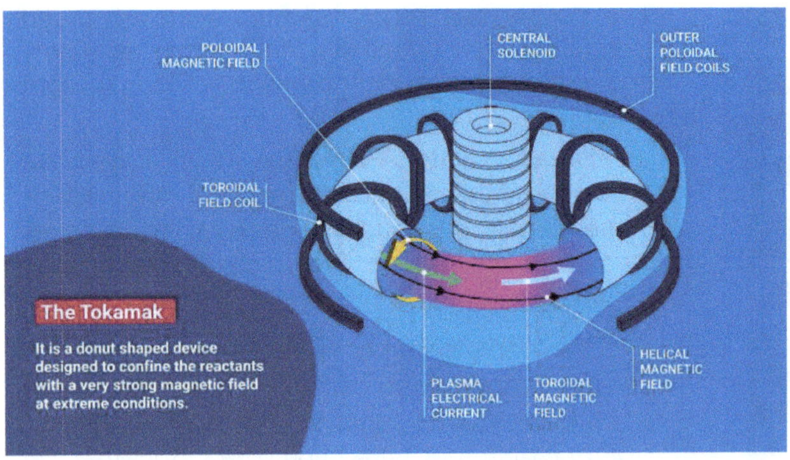

The world's largest fission reactor prototype, "Tokamak," was built in the South of France with funding from 35 countries.

Nuclear Fusion Plant

Nuclear fusion reactors are proposed to simulate a process in stars when two or more light hydrogen nuclei fuse together under enormous pressure to form a single heavy helium isotope nucleus creating a mass difference that converts into energy.

- Although not yet demonstrated at economically practical scale, nuclear fusion technology has a theoretical capacity to release nearly four times as much energy as nuclear fission.
  - Fuels that can be used for fusion are widely available and can be derived from water and lithium.
  - Nuclear fusion also has a particularly attractive advantage over fission in avoiding long-lived nuclear waste.

595. Green car congress

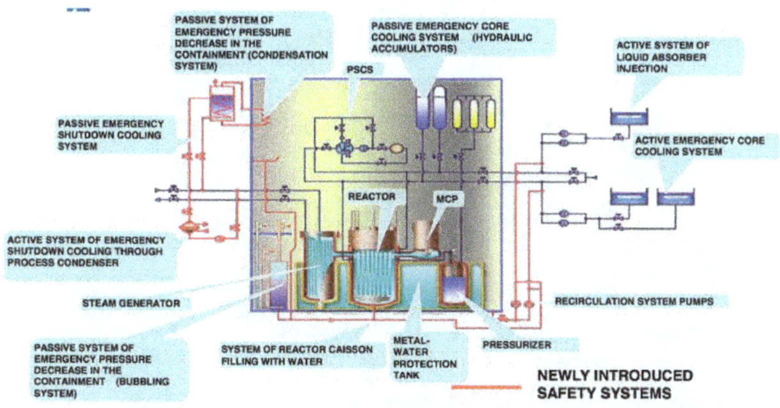

The KLT-40S is one of the two operating SMR in Russian small modular pressurized water reactor (PWR) used in the Akademik Lomonosov floating nuclear power plant, generating 35 MWe of electricity and 150 MW of thermal power.

SMR Nuclear Reactor

New generation small modular reactors (SMRs) typically produce about 300 MW(e) of electricity or less, with each megawatt capable of powering approximately 1,000 window air conditioners or a Walmart Supercenter.

- SMRs have advantages of much simpler construction than traditional nuclear reactors utilizing components that can be built in factories and shipped.

- Operating at lower power and pressure, SMR designs offer high safety with lower maintenance than traditional nuclear reactors.
- In addition, SMRs require less frequent refuelling than conventional plants (3-7-years vs. 1-2-years), with some designed to operate without refuelling for up to 30 years.

596. Energy Matters

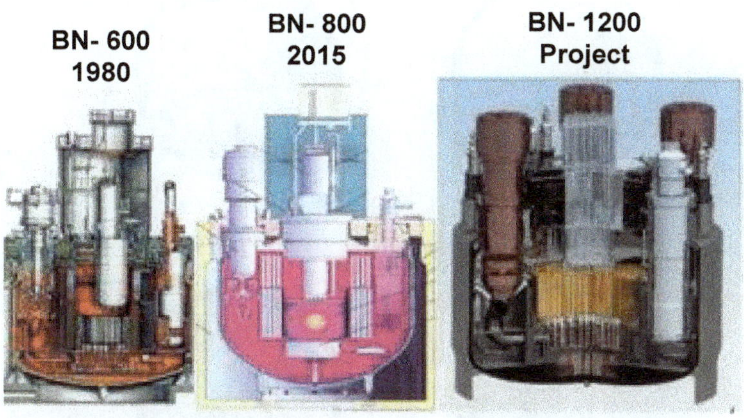

As of 2024, Russia operates the only commercial-scale fast breeder reactors, the BN-600 and BN-800, while other countries like India and China have fast breeder reactors in various stages of development or operation.

Breeder Nuclear Reactor

Breeder reactors produce more fissile energy materials than they consume, enabling these fissile products to be reintroduced back into the energy production process to reduce fuel costs.

- The reactors can be fueled with isotopes of uranium such as U-238 and thorium-232, whereby thorium reactors are natural breeders:

- Fast breeder reactors (FBRs) typically use U-238 to breed fissile plutonium, or if desired, can breed U-233 from thorium.

- Thermal breeder reactors are believed to be commercially feasible only with thorium fuel to breed U-233.

While no current or emerging technology offers a perfect or total solution to pressing conservation challenges, we cannot afford to wait for an ideal panacea to materialize.

- We have no other real choice but to combine and optimize all available options:
  - Although biofuel production consumes energy and displaces food crops, and its use releases greenhouse gases, biotechnology and engine advancements may enhance benefits.
  - Wind, solar, geothermal and hydropower aren't always abundant wherever and whenever needed, but they can be valuable.
  - Hydrogen is energy-expensive and presents transport and storage problems, but is clean if processed using clean power.
  - Nuclear presents waste storage and safety/security issues, but is otherwise clean and substantial, and technologies have improved.
  - Fossil fuels are limited and versatile natural treasures that can be made cleaner and must be carefully conserved.

597. Larry Bell

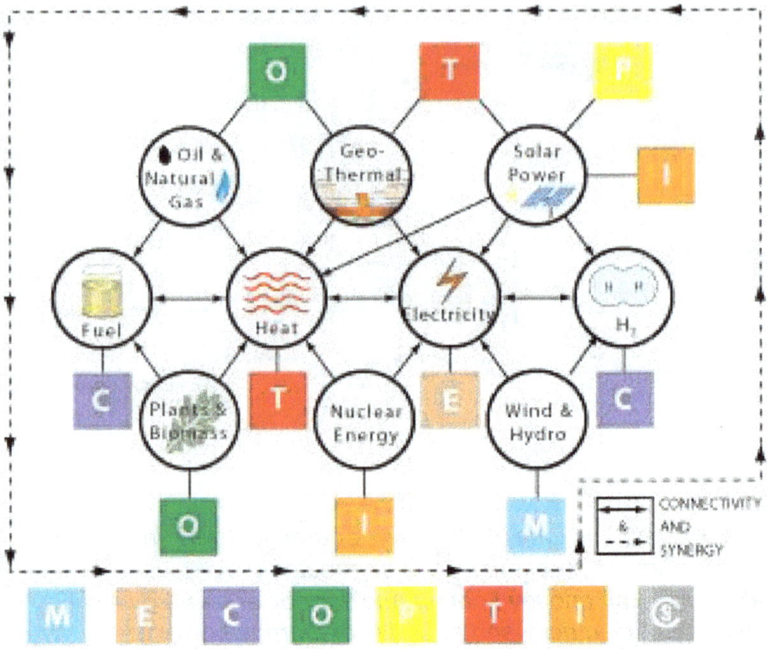

In Nature, and in natural conservation, everything is important, everything connects, and everything works together.

MECOPTICS in Practice

598. Larry Bell

599. Larry Bell

Biofuels

Advantages:

- Offer alternatives to extend or replace fossil fuels and dependence upon imports.
- Can produce lower greenhouse gas emissions than petroleum fuels.
- Production uses conventional processes and technologies.

Disadvantages:

- Low product life cycle energy efficiencies with plant sources that also compete for land with food crops.
- Cellulosic processing is currently difficult / expensive.

Hydrogen Fuel Cells

Advantages:

- Provide a clean fuel source if hydrogen is processes using clean energy for electrolysis.
- Can provide heat as well as electricity through CHP applications.

Disadvantages:

- Requires more energy to produce hydrogen than it yields.
- Is difficult to store and transport, requiring energy for compression or liquification.
- Is an explosive gas that tends to leak out of containers.

Nuclear Power

Advantages:

- Power production releases only water vapor rather than greenhouse gases.
- Capable of producing large amounts of electricity at competitive costs.
- Good safety over decades, and technologies have advanced.

Disadvantages:

- Storage of wastes remains a controversial issue.
- Accidents and terrorism threats require aggressive safeguards; breeder reactors present weapons concerns.

Fossil Fuels

Advantages:

- Provide highly versatile uses, including energy for power, heating and transportation, and feedstocks for plastics and other products..
- Have relatively high energy densities compared with biofuels and hydrogen.

Disadvantages:

- Produce CO_2 and other greenhouse gases / particulates.
- Supplies of natural gas and oil are rapidly being depleted.
- Dependence on foreign oil creates geopolitical vulnerability.

Wind Power

Advantages:

- Wind farms can produce substantial energy to localities and broader regional power grids.
- Large systems require high capital investments, but operational costs are low.
- Small, inexpensive turbines can provide supplementary power for a variety of uses.

Disadvantages:

- Wind is unpredictable, depending upon seasons, weather conditions, and time of day.

Solar Power

Advantages:

- Large and small systems can power electrolysis for hydrogen production.
- Ground-mounted modules and arrays can provide supplementary power for off-grid facilities / devices.
- Roof mounted modules can power residential and commercial water heating.

Disadvantages:

- Operations are daylight only and influenced by seasonal and daily weather / atmospheric conditions.

Hydropower

Advantages:

- Major installations supply large amounts of clean and dependable power.
- Very small, inexpensive turbines in flowing streams can provide supplementary power for remote applications.

Disadvantages:

- Operations can kill fish and damage sensitive marine and marsh ecosystems.
- Tides are predictable, but short periods, and waves are unpredictable depending upon seasonal / daily conditions.

Geothermal Energy

Advantages:

- A very dependable source which can provide heating / cooling as well as conversion to electrical power.
- Low-to-moderate temperature reservoirs provide heat for direct use or electricity in many parts of the world..
- Flash steam plants provide electrical power for hydrogen electrolysis and other commercial purposes.

Disadvantages:

- Appropriate supply sites are very limited and are often legally restricted.

While many new and emerging technologies do provide encouraging means to advance conservation goals, it is evident that such innovations alone cannot resolve imminent energy shortages and environmental crises.

- Government and industry can provide incentives and tools, but real progress can only be realized through responsible choices involving large numbers of private citizen, business, and community stakeholders:
  - We must reduce consumption by choosing to purchase and apply efficient and environmentally-benign technologies, and avoid wasteful lifestyles.
  - We must choose to regard discarded by-products of daily activities as valuable resources that can be recycled.
  - We must exercise choices to support public legislation and programs that conserve, reprocess, and reuse resources in a safe and sustainable manner.

Source Reduction

Minimize consumption and waste through responsible technology selection and utilization.

Recycling

Recover, sort, and treat waste materials so that they can be reclaimed for useful purposes.

Reuse

Reapply waste and discarded products in their current or in altered form.

Primary goals of responsible conservation practices are to extract maximum resource benefits while minimizing wastes and hazards.

Source reduction, the most important conservation strategy of all, can be practiced by everyone to minimize the amount of energy and material used and the quantity of waste created.

- Doing more with less makes good economical and moral sense, and can often be accomplished without serious lifestyle compromises:
  - Passive solar building design is a simple and effective example that when properly applied, can greatly reduce energy consumption and pollution impacts.
  - Purchasing energy-efficient cars and use of public transportation expands market and service demands that drive costs down even more.
  - Combined heat and power-generating systems and inexpensive solar water and space heating devices, can save money and fuel.

600. Chevron Corporation

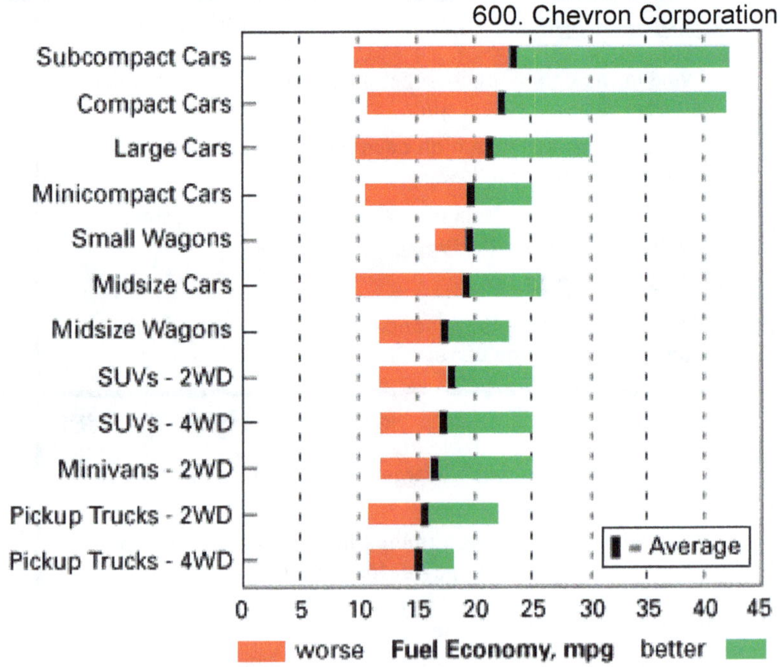

Vehicle Fuel Economy Comparisons

601. Auckland City Council

Former landfill converted into a public park in Auckland, New Zealand.

Landfill Methane Recovery

Methane recovery from landfills is gaining commercial interest as an energy source to supplement dwindling natural gas supplies.

- While public concerns about objectionable odors, wind-blown litter, vermin and leachate contamination of groundwater persist, safeguards are addressing these problems:

- Separation and recycling programs are reducing necessary landfill sizes, and are also much more labor- and cost-effective when accomplished prior to disposal.

- Groundseal barriers and modern gas extraction technologies are also making methane odor control and collection more efficient and productive.

- Many landfills are being developed by operators into public parks and sports areas as added benefits.

Recycling fundamentally involves recovery of discarded products and materials that would otherwise be considered waste.

- "Secondary resource recovery" is increasingly being practiced in metropolitan areas world-wide, particularly where space for new landfills is becoming scarce:

- There is a growing awareness that simply disposing of waste materials such as metals is unsustainable in the long-term due to diminishing raw sources, and some people believe current landfills may be revisited as "mines of the future".

602. Earth Works Recycling, Inc.
604. Rochester Environment
603. Truman State U
605. State of South Dakota

New recovery methods and technologies are constantly being developed, creating economical business opportunities for equipment developers and product reclaimers.

Recycling Useful Resources

606. Science Daily
607. JCC. S. nevada
608. Marina del Ray, CA
609. Carla Perrotti
610. Nature Conservatory
611. Fox Island
612. Rec. Travel USA-Canada
613. Nature Products Network
614. Dalhousie University
615. Wmdstar

PART IV: CHALLENGES/POSSIBILITIES

Fuel crises, unless averted through preemptive planning and actions, can provoke hasty stop-gap responses that are costly and ineffective.

- Prevention interventions will not be simple or easy, and must simultaneously address daunting issues at many levels of scale:

- International policies influence geopolitical cooperation associated with joint resource and technology development, imports/exports and responses to terrorist threats.
- Federal, state and local programs influence domestic production capacity and consumption through regulatory incentives related to technology investment and user costs.
- Industries, businesses and individuals influence lifestyles and demands that determine which technologies will succeed in the marketplace, and whether conservation will be broadly practiced.

616. Brad DeLong

1970s Fuel Crisis

617. Larry Bell

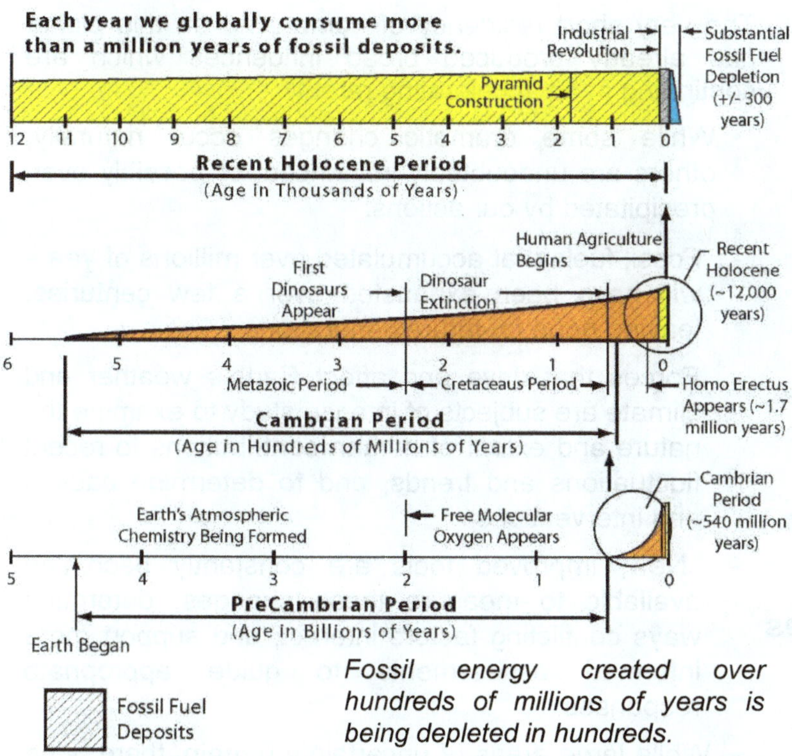

Fossil energy created over hundreds of millions of years is being depleted in hundreds.

Earth's Fossil History

Eliminating dependence upon fossil fuels is not a choice, but rather, is an imperative.

- These resources which now account for most of our energy needs are being depleted at an incredible rate, and deposits which accumulated over billions of years will have been largely consumed over hundreds:

- Free molecular oxygen wasn't present until about 2 billion years ago, and photosynthetic plants later evolved over millions to billions of years.
- The Cambrian period (most recent 540 million years or so) witnessed development of life forms which evolved and diversified rapidly, leading to our Homo Erectus ancestors who disappeared about 1.7 million years ago.
- The Recent Holocene period saw the beginnings of organized human agriculture about 12,000 years ago, exploding population growth, and the very recent Industrial Revolution which has produced large-scale global impacts.

The trend over the last two centuries for world populations to have switched from biomass energy to depend predominantly upon fossils for heat and power is not a sustainable development.

- While prominent experts disagree about how long it will be before accessible fossil supplies will be exhausted, predictions typically vary at most tens of years for oil and gas, and hundreds for coal:
  - As liquid and gas reserves play out, increased dependence upon coal can be expected to drive consumption and cost rates higher.
  - Rising prices of all fossils will make alternatives more attractive, but will also increase global competition which is likely to produce shortages, crises and conflicts.

619. Larry Bell

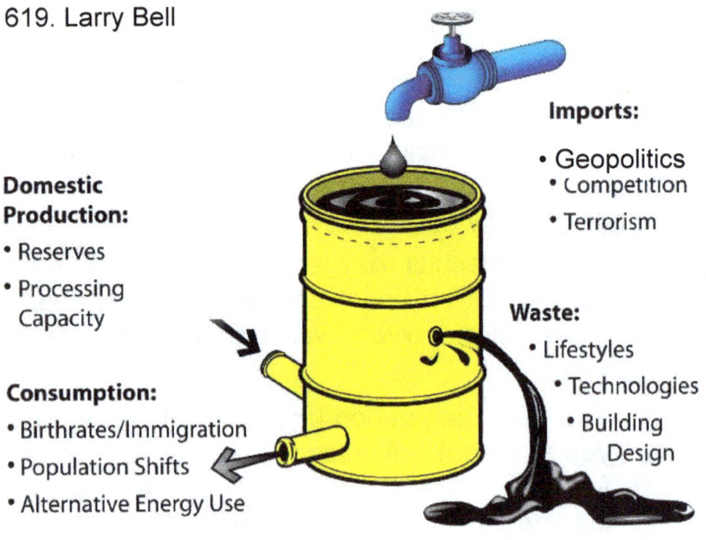

Oil Dependencies

618. Association for the Study of Peak Oil

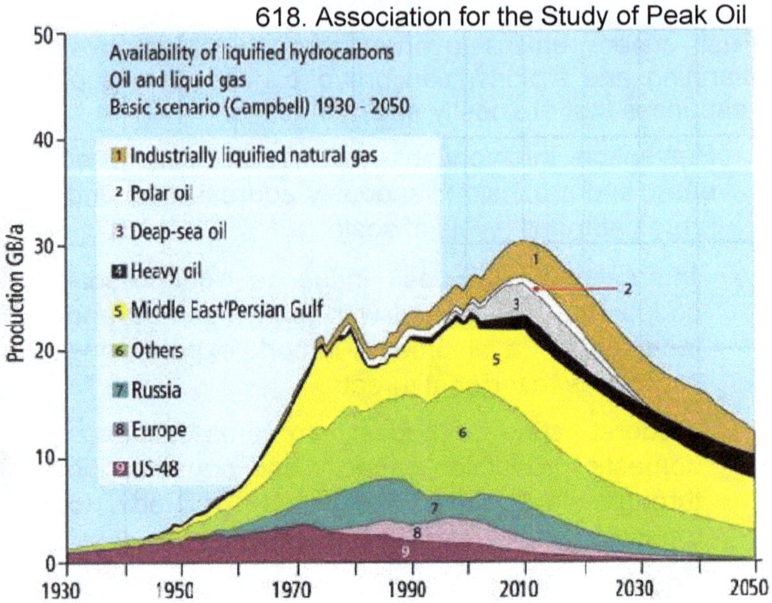

ASPO Oil and Liquid Gas Forecast

The Association for the Study of the Peak Oil and Gas (ASPO), a group of oil geologists, has forecast that oil production will increase roughly until 2010, and then production from new fuels will no longer be able to offset declines from old ones.

Running on Empty

The very short residency of humankind on this planet has already produced broad influences which are continuing at an accelerating rate.

- While some dramatic changes occur naturally, others are undoubtedly influenced, or possibly even precipitated by our actions:
  - Fossil fuels that accumulated over millions of years will have been exhausted over a few centuries, leaving none for future generations.
  - Forces that drive and affect Earth's weather and climate are subjects of intense study to examine the nature and extent of human contributions to recent fluctuations and trends, and to determine causes and interventions.
  - New, improved tools are constantly becoming available to measure these changes, determine ways conflicting factors interact, and support more informed assessments to guide appropriate responses.
  - While large areas of uncertainty remain, there is no doubt that our current human course is unsustainable, and that we must either alter our ways, or expect unacceptable consequences.

Rapid transitioning away from fossil fuels is urgent.

- Rising numbers of people, increasing demands for resources and alteration of Earth's atmospheric composition through combustion of fossils are influencing both the rate and magnitude of global impacts:
  - Atmospheric CO_2 has increased by 32% since measurements began in 1958.
  - Based upon samples of gas bubbles trapped in glacial Antarctic ice, CO_2 has increased by about 45% over the previous two centuries.
  - Some believe that the only comparable influences upon Earth's atmosphere and climate were caused by catastrophic collisions of huge meteorites and / or comets that led to large-scale extinctions of many species, including dinosaurs.

620. Global Monitoring Laboratory

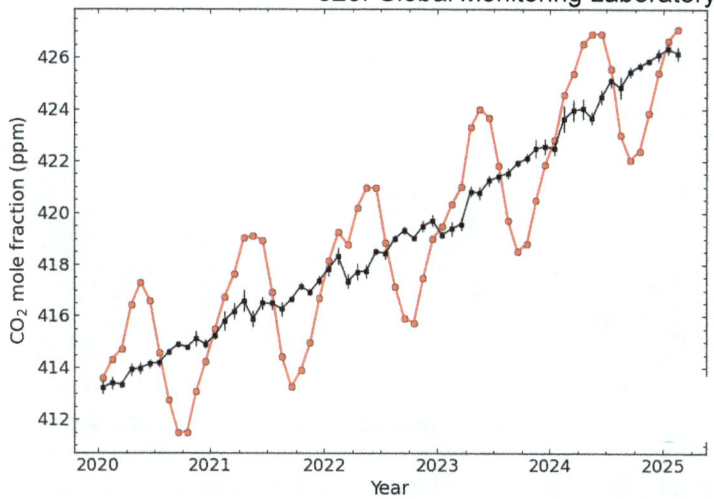

There appears to be a clear, yet inexact relationship between temperature changes and atmospheric greenhouse gas concentrations (carbon dioxide and methane in particular), all of which have risen since the Industrial Revolution.

Greenhouse Gas

Ironically, much research suggests that the Northern Hemisphere of our planet is ultimately heading towards a protracted new glacial age.

- During the past million years Earth has experienced a major ice age about every 100,000 years, interrupted briefly by warmer "interglacial" periods like our current Recent Holocene that last about 12,000 years or so:
  - About 21,000 years ago, North America and northern Europe were covered by ice sheets estimated to be about 1.9 miles (3km) thick.
  - More recently, about 6,000 years ago, a thaw caused ice sheets to retreat to Greenland and Antarctica and stream into the sea.
  - Although the last thousand years have been generally mild, substantial fluctuations have occurred, such as the "Little Ice Age" which brought bitterly cold weather to many parts of the world during the 17th and 18th centuries.

621. WFLA

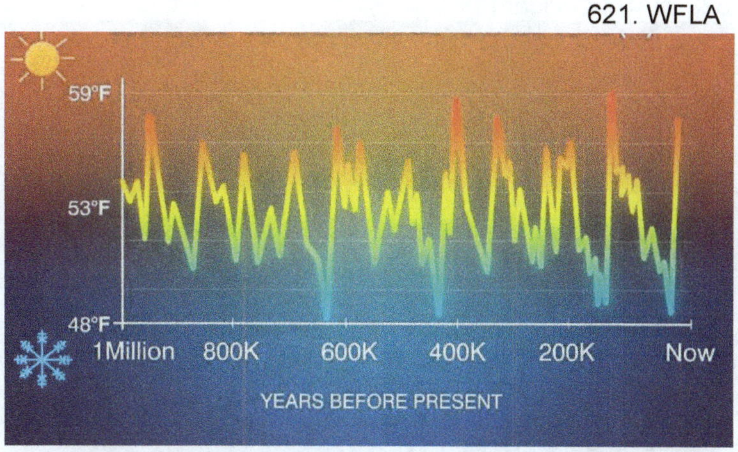

The difference between an ice age and a warm interglacial period like today is not very much. Earth's average temperature 20,000 years ago was 46 degrees and today it's 57. A difference of just 11 degrees.

Previous Ice Ages

622. Hendrick Avercamp *

Scene inspired by the harsh winter of 1608 in The Netherlands during the Little Ice Age

Rather than continued global warming, our planet may quite likely be near the end of a relatively short period between major ice ages. Periods such as these are characterized by abrupt, vacillating, and sometimes extreme climate changes, ranging from cold to hot, and wet to dry. An example during this present interglacial occurred in the Northern Hemisphere between the 16th and 19th centuries.

Little Ice Age Cooling Paradox

If Earth climate patterns hold true to recorded schedules, a predicted ice age may present even more serious adaptation challenges than global warming, creating both food and energy shortages.

- Results of the Little Ice Age (not a real glacial ice age) can offer some examples of such consequences:

- In 1250, Atlantic pack ice began to grow; warm summers stopped being dependable in Northern Europe in 1300; torrential rains followed, and the Great Famine of 1315-1317 required crop practices to be altered in order to adapt to shortened, unreliable growing seasons.
- Later, in the mid-17th century, glaciers in the Swiss Alps advanced to gradually engulf farms and villages; the Thames River and canals of The Netherlands froze over (as did New York Harbor by 1780); and sea ice closed shipping harbors in Iceland.
- The climate began to warm again beginning in about 1850 to more recent conditions.

623. NASA Goddard

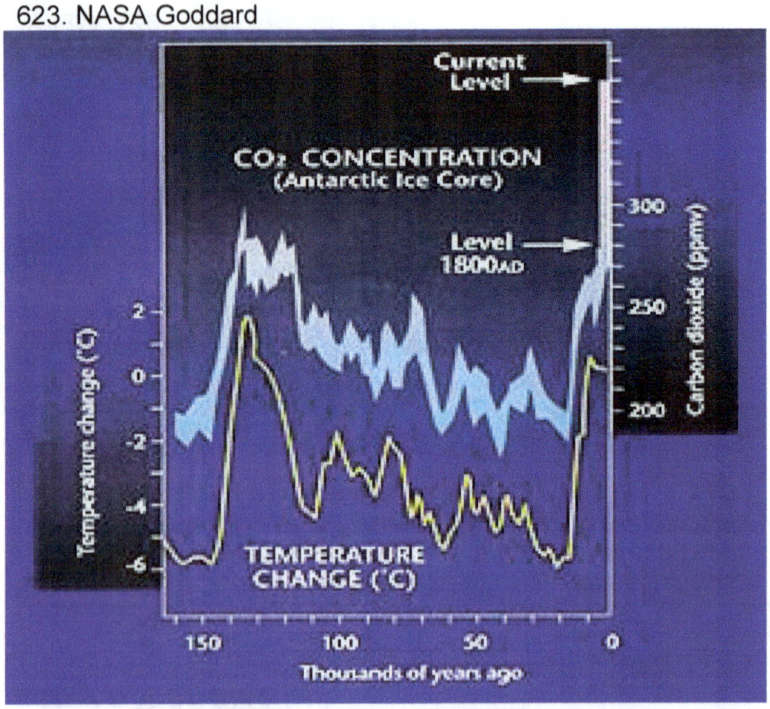

Early Climate Changes (Past 150,000 Years)

Except for relatively brief interglacial episodes such as the "Eemian" that peaked about 125,000 years ago and our present one, much of the Earth's surface has been under a siege of ice over most of the past 160,000 years.

- Occurrences and impacts of these and other glacial periods has been verified by three types of methods:

- Geological studies reveal evidence in the form of rock scouring and scratching, glacial moraines, valley cutting and other surface effects.
- Chemical analysis of isotope variations in sedimentary rocks, ocean sediment and ice cores reveal biological changes and extinctions.
- Paleontological surveys of fossil distributions indicate when and where cold-adapted organisms spread into higher latitudes, and others requiring warmer conditions became extinct or migrated into lower latitudes.

* "A Scene on the Ice" (Painting, circa 1600s)

Analyses of ocean sediment cores drilled off the African coast indicate that climate shifts may have dramatically influenced the course of human evolution.

- These recent studies reveal occurrences of three major climate cycles that coincide with developments evident in fossil records:
  - Brief, oscillating warm/wet and cold/dry climate cycles about 2.8 million years ago became progressively colder and longer, killing off fruit trees, rainforests and species that depended upon them, while grassland species thrived.
  - After that period Australopithecus Afarensis (known as "Lucy") evolved and divided into at least two separate lineages, the genus Paranthropus and genus Homo, and the first stone tools appeared.

624. Content Answers

***Lucy* About 2.8 Million Years Ago**

625. Archaeology Info

***Homo Erectus* About 1.7 Million Years Ago**

626. Michigan State U.

***Homo Sapien* About 130,000 Years Ago**

Another period of colder, drier African cycles about 1.7 million years ago caused Homo Habilus to die out, and a larger-brained Homo Erectus first appeared along with more sophisticated tools used by our human ancestors.

About 1 million years ago when cold, dry periods again became extreme, the genus Paranthropus became extinct, and Homo Erectus migrated from Africa to eventually evolve into modern humans, the Homo Sapiens.

Influences on Human Evolution

No one can really predict how long our current interglacial period will last before the next glacial period arrives, and if the term "ice age" is used to refer to a long generally cool interval when glaciers advance and retreat, some can argue that we are still in one today.

- Although a "typical" interglacial may be estimated to last about 12,000 years, this is difficult to substantiate from ice core records, and our current one may be more analogous to a previous interglacial estimated to exist for 28,000 years:
  - The Earth has seen at least four major ice ages, with many glaciation cycles of advancing and retreating ice sheets within and between them.
  - These warming and cooling periods seem to occur in small-scale cycles of about 40 years, within larger-scale 400 year cycles such as the Little Ice Age, within still larger 20,000 year cycles, and so on.

627. Maria Koshland Museum of the National Academy of Sciences

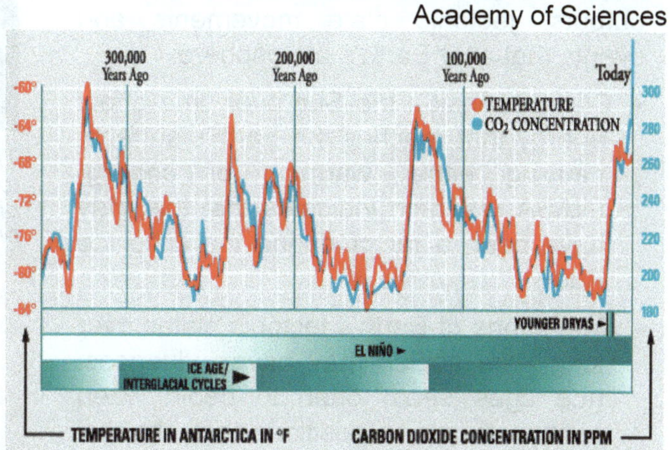

Over more than the past 400,000 years of Earth's history, ice ages have occurred at regular intervals of about 100,000 years each. These cycles have witness changes when huge glaciers have destroyed entire regional ecosystems and altered topographical surface features.

Glacial and Interglacial Cycles

Although causes of major glacial events are not confidently understood, changes in Earth's orbit are often considered to be important factors.

- Variations in the shape (or eccentricity) of Earth's orbit around the Sun (called "Milankovitch" cycles) and in the planet's tilt axis both affect the amount of sunlight received on the surface, influencing periods of warming and cooling:
  - Tilt is usually considered to be the most important of these in terms of glacial and interglacial periods.
  - Distance changes as Earth processes elliptically around the Sun are expected to have contributing effects, particularly with regard to climate shifts within an interglacial period.

628. University Corporation for Atmospheric Research

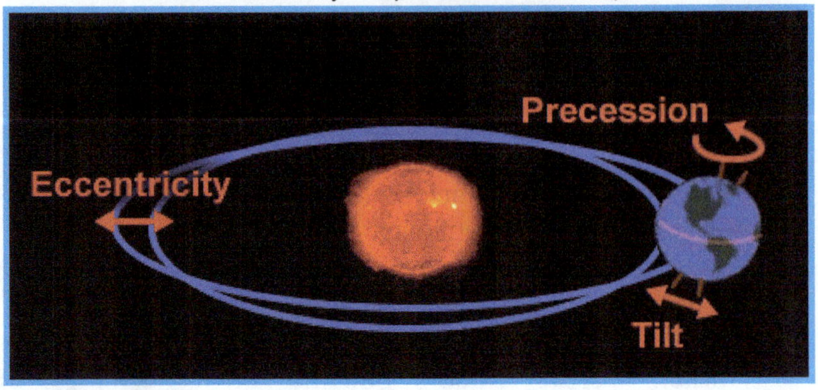

Changes in the Earth's orbit eccentricity and its axis tilt angle can produce changes over thousands of years. Its axis is offset from the perpendicular by an angle that "wobbles" from a minimum of 22.5° to a maximum of 24° over about a 40,000-year cycle. It is currently tilted at 23.5°. Approximate 100,000-year dominant patterns of glacial periods roughly correspond with both types of changes.

Influences of Earth's Orbit

Other naturally-occurring conditions are also theoretically linked to major climate changes.

- Key among these are variations in solar outputs, tectonic plate movements and events that alter Earth's atmosphere:
  - Changes inside the Sun over time affect energy that reaches the Earth's surface, causing either warming or cooling. (NASA research indicates that this may have been a factor during the Little Ice Age.)
  - Migrations of Earth's tectonic plates over long time scales may have influences. (Ice ages seem often to begin when continents are in positions that obstruct warm water flow from the equator to the poles, causing ice sheets to form.)
  - Volcanoes and oceans release greenhouse gases that can produce warming effects, and large meteorite strikes release dust that block sunlight (a cause cited for dinosaur extinctions).

629. NASA
631. Isaac Newton Group
630. NASA
632. Orbit @ Home

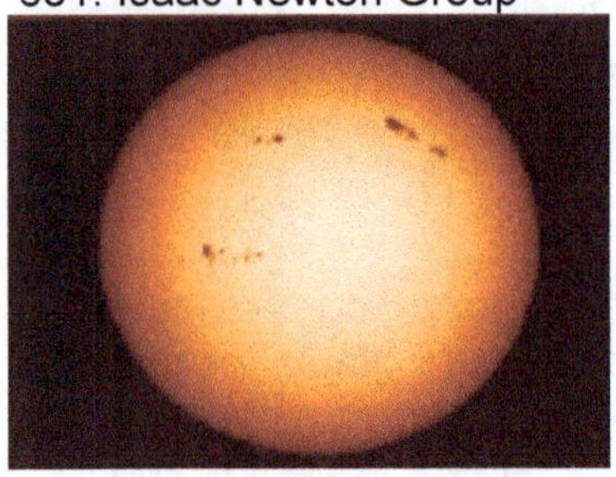

The Sun's output is estimated to increase about 10% each billion years, and sunspots produce short-term effects.

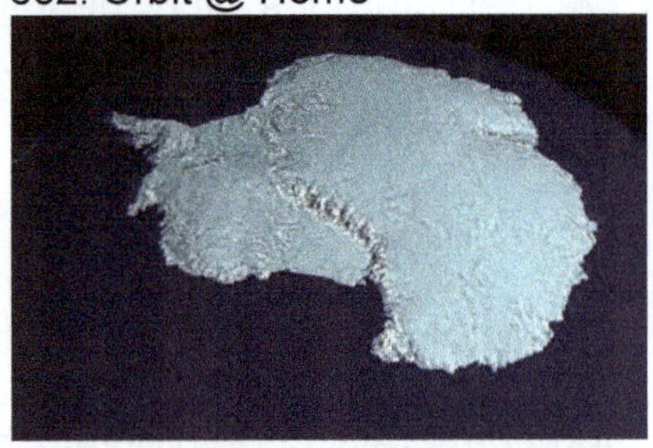

A continent at the top of a pole (as Antarctica is presently) can block warm water flow.

Volcanoes produce dust and aerosols that block sunlight, and also release warming greenhouse gases.

Dust ejected by large meteor strikes may have blocked sunlight and cooled some regional environments in the past.

Other Potential Influences

Geological evidence shows that over the past 400,000 years CO_2 levels and temperatures have greatly fluctuated at regular intervals, with temperature rises preceding CO_2 increases.

- CO_2 levels today are similar to the Eemian interglacial period that occurred between 120,000-140,000 years ago which was followed by a full-fledged ice age immediately afterwards:
  - Atmospheric CO_2 has been rising steadily since about 18,000 years ago when Earth began to warm its way out of the Pleistocene ice age that covered much of North America, Europe and Asia with glaciers.
  - Records indicate that CO_2 levels fall at the start of ice ages when more of the gas is absorbed by the colder oceans, and levels rise during glacial retreats when the processes reverse.

633. Petit et al. Nature, 1999

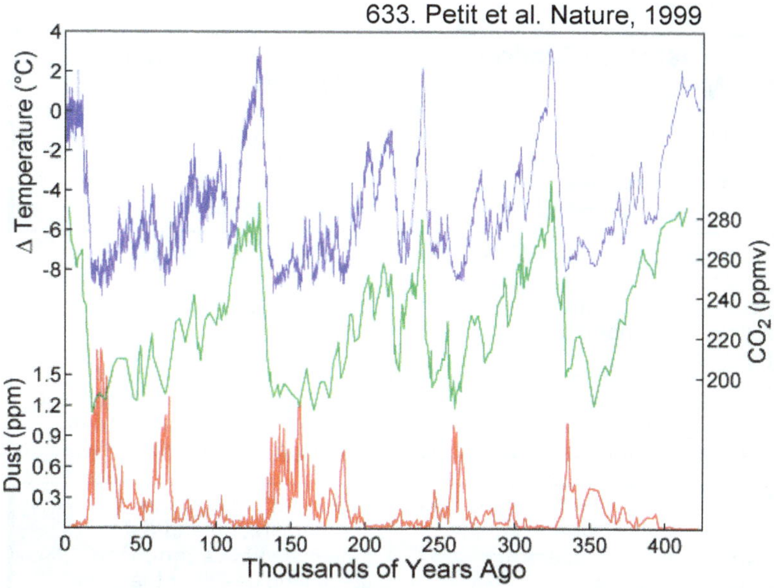

Core samples taken from Vostok ice show direct correlations between CO_2, temperatures, and dust over the last 400,000 years. Warmer global temperatures release higher CO_2 levels from oceans, events that may be triggered by Earth's orbit changes and/or other causes.

Correlations with Atmospheric Dust and CO_2 Levels

634. W Broecker Columbia University

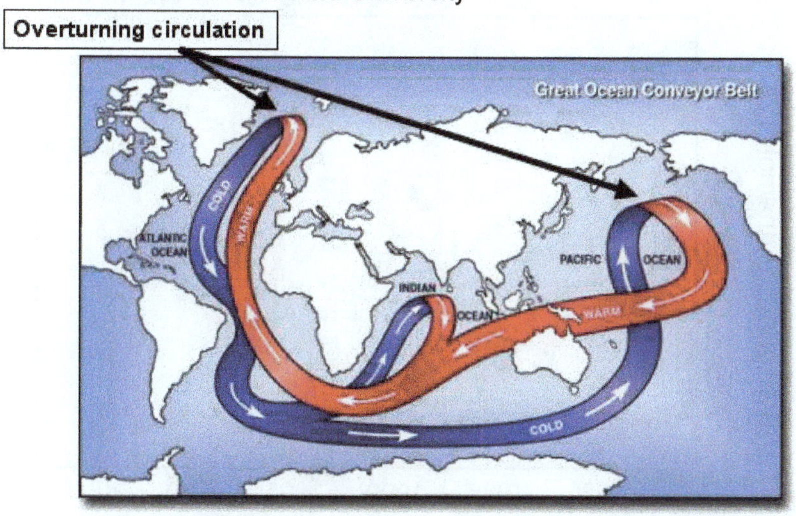

Thermoline Circulation in the North Atlantic

Terrence Joyce, former chair of the Woods Hole Physical Oceanographic Department in Cape Code, Massachusetts believes that freshwater infiltration into the North Atlantic from melting Arctic ice could shut down thermoline circulation that powers heat transfer from the tropics to Europe in winter. Such a shutdown might possibly precipitate a sudden drop in North Atlantic water temperatures and bring colder temperatures to land masses on both sides of the ocean.

Warming might trigger Mini-Ice Age

At least one theory suggests that global warming may actually trigger a mini-ice age.

- This possibility is supported by the appearance over the past 30 years of huge freshwater "rivers" in the salty North Atlantic, along with a recent drop of salinity in the Labrador Sea between Canada and Greenland adjoining the Atlantic:
  - It is suspected that these events may have been caused by melting Arctic ice that might subvert northern movement of heat-laden tropical Gulf Stream waters.
  - Instead of transferring heat to eastward winds that warm Europe in winter and then sinking as it cools and becomes denser, the North Atlantic fills with fresh water that is less dense and doesn't sink, remaining on top of the ocean like a big thermal blanket

635. Larry Bell

Ground Station Measurement Critics	Balloon/Satellite Measurement Critics
If monitoring stations are located in urban areas, they may become heat islands that skew results upward.	Measurements show a lack of observed differences between urban and rural areas, indicating no bias.
Suitable measurements sample less than 30% of the global surface. Localities exhibit different temperature trends.	Evidence of warming is apparent in glacier meltings and effects upon ocean salinity levels.
Satellite measurements have global coverage, and can precisely measure tropospheric and stratospheric temperatures.	Short measurement periods, drifts over local solar times, and ozone depletion cooling of the stratosphere skew results*
Balloon radiosonde records are longer and provide accurate tropospheric temperature measurements.	Significant changes in instrumentation and data processing have produced discontinuities over the years.

Temperature Measurement Process Arguments

Broad alarm during the mid-1980s about indications of global cooling portending the arrival of a new ice age reversed to predictions of cataclysmic warming threats by the later part of that century.

- The existence, magnitude and causes of this potential threat are subjects of contentious scientific and public policy debate:

- Measurements taken after about 1940 did seem to indicate a cooling trend, although declines slowed to halt in the late 1970s.
- Observations since the 1980s have recorded slow but steady increases in global near-surface temperatures in combination with rising amounts of atmospheric CO_2.
- Atmospheric balloon and satellite instrument measurements since 1979 have seemed to indicate smaller temperature increases than ground stations, although differences appear relatively minor and uncertain.

There appears to be a strong consensus among climatological experts that global warming is in fact occurring, and many, but not all, agree that human activities are an important contributing factor.

- An Intergovernmental Panel on Climate Change (IPCC) established in 1988 by two United Nations organizations, the World Meterological Organization (WMO) and the UN Environmental Program (UNEP) supported global warming assertions in a 2001 summary report:

- After much debate, the IPCC concluded that *"In the light of new evidence, and having taken into account the remaining uncertainties, most of the observed warming over the past 50 years is likely to have been due to increases in greenhouse gas concentrations."*
- The Panel also concluded that they expect human activities to continue to change atmospheric composition throughout this century.

636. Rober A. Rohde, Global Warming Art

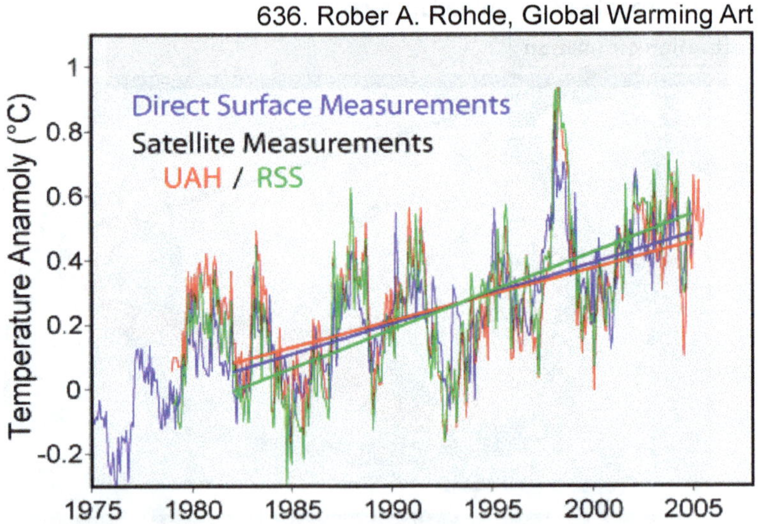

*Surface ground stations have recorded temperature increases of about +0.11°F (+0.06°C) / decade and +0.31°F (+0.17°C) / decade since 1979, while Remote Sensing Systems (RSS) taken by the U of Alabama Huntsville since 1979 show increases of about +0.063°F (+0.035°C) / decade. Removing RSS measurements of the lower troposphere only to eliminate possible stratospheric cooling effects, the surface and atmospheric measurements may be closer.***

Reconciling Ground and Atmospheric Temperature Data

* See Appendix: 3
** See Appendix: 4

637. MDPI

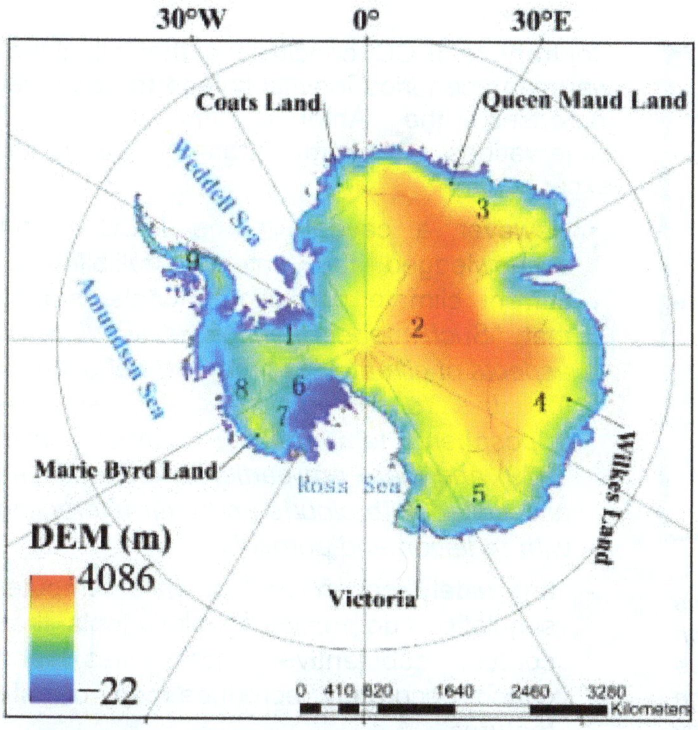

Elevation of Antarctic Ice sheet, 2017

In addition to projecting continued average global temperature increases, the IPCC also predicted that sea levels will rise "under all scenarios".

- Given that rising sea levels are principally associated with anticipated Arctic and Antarctic ice melts, measurements of snow accumulations in Antarctica may offer an alternative picture:
  - Satellite altimeter measurements made between 2003-2017 indicate, on average, the elevation of about 3.2 million square miles (8.5 million square kilometers) of the Antarctic interior has experienced a net snow gain of about 50-70 billion tons/year, potentially enough to slightly lower the surrounding sea level.
  - These observed snowfall increases covering about 70% of the total ice sheet area are likely to be linked to warming of southern oceans from either natural or human-influenced causes.

Snow Accumulation in the Antarctic Interior

European Space Agency satellite altimeter measurements of the Greenland Ice Sheet since 1992 record recent changes similar to those observed in Antarctica, with thickness growth in higher interior areas, and thinning at coastal margins.

- While total mass balance gains or losses are yet unknown, spatial averaging indicates a net increase of 2.1 in (5.4cm) per year, with thickness increases of about 2.5 in (6.4cm) per year in interior areas above 4,900 feet (1,500 meters), and 0.79 in (2 cm) per year reductions below that elevation:
  - The recording period is too short to know if these observations represent a long-term trend, since Arctic temperatures appear to naturally fluctuate in 60-80 year time cycles.
  - Annual snowfall increases and melts have been observed to be strongly linked to variability in regional atmospheric circulation (the North Atlantic Oscillation), along with possible smaller influences associated with temperatures.

638. NSIDS

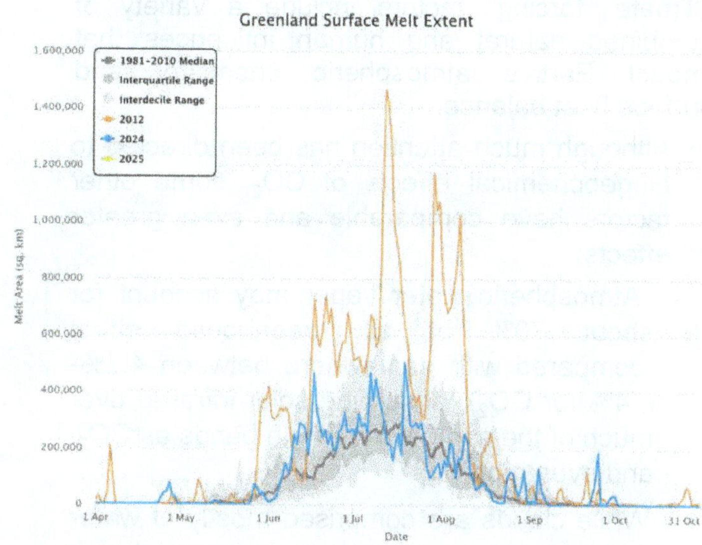

*Annual Greenland Ice Sheet melt changes revealed by satellite data since 1978 are believed to be largely influenced by regional atmospheric circulation patterns. Warming (from natural and human-enhanced influences) may be a contributing factor. The lower-than-average 1992 melt appears to be linked to a large Mt Pinatubo eruption that released aerosols.**

*** See Appendix: 5**

639. CSIRO Marine and Atmospheric Research

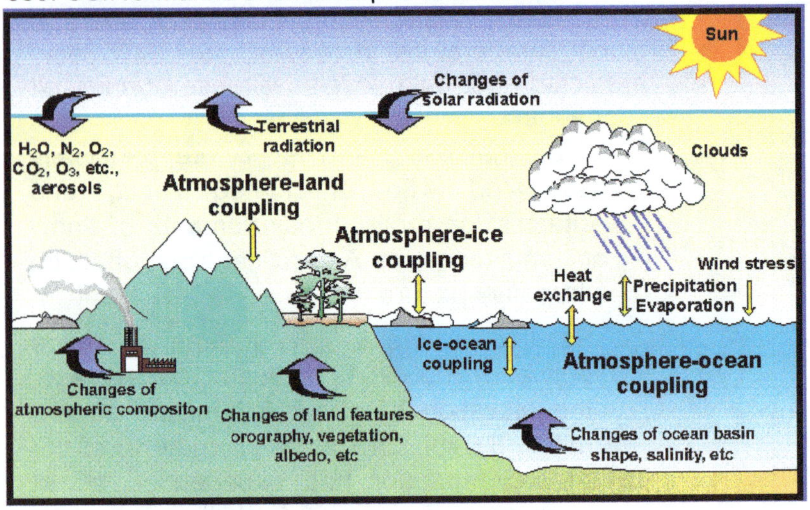

Weather is influenced by many interactive day-to-day conditions and events in a locality that are often difficult to predict with confidence. Global climate is conventionally thought to be comprised of interactive local weather conditions over large scales measured over 30 years or more, adding enormously greater complexities and uncertainties.

Climate Modeling Challenges

Contrary to IPCC conclusions that all global warming scenarios inevitably lead to sea level increases, the Antarctic and Greenland observations demonstrate apparent exceptions.

- However a caveat in the IPCC report acknowledged the potential fallibility of current climate prediction models, stating that: "*Such models cannot yet simulate all aspects of climate (e.g. they cannot account for the observed trend in surface-troposphere temperature changes since 1979) and there are particular uncertainties associated with clouds and their interaction with radiation and aerosols*":

- It is widely recognized that while computer simulations do provide scholarly tools, they contain substantive uncertainties, and cannot accurately reproduce many climate features.

Climate "forcing" factors include a variety of combined natural and human influences that impact Earth's atmospheric chemistry and surface heat balance.

- Although much attention has been directed to biogeochemical effects of CO_2, some other factors have comparable and even greater effects:

- Atmospheric water vapor may account for about 70% of all greenhouse effect (compared with somewhere between 4.2%-8.4% for CO_2), absorbing solar infrared over much of the same wavelength bands as CO_2, and even more.

- While clouds are comprised mostly of water droplets (not greenhouse gases) they absorb about one-fifth of the long-wave radiation from Earth, although their influences upon climate mechanisms are not fully understood.

640. NASA Goddard

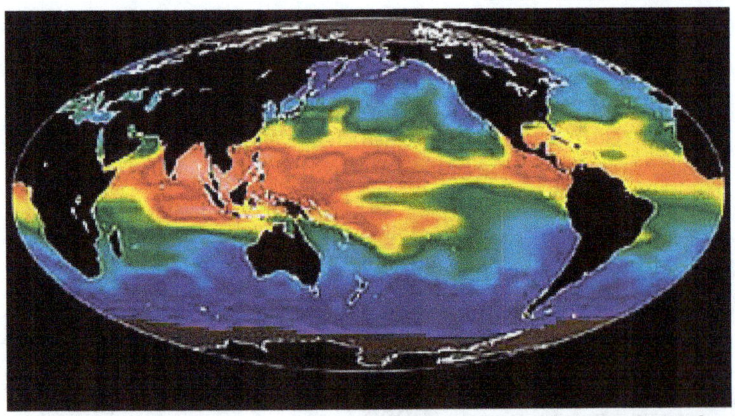

Satellite Water Vapor Imaging

*Although water vapor is a primary greenhouse gas, upper atmosphere satellite measurements at a 6-9 mile (9.5-14 km) altitude show less positive feedback correlation between vapor concentration and Earth warming than previous computer models predicted.***

*Separate satellite measurements of global surface and ocean precipitation between 1979-2019 also reveal virtually no net vapor concentration increases due to climate warming evaporation effects.****

Water Vapor Influences

Climate Forcing Factors

** See Appendix: 2
*** See Appendix: 6

While it is evident that atmospheric CO_2 levels have increased about 110 ppm from a pre-Industrial Revolution level of 280 ppm to about 360-380 ppm now, the actual influences of this trend upon global warming or its future consequences are not nearly as clear.

- Although CO_2 represents only a very tiny amount of total atmospheric gas content (less than 4/100ths of 1% which is much lower than water vapor per volume and weight), steady increases over the past 50 years are raising international concerns:

- Much or most of this increase is generally attributed to human causes, yet computer models cannot yet reliably predict climate change impacts.

- Key areas of uncertainty include heating and cooling influences of clouds upon the atmosphere, roles played by oceans, and land influences such as deforestation, agriculture, and seasonal changes.*

641. US Geological Survey

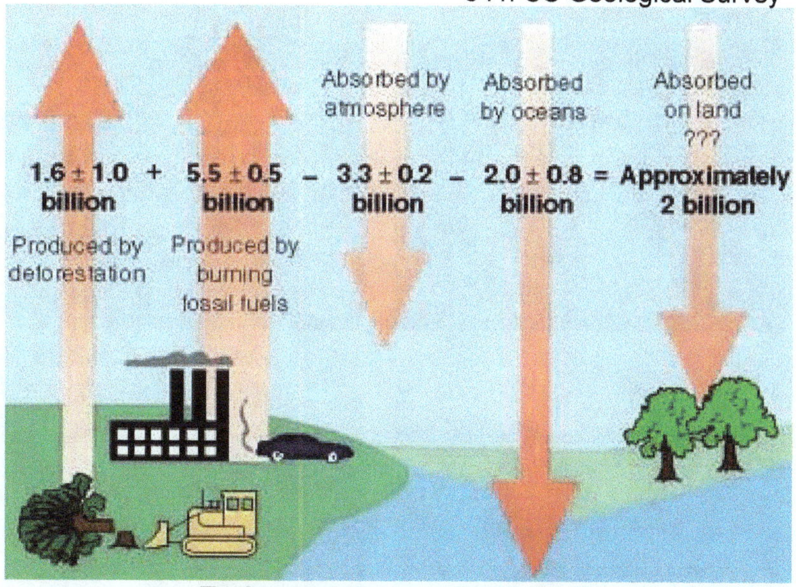

Estimated Human Contributions

Atmospheric CO_2 is produced by a variety of human and natural sources, with much ultimately absorbed by the atmosphere, oceans, and vegetation. Many researchers believe that influences upon Earth temperature changes may be well within natural climate variability. **

Atmospheric CO_2 Influences

Aerosols include atmospheric dust particles and sulfate droplets released from natural sources such as volcanoes, sea spray and land wind erosion in combination with human activities such as fossil burning from industry and forest clearing for agriculture.

- Current models for distinguishing concentrations from natural vs. human sources, or assessing climate effects, are unreliable and often contradictory due to issue and calculation complexities:

- It is generally believed that unlike greenhouse gases, sulfate aerosols tend to cause cooling rather than heating by scattering radiation at the top of the atmosphere. This enhances reflectivity that reduces cloud lifetimes by increasing the number of droplet collisions needed to produce rain.

- Dust particles do this to some extent, but also absorb some sunlight to produce warming effects.

642. US Climate Change Science Program

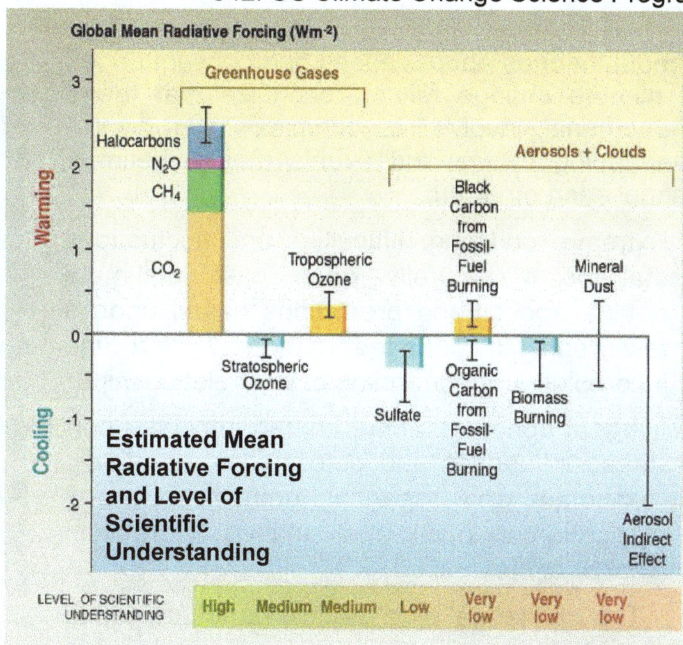

Aerosols tend to create cooling rather than warming effects. They remain in the atmosphere for only days or weeks before they fall out or are washed out by rain. These short periods do not allow enough time for mixing to occur that would enable global transport associated with greenhouse gases. ***

Estimated Aerosol / Greenhouse Influences

* See Appendix: 7
** See Appendix: 8
*** See Appendix: 9

643. UCAR
645. Exec. Action Team
644. U. of Missouri
646. Julian Pye

Human land use influences upon weather and climate are associated with a number of factors, including growth of farming and industry, population size and distribution, and types of technologies used in each region. The combined effects of these impacts can, in turn, affect the suitability of the land and land cover for desired future uses.

Land Use Influences

Human influences upon land uses and vegetation can have a variety of complex links to weather and climate mechanisms, including surface-atmosphere greenhouse gas exchanges, reflected and absorbed radiation and thermal balances, and wind interactions with surface topologies.

- Growth of agriculture has had particularly significant land use influences over much of the world:

- Although global impacts are not well understood, replacement of dark forests with croplands is known to affect local conditions, increasing surface albedo levels to cool some regions, while denying shade to other heavily-irradiated areas to raise ground temperatures.

- Large-scale crop irrigation can also produce measurable regional cooling effects, as has been demonstrated in the Great Plains of North America since 1945.****

Simple theories about causes and consequences of climate change fail to recognize that these phenomena involve a complex interplay of overlapping cycles and events that sometimes cancel each other out.

- Extreme modeling difficulties and fluctuations that occur naturally often cast legitimate doubts about trend predictions based upon a few years or decades of limited data and incomplete understanding of interrelationships:

- Urgent concerns about human impacts upon these influences and outcomes are certainly legitimate and essential, both as a moral obligation to our planet and as a practical matter of survival.

- The challenge is to avoid premature, faulty conclusions and policies driven by inadequate models, unsubstantiated hypotheses, and alarmist observations based upon both.

647. Solcomhouse

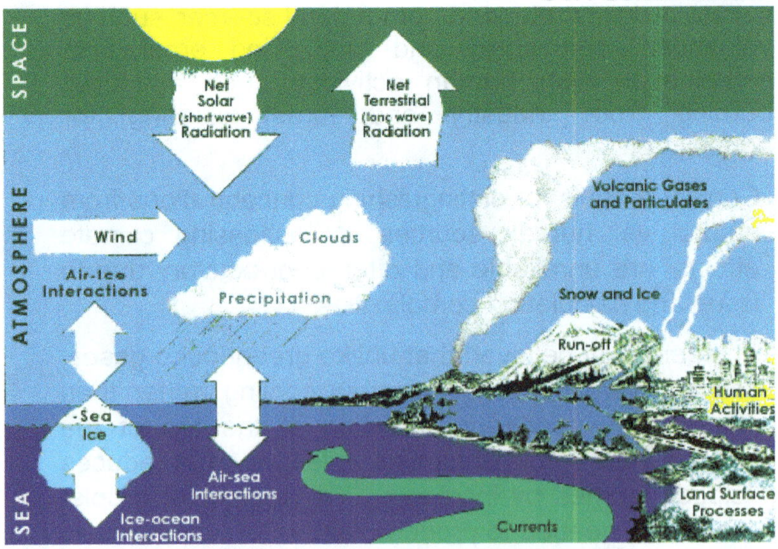

Climate and temperature changes are influenced by many highly interactive forcing factors that are not well understood as a basis for developing accurate or reliable models.

Climate and Temperature Influences

****See Appendix: 10

Important issues and controversies concerning human impacts upon climate revolve around questions regarding which observed trends and fluctuations are natural or "normal".

- In regard to global temperatures it can be argued that normal can be characterized by ice ages that have dominated Earth's history:
  - Except during brief interglacial periods, our planet's climate has been invariably glacial, with more than half of the surface covered with ice and arid deserts.
  - Present global warming trends began about 18,000 years ago near the end of the Pleistocene Ice Age when much of North America, Europe and Asia began to thaw out from a very long deep freeze.

648. Australian Academy of Science

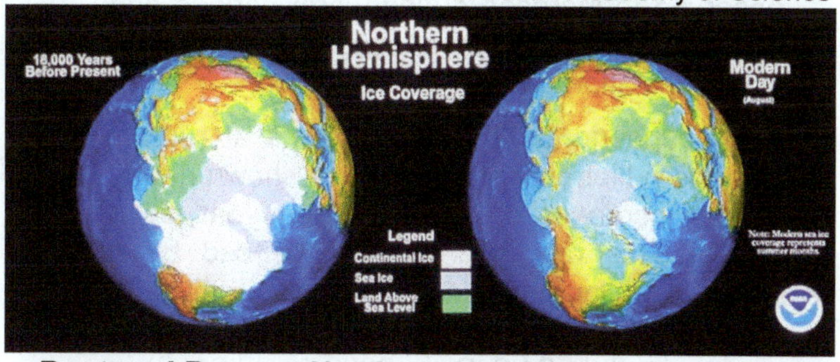

Past and Present Northern Hemisphere Ice Coverage

Earth has been relatively ice-free even at high latitudes, only during relatively short interglacial periods such as our current one. From a historic perspective our climate is, without doubt, "abnormally warm".

Past and Present Northern Hemisphere Ice Coverage

Interglacial periods, such as ours, appear to be characterized by great climate variability.

- Contemporary observations reveal constant fluctuations from month-to-month, season-to-season, year-to-year, decade-to-decade, and century-to-century:
  - The 19th century ended the well-documented, centuries-long Little Ice Age that brought "abnormally cold" temperatures to many parts of the world, and may have biased observations commencing during the second half of that century as "normal", and the 20th century as "abnormally warm".
  - Today's climate is generally comparable to that of the Medieval Warm Period (believed to extend from the 9th-13th centuries AD) which preceded the Little Ice Age.
  - Dramatic, but very temporary changes were also witnessed as influences of tropical Pacific conditions that produced cooling during "triple-dip" La Niña from 2020-2022 and during the warming El Niño in 2023-2024.

649. Ljungqvist (2010)

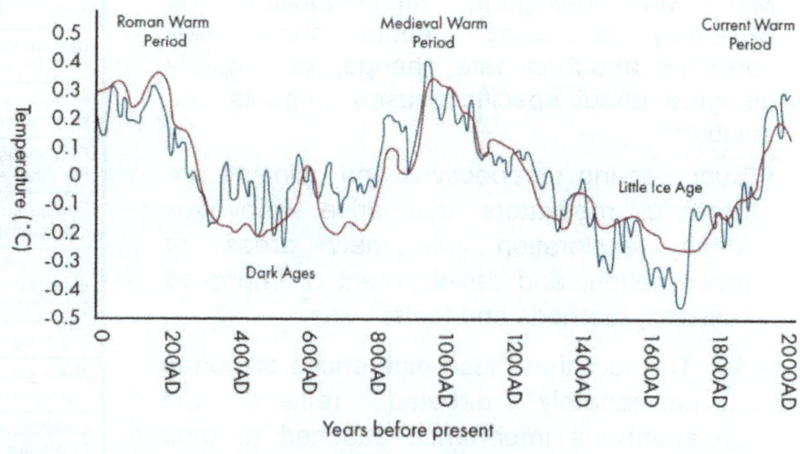

*The Northern Hemisphere has experienced significant climate shifts during the past 2,000 years, ranging from the Roman Warm Period to the Modern warm period we are currently experiencing.**

Northern Hemisphere Climate Shifts

***See Appendix: 11**

Influences of atmospheric CO_2 concentrations upon climate change and human-induced enhancements in particular, are issues of continued uncertainty and controversy.

- Although CO_2 levels have generally been observed to increase during warm periods and fall with colder temperatures, it is not clear which condition causes the other:
  - Atmospheric CO2 measurements at remote Northern Hemisphere sites have shown nearly linear increases of about 2.5 ppm/year since 2000. This rate has been influenced by a significant increase in fossil fuel combustion emissions, which have risen from approximately 24.8 Gt in 2000 to 37.01 Gt in 2023
  - Previous climate shifts between the Medieval Warm Period and Little Ice Age also raise questions, having occurred long before the Industrial Revolution or invention of the internal combustion engine.

650. NOAA

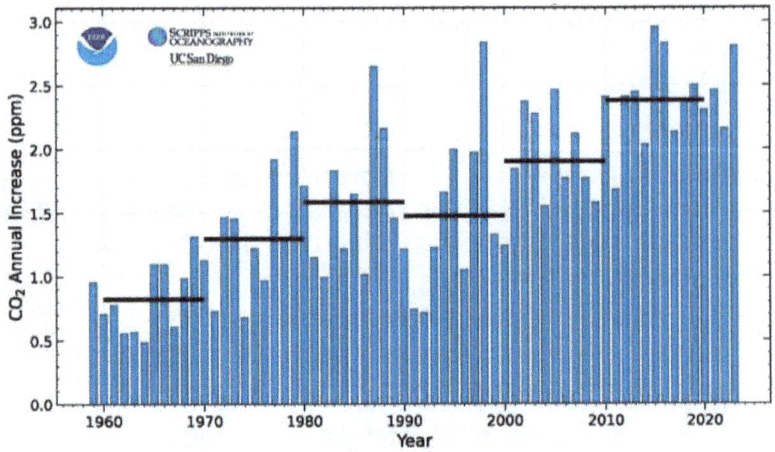

*On average, only about half of CO_2 emissions released by fossil fuel combustion remain in the atmosphere, with inter-annual fluctuations due to variations in natural sources and sinks.***

Annual Global increase of CO_2, 2023

Many very intelligent, highly-educated and extremely dedicated people share real concerns about climate change, yet sharply disagree about specific causes, impacts and solutions.

- Such varying perspectives and opinions are powerful motivators that drive innovative ideas, exploration into new areas of investigation, and development of improved research methods and tools:
  - The fact that these differences are often passionately debated reflects the enormous importance attached to their concerns, and deserves to be honored with appreciation and respect.
  - Real progress demands that theories be regarded as unproven possibilities, and "established facts" be constantly tested as newer, more comprehensive information is revealed.

651. Larry Bell

Less Concerned	More Concerned
Climate models have proven faulty and don't include major influences.	Humankind is performing a dangerous geophysical experiment.
Human vs. natural contributions to atmospheric CO_2 are very small.	A precautionary principle requires action to prevent crises later.
CO_2 levels follow a temperature rise, not the other way around.	Temperatures and CO_2 have risen rapidly over the past 50 years.
Temperatures always rise and fall during interglacial periods.	Most scientists today share global warming concerns.
Issues are politicized; climatologists who disagree are afraid to speak out.	Most scientists agree that human activities contribute to warming.
CO_2 is a very minor greenhouse gas, and most comes from natural sources.	CO_2-water vapor interactions can have a compounding effect.
Natural ocean and land CO_2 sinks rapidly reduce accumulations.	CO_2 has a long average atmospheric lifetime.
Solar activity is a major factor: present levels are highest in 8,000 years.	Without greenhouse forcing, solar activity doesn't account for changes.
Greenland/Antarctica melts are balanced by interior snow accumulations.	Greenland/Antarctica melts may flood global coastlines.
Global warming may have prevented an overdue new ice age.	Greenland melts may trigger a Northern Hemisphere glacial event.

How Concerned Should We Be About Global Warming/CO_2?

Honoring Disagreements

**See Appendix: 12

652. Larry Bell

Causes:
- Do CO_2 changes lead or follow temperature changes?
- How important is CO_2 as greenhouse gas?
- What are other forcing factors and interrelationships?
- How significant are fossil fuel combustion influences?
- Will switches to biofuels make measurable differences?
- How important are land use change influences?

Effects:
- How warm is ultimately "too warm"?
- Is warming a micro-trend within this present interglacial?
- When is the next ice age "scheduled" to arrive?
- Are net Arctic / Antarctic water mass balances changing?
- Will Northern Atlantic salinity dilution trigger cooling?
- What human greenhouse reductions are possible/ realistic?

Global Warming Questions

Wide-spread concerns about global cooling trends during the 1970s and warming in the present may have a solid basis (possibly in reverse order), but even then do not necessarily warrant dire catastrophic forecasts.

- There are no conclusive reasons to believe that slightly lower temperatures are preferable to slightly higher ones, or that modest warming trends will trigger runaway greenhouse events:

- Recognizing that atmospheric and surface climate mechanisms may well be sensitive to relatively small influences, CO_2-water vapor interactions linked to greenhouse feedback loops do appear to be less active than climate models formerly suggested.
- It is also apparent that much of the CO_2 released from human and natural processes is absorbed by the oceans and terrestrial biosphere, providing an essential nutrient for agriculture, forests and all plant-based ecosystems.

It is rational to recognize global warming as a current reality that can possibly precipitate non-linear sudden climate shifts.

- A 2002 report titled "Abrupt Climate Change: Inevitable Surprises" issued jointly by the Ocean Studies Board, Polar Research Board, and Board on Atmospheric Sciences and Climate of the National Research Council advocated preparation without panic:

- *"The climate record for the past 100,000 years clearly indicates that the climate system has undergone periodic and often extreme shifts, sometimes in little as a decade or less"... "Societies have faced both gradual and abrupt changes for millennia and have learned to adapt..."*
- *"It is important not to be fatalistic about the threats of abrupt climate change"... "Nevertheless, because climate change will likely continue in the coming decades, denying the likelihood or downplaying the relevance of abrupt changes could be costly."*

653. Larry Bell

Complacency, possibly rooted in denial or apathy, promote hazardous tendencies to assume that "everything is really fine"; or "someone else will solve the problem"; or it won't happen in my lifetime or anytime soon, so why worry?"

Hysteria, potentially fueled by simplistic assumptions and sensationalism can promote desperate, misguided and costly decisions at the expense of seeking better understanding of underlying issues and pursuing the most effective intervention options.

Dangerous Extremes

*The "Abrupt Climate Change: Inevitable Surprises Report" issued by the National Academy of Sciences in 2002 emphasizes that "Increased knowledge is the best way to improve the effectiveness of response; research into causes, patterns, and likelihood of abrupt climate change can help reduce vulnerabilities and incr*ease our ability to react."*

***See Appendix: 13**

The fact that theories and models do not always match measured observations does not necessarily indicate that basic underlying premises are wrong, that predicted outcomes are invalid, or that reasoned interventions should be delayed until all conditions are perfectly understood.

- There is inescapable evidence that human activities are changing Earth's environment, including its climate, in ways that pose unacceptable perils to all life:

- Air, water and land pollution is an expanding global reality that must be rapidly reversed before the impacts are irreversible.
- Fossil fuel depletion is a near-term certainty that demands a greatly accelerated adjustment of consumption habits and transition to clean, sustainable alternatives.

655. NOAA / NASA

NOAA Polar-Orbiting Operational Environmental (NOAA-K) Satellite

Information Advancements

654. Larry Bell

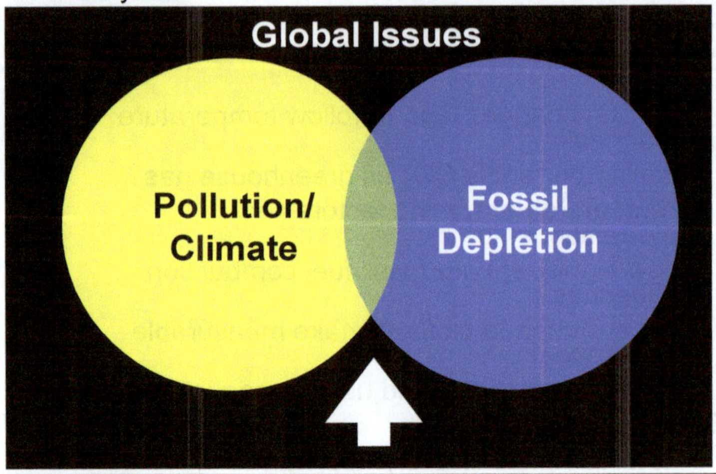

Information: Determine interrelationships between causes / effects and human contributions / interventions.	**Innovation:** Develop responsive technologies, strategies, and programs that address short- and long-term needs.
Education: Disseminate accurate information to advance public, industry, and government awareness.	**Application:** Put responsive plans and technology infrastructures into action through national and global initiatives.

Responsive Priorities

International researchers from diverse disciplines using a variety of new tools are making real progress towards achieving a better understanding of natural and human-enhanced influences upon Earth's environment and climate.

- Increasingly precise methods are being used and improved to measure distant and recent climate and temperature fluctuations, and to assess observed and potential influences upon atmospheric and surface phenomena and their interactions:

- Expedited advancement and expanded application of these capabilities is vitally important to construct more comprehensive and reliable climate models, both to examine current forcing conditions, and to predict future trends and consequences.
- Rapid advancements in computing technologies are also encouraging, and are critically important to study enormously complex interrelationships between forcing influences that are continuously being revealed.

Information technology is but one of many tools afforded to guide better understanding of Nature's complexities and intricacies, our influences upon these phenomena, and lessons we can apply to be more positive contributors.

- Other important tools are our natural gifts of human curiosity, reasoning, creativity and the capability to recognize our dependence upon the wellbeing and stability of other creatures, habitats and support systems that our actions impact:

- As students of Nature and Earth's history we learn that very large events and consequences can precipitate from small compounding influences over relatively short time frames.
- We can also readily observe that pervasive human influences upon this plant's fragile lands, oceans and atmosphere are neither small nor sustainable.

656. Asel Electronic 657. RST3
658. NOAA 659. Welfare Service Club

Lessons from Experience

660. Larry Bell

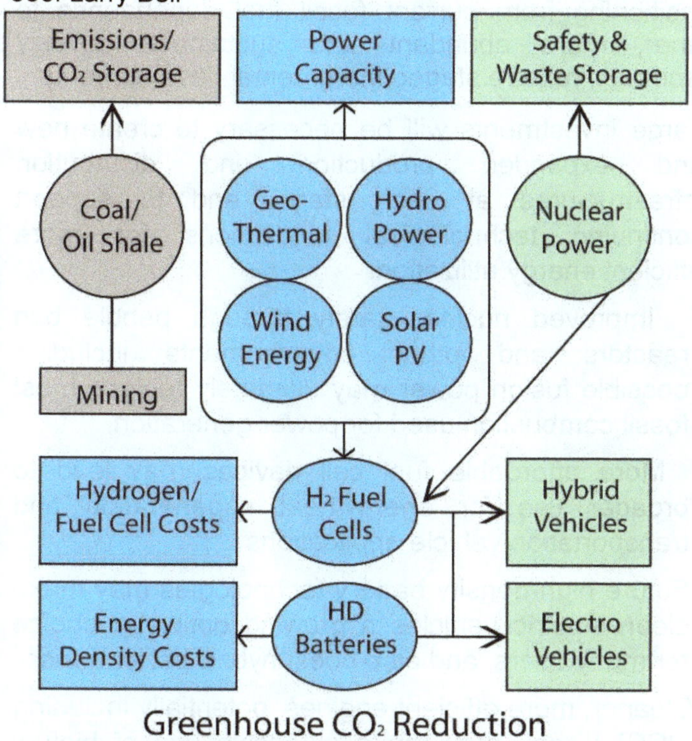

Power / Fuel Innovation Challenges

Recent and emerging technologies reflect progress in response to serious environment and resource depletion issues, but none yet offer perfect or complete solutions.

- Continued innovation will be needed to improve or replace many of these approaches:

- More efficient and effective clean coal and CO_2 sequestration methods must be implemented, ultimately transitioning to non-combustion processes.
- Nuclear energy is potentially a major energy source, but safety and hazardous waste storage concerns must be conclusively addressed.
- "Alternative" geothermal, hydropower, wind and solar approaches can provide clean hydrogen processing for fuel cells (at the expense of other electricity uses), but fuel cell costs must also be reduced.
- Development of more powerful, yet affordable high-density batteries is vital to improve performance and expanded use of clean electric vehicles.

661. Larry Bell

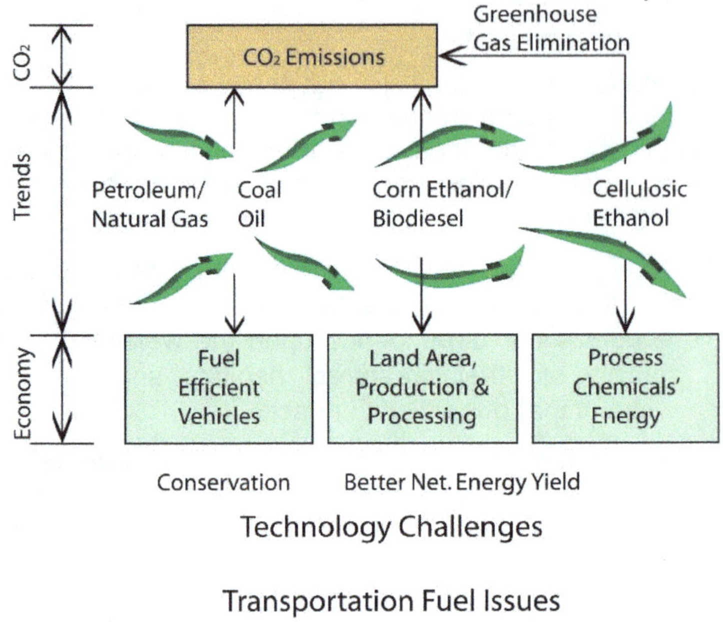

Fuel Transition Challenges

Transitioning to organic and coal-derived liquid fuels to extend dwindling crude oil reserves appears to be a useful but limited longer-term solution.

- All current hydrocarbon combustion processes release CO_2 and other pollutant emissions and consume substantial amounts of production energy relative to outputs:

- Corn and other grain ethanol and biodiesel fuel may become more economical provided that bioengineering developments can improve average yields, and ways can be found to reduce farming and processing energy budgets.
- Cellulosic bioethanol may gain advantages if difficult chemical digestion and process energy problems can be solved.
- Petroleum derived from coal can be expected to become increasingly attractive as oil costs continue to rise provided that current experimental technologies prove to be scalable and effective.

662. Larry Bell

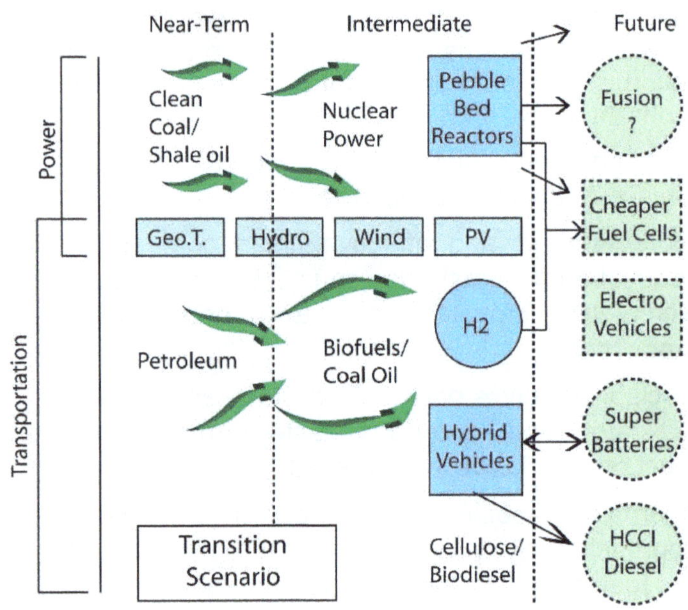

Applying Alternatives

Transitioning from current fossil fuel dependence to cleaner, more abundant and sustainable energy sources will require staged incremental developments.

- Large investments will be necessary to create new and expanded production and distribution infrastructures at each stage, and to support continuing technological innovations for more efficient energy utilization:

- Improved nuclear safety through pebble bed reactors and other advancements including possible fusion power may ultimately replace most fossil combustion used for power generation.
- More affordable fuel cell devices may lead to broader use for energy/heat cogeneration and transportation vehicle applications.
- Future high-density battery technologies may make clean electric vehicles a growing option of choice for many users, and also boost hybrid performance.
- Cleaner, more efficient engines, potentially including HCCI diesel, may make extensive use of biofuel production and processing advancements.

Public attention and importance attached to resource depletion, energy independence and the health and stability of our planet are powerful incentives for innovation and change.

- Government, industry and academic organizations along with citizens that support them recognize that better solutions are essential for many reasons:

- Because they care about our children, grandchildren and the future of all generations who will follow.
- Because they are concerned about the quality of the natural environment and fragile ecosystems that are endangered.
- Because new and better solutions make good economic sense to create markets, provide employment and produce operating economies and profits needed for progress.
- Because change, whether forced or elective, is a necessity, not an option.

663. Children's Research Center
665. US Dept. of State
664. National Park Service
666. Salem Cnty. AVA Center

Incentives for Progress

667. Chemistryland
669. NASA
668. Chemistryland
670. NASA

Applying Knowledge and Tools

The emergence of Homo Sapiens from the Late Pleistocene glacial period about 13,000 years ago demonstrated our remarkable human capacity to adapt and survive.

- Since that time humankind has achieved much, beginning with stone tools, and ultimately returning similar rocks during expeditions to the Moon:
  - Our genus has learned not only how to adapt to the environment, but through agriculture, industry and living, to adapt the planet to serve human purposes.
  - We are now realizing that our continued survival will depend upon learning how to protect the planet from our impacts, probably the greatest adaptation challenge of all.

Our human capacity to gain knowledge about changes we are imposing upon our planet provides opportunities to adapt our living habits, industries and technologies to prevent events unscheduled by Nature from creating unfortunate and avoidable surprises.

- Such events, including resource exhaustion, ecosystem destruction and severe climate changes can have harsh consequences effecting all life:
  - Rapid changes, even lasting over relatively short periods, can be more hazardous and disruptive than slow ones, allowing less time to respond.
  - Unlike our early ancestors, we won't be able to simply "pack up and leave", since there may be few places to escape to, and it may be impossible to relocate present large populations to more favorable areas, particularly across restricted national borders.

671. Scientific American
672. European Space Agency
673. United Nations Magazine
674. Vestal Design

Penalties for Failures

675. Regional Municipality of Wood Buffallo

Learning from Mistakes

There is little disagreement within our global community that we must change the ways we treat our natural environment before it retaliates in ways we cannot correct.

- Whether these repercussions occur rapidly or over longer time frames, present trends are unsustainable, promising unthinkable consequences that transcend geographic boundaries:
  - Citizens, children and future generations throughout the world are equal stakeholders in the outcomes of our actions.
  - Industrialized and developing nations enjoy special responsibilities and opportunities to provide essential solutions and examples.
  - Our decisions and commitments may well determine the ultimate destiny and natural legacy of human civilization.

There is no escape from a reality that we are all members of a global civilization, nor from the impacts that our common community create upon our planet and our future.

- Pollutants from all parts of the world rapidly circumnavigate through the atmosphere and oceans to distant locales, and weather and climate influences recognize no geopolitical boundaries:
  - Environmental abuses everywhere are ultimately visited upon us as our problems, as are those we export to others.
  - Interventions we enact, while important, offer only marginal protection from cumulative impacts of other nations, including contamination of marine ecosystems we share and the air we breathe.
  - It is clear that just as the problems are global, real solutions must be also.

677. Sydney Morning Herald 678. Space and Motion
679. Planetsave 680. NASA

Los Angeles

India

China

Mexico City

Pursuing International Solutions

676. NASA

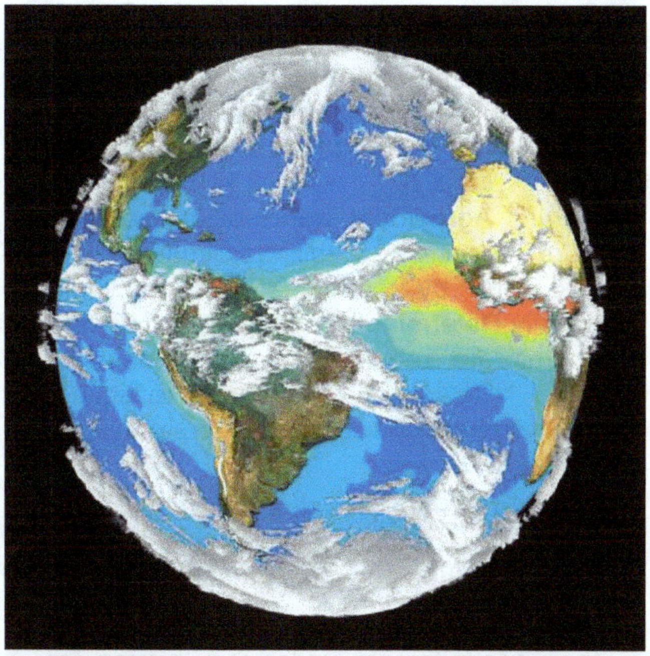

Image created using data from four satellites, showing fires burning on land areas, a large aerosol cloud over the Atlantic Ocean formed by biomass burning in Africa, and a dust cloud surrounding the globe.

Establishment and enforcement of universal international environmental protection standards has thus far proven to be an elusive goal.

- The Paris Agreement, which succeeded the Kyoto Protocol, has been ratified by 195 parties (194 states and the European Union), representing over 98% of global greenhouse gas emissions:
  - Under the Paris Agreement, all countries have set their own targets (Nationally Determined Contributions) to reduce emissions.
  - The US joined the Agreement in 2016, withdrew in 2017, rejoined in 2021, and withdrew again in 2025.
  - India aims to reduce the emissions intensity of its GDP by 33-35% below 2005 levels by 2030 and achieve 50% cumulative electric power installed capacity from non-fossil fuel-based energy resources by 2030.
  - Germany, as part of the EU, aims to achieve the EU's target of reducing emissions by at least 55% below 1990 levels by 2030.

The Paris Agreement, adopted in 2015 and entered into force in 2016, is a landmark international treaty on climate change that aims to limit global warming to well below 2°C, preferably to 1.5°C, compared to pre-industrial levels.

- The Paris Agreement works on a five-year cycle of increasingly ambitious climate action carried out by countries through their Nationally Determined Contributions (NDCs). Countries are required to submit new or updated NDCs every five years, reflecting higher ambition compared to previous version:
  - The agreement provides a framework for financial, technical, and capacity-building support to those countries that need it.
  - Countries continue to work towards fulfilling their commitments under this, aiming for significant reductions in greenhouse gas emissions.

681. Our World

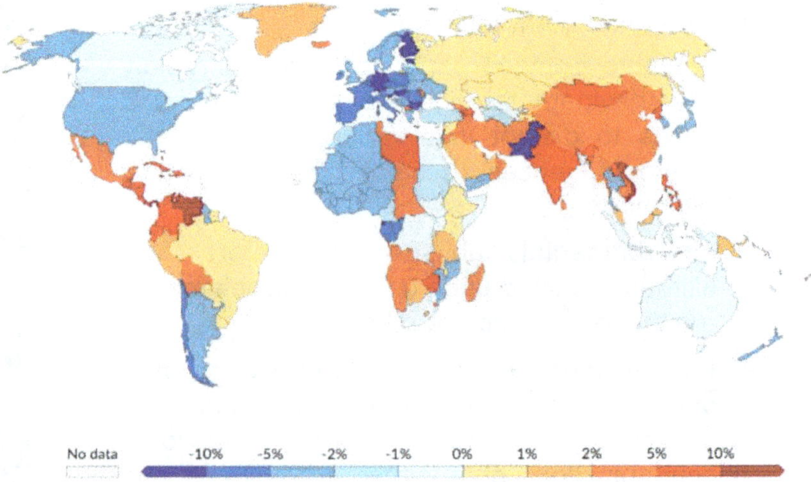

World CO2 Emission, 2023

682. Larry Bell

- Establish a responsible balance between state/regional/national economic and environmental interests by ensuring that both priorities are fully and effectively represented in policy and finance decisions.
- Provide business incentives and regulatory mechanisms that promote clean industries without imposing overly burdensome costs and restrictions that send production and jobs overseas to places that are cheaper in part due to lack of equivalent standards.
- Protect national economic competitiveness in world markets through international agreements that insist upon fairness, while also demonstrating that global cooperation benefits everyone.
- Recognize responsibilities of all nations to support less developed ones through technology funding and transfer for climate-related studies and projects.

Economic Policy Priorities and Challenges

Interjections of economic considerations into national and international environmental policy decisions can be regarded by some as an inevitable compromise of principles.

- Recognizing that all-too-often these separate priorities have produced conflicts, with environment the loser, it is also important to realize ways the underlying objectives of both are interdependent:
  - Commercial investment, production and profits provide employment essential to support families of all income levels, finance education and public services, and create safe and affordable energy and food used in daily life.
  - Businesses that fail, or are forced to relocate to countries with lower environmental standards, compound economic stress with additional global environment impacts.
  - Economic prosperity reinvested by industry provides cleaner, more efficient production methods, products and advancements.

Each of us affects courses of human and natural events through our choices and actions.

- Individually and collectively, we change the world for better or worse through daily decisions:
  - We choose which information sources we seek and heed in determining who we will trust to lead us and implement policies we believe are important.
  - We determine which business will be influential and which products will be successful in the marketplace through our purchasing power.
  - We decide how many resources we will consume, how much waste will be created, and where it will go based upon values that guide how we live.
  - We influence our children and others around us through our conservation outlooks and examples we put into practice.

683. Missouri Secretary of State 684. URFAN
685. Star Bulletin 686. City of Cambridge

687. NASA

Viewing our tiny, fragile planet from space we can more fully realize that there is really no “them”, only us, and that we must learn to live and work within that sphere of reality.

- Our world, shared with other creatures, is a place of overwhelming, magnificent complexity:
  - It is a place within an infinitely larger Universe where marvelous, often poorly-understood natural forces interact and exchange energy at unimaginably large and small levels of scale.
  - It is place where global and microscopic events occurring over millennia and microseconds cause some creatures to evolve, while others unable to adapt perish.
  - It is a place where societies bound together by common purposes and dreams sometimes become separated by conflicts that impact our lives, future and planet.

688. NASA

Nature's incomprehensible scale, diversity, intricacies and complex interconnections are impossible to grasp by intellect alone, leaving only spiritual consciousness to embrace its full magnitude.

From this vantage point we can perhaps be inspired by the wonders about us, rather than intimidated by a sense of insignificance and futility:

- Viewing ourselves as part of Nature's majesty can empower us to recognize that we are creatures constructed of cosmic potentials.
- We can appreciate that while we are small, everything, including Earth's Sun is tiny compared with the Universe, and that Nature makes no size distinction in assigning importance.
- Ultimately we may realize that in Nature's grand model everything is truly part of everything, including you.

The challenges that we share in understanding and addressing fossil depletion, ecosystem damage, climate issues and other human-induced conditions that will influence our lives, future generations and our planet are extremely complex and daunting. Those who read this book seeking simple solutions or reassurances that everything will resolve itself will be disappointed. Others who accept dire predictions or believe that singular answers hold the key may likewise be dissatisfied that contrary views are presented. Such matters warrant serious concern by all caring people; concerns that both unite and polarize passionately-held interests and opinions.

This book was written by someone holding no special claims of expertise concerning many or even most of the issues discussed, the research that underlies myriad observations addressed, or the numerous technologies that are highlighted. Yet as an architect with some experience in studying complex phenomena both on Earth and in space, I hope that this effort to put important factors into more comprehensible perspectives will offer useful insights. As a professional who participates in the planning and design of habitats and as a fellow stakeholder with a family, I share your hopes and frustrations.

It may be of interest to note that this project began with absolutely no thought of producing a book, but was entirely motivated only to explore and reconcile information that appeared ambiguous, confusing and often contradictory. This process has been illuminating for me, and while continuing to raise more questions than answers, has altered my outlook on many topics. Clearly most importantly, it has caused me to much more fully appreciate the wonderments of Nature, to be inspired by the nature of human innovation, and to be heartened by the dedication of so many people who are advancing knowledge and progress so desperately needed.

The amount of information readily available to all of us through the internet is absolutely revolutionary and remarkable, and this book is truly a “Google” product. Yet while data and viewpoints are amazingly abundant, large challenges are to objectively vet sources, accurately present facts and positions, maintain a fair balance where disagreements and controversies exist, and summarize contents as simply as possible without being simplistic. This quest has also led to certain ideas, such as the MECOPTICS framework concept that represents an attempt to organize some basic principles in a way that hopefully provides another reference for consideration.

Readers may observe that up to this point the word "science" does not appear in the text other than where it is used in a cited quotation by others. This omission is entirely intentional, but certainly not due to any lack of respect for its vital importance. Instead, this decision is an attempt to discourage any tendencies to assume that the book is about "things that specialized, technically-oriented people do", rather than putting emphasis upon observing Nature more holistically through our own eyes and discovery. All-too-often what we regard to be guided by "higher education" is bounded by disciplines that may obscure Nature's connected, synergetic realities within larger, more important contexts.

Viewed more broadly by mind and spirit we may realize that Nature is unbounded by descriptions of systems we apply to characterize parts and relationships we find most interesting and easy to measure. Instead, we can observe that Nature simultaneously mixes things together across categories we define, making causes and effects difficult or impossible to isolate.

A broad view also reveals that our natural environment is fundamentally dynamic. Earth's atmosphere, for example, is comprised of elements and events that are constantly moving and changing, with substances entering, leaving and forming different compounds at different times and places, often to be distributed world-wide.

If there is a positive side to urgent challenges facing us, it must be a growing general awareness that we are all responsible for solutions. Progress in our recognition of issues is being revealed through interdisciplinary activities involving combined and coordinated efforts of paleontologists, climatologists, physicists, space and atmospheric researchers, and others using ever-improving methods and tools. Government, university and industry programs are working across scientific and business fields to develop new technologies, products and infrastructures. Architects, developers, investors and building owners are emphasizing green design to conserve energy and reduce costs.

We are also learning that a "Spaceship Earth" analogy is entirely realistic. Our natural environment cannot be compartmentalized into isolated divisions defined according to cultural, economic or political priorities. Municipalities, states and nations are beginning to realize this fact, and are seeking, developing and implementing responsive policies and initiatives. It has become clear that focusing upon differences, competitive economic advantages and blame-placing will not serve long-term interests of anyone. And while cooperation often imposes difficulties that test patience, and goodwill, there is really no alternative. We are all in that tiny, fragile spacecraft together, depend upon the same life support systems, and share a vital mission.

1. "Greenhouse Effect"

This "greenhouse effect" is a term applied to warming that occurs when the atmosphere traps heat radiating from Earth towards space. The analogy is to certain gases in the atmosphere that act roughly similar to a glass greenhouse structure that allows sunlight to pass inside, but blocks Earth's heat from escaping into space. As a result, on average, the Earth becomes warmer, with some regions welcoming this change, and others not. The warmer conditions generally lead to more evaporation and precipitation overall, but individual regions vary, with some becoming wetter, and others dryer.

Although water vapor is by far the most abundant atmospheric greenhouse gas, CO_2 is cited by many as an important contributor also. Some crops respond favorably to increased CO_2, growing more vigorously and using water more efficiently. Shifting climate patterns that result may also change areas where certain crops grow best.

2. Water Vapor as a Greenhouse Gas (Feedback Observations)

A NASA-funded study found that some climate models may be overestimating the amount of water vapor entering the atmosphere as the Earth warms to trap greenhouse gases. The study was conducted by Ken Minchwaner, a physicist at the New Mexico Institute of Mining and Technology and Andrew Dressler at the University of Maryland and the NASA Goddard Space Flight Center in Greenbelt, Maryland. The study was published in the March 15, 2004 issue of the American Meteorological Society's *Journal of Climate*. It used data on water vapor in the upper atmosphere at a 6-9 mile altitude from NASA's Upper Atmosphere Research Satellite (UARS). This data measured specific humidity (actual amount of water vapor in the air) and relative humidity (amount of water vapor divided by the amount of water air is capable of handling at a given temperature). As the Earth warms, more water evaporates from oceans to increase water vapor (the primary greenhouse gas) leading to an increase in surface temperatures (a "positive feedback"). The existence and size of this feedback has been argued for many years. The study confirmed that water vapor does increase with surface warming, but not as much as many climate-forecasting models had previously assumed.

3. Stratospheric Ozone Depletion

Ozone (O_3) in the Earth's stratosphere between about 11-50 km altitude is essential to life because it absorbs a portion of incoming ultraviolet solar radiation with wavelengths between 210-310mm. Excessive UV radiation can damage immune systems, and cause genetic mutations and skin cancers in humans. It can also reduce photosynthetic capabilities in plants, including marine algae and phytoplankton which are at the base of the ocean food chain.

Ozone exists in a delicate balance with stratospheric oxygen. UV from the Sun breaks down molecules of O_2 into separate atoms which are free to combine with other O_2 to create O_3 (again absorbing UV). This balance can be disrupted by compounds such as chlorofluorocarbons (CFCs) released by human activities. The CFCs are broken down by UV, releasing chlorine (Cl) which so aggressively destroys ozone molecules that it is estimated that one Cl atom can affect 100,000 ozone molecules before being removed from the stratosphere by other chemical reactions. Although ozone is constantly being created, many believe that the rate of destruction currently exceeds the rate of production, possibly by as much as 1% every 3 years over the Northern Hemisphere.

4. Atmospheric vs. Surface Measurements/Assessments

In-situ radiosonde balloon sensors have been measuring lower atmosphere temperatures since 1958, and satellite Microwave Sounding Units (MSUs) have been sensing tropospheric temperatures (up to about 5 km above the surface) since 1979. While satellite measurements have the advantage of global coverage, they have been recorded over too short a period to provide a good basis for trend predictions. Radiosonde balloon measurements have occurred longer, but extensive instrumentation changes over their period of use may also account for possible errors in assessing temperature changes over time.

The MSUs on NOAA polar-orbiting platforms have measured the intensity of upwelling microwave radiation from atmospheric oxygen at different frequencies that sample "weighted ranges" associated with different altitudes. Records have been merged from nine different MSUs, each with special "peculiarities" calculated and removed to correct for errors (e.g., time drift of the spacecraft relative to local solar time).

The process of constructing temperature records from satellite data is complex. For some time, measurements analyzed at the University of Alabama Huntsville (UAH) have appeared to contradict surface measurement data at ground stations that have indicated a warming rate estimated by climate models. These earlier atmospheric measurements seemed to suggest that warming was not occurring or was occurring at rates below model forecasts. Subsequent corrections for possible errors in the analyses of satellite data have brought these two datasets more closely in line with each other and with the general conclusions of the UN Intergovernmental Panel on Climate Change (IPCC).

During the late 1990s the National Research Council observed that earlier record assessments contradicted predictions that temperatures should increase in the upper air as well as on the surface if increased GHG concentrations were the cause of the warming. Still, even at that time their report concluded that "*the warming trend in global-mean surface temperature observations during the past 20 years is undoubtedly real and is substantially greater than the average rate of warming during the 20th century. The disparity between surface and upper air trends in no way invalidates the conclusion that surface temperatures have been rising.*"

General theories suggest that the stratosphere should cool while the troposphere warms. Lower observed stratospheric records may be explained by other influences, such as effects of ozone depletion which may have cooled this region.

5. Greenland Ice Sheet

In the past decade, the Greenland Ice Sheet has experienced significant changes, marked by rapid thinning along its edges, faster-moving outlet glaciers, and a general increase in mass loss. Given its global significance, the state of the Greenland Ice Sheet has been included in the CCI Programme and is monitored through various Essential Climate Variables (ECVs).

Between 2002 and 2021, data from the Gravity Recovery and Climate Experiment (GRACE) satellites (2002-2017) and GRACE Follow-On (from 2018 onward) indicate that Greenland lost approximately 269 gigatons of ice per year, leading to a global sea level rise of 0.03 inches (0.8 millimeters) annually. In general, higher-elevation areas near Greenland's center experienced minimal to no ice loss, while lower-elevation and coastal regions experienced a significant reduction in ice mass, exceeding 16.4 feet (5 meters) over the 22-year period. The most significant ice loss was observed along the West Greenland coastline.

The satellite measurements provide records of Earth's oceans and land, as well as ice fields, and offer important data for assessments of whether ice sheets are growing or shrinking. While there is still no consensus regarding Greenland's overall ice sheet mass balance, there is evidence of melting and thinning in coastal marginal areas in recent years, and indications that outlet glaciers can surge, possibly in response to climate variations.

Many researchers attribute Greenland's interior elevation growth to increased snowfall linked to variability in regional atmospheric circulation known as the North Atlantic Oscillation (NAO) first discovered in the 1920s. This acts in a way similar to the El Niño phenomenon in the Pacific that contributes to climate fluctuations across the North Atlantic and Europe. There appears to be a direct ice sheet elevation change with strong positive and negative NAO phases during winter which largely controls temperature and precipitation patterns across Greenland.

In the past two decades, Greenland ice sheets have lost an average of 267 gigatons per year, or more than 5,400 gigatons in total It is estimated that global sea level rises by one millimeter for every 360 gigatons of land ice lost contributing to a combination of higher temperatures and altered, stormy wind fields that tend to break up ice cover. 1992 had a less-than-average melt attributed in large part to a Mt. Pinatubo volcano eruption that sent cooling aerosols into the atmosphere.

Researcher caution that the altimeter survey data does not necessarily reflect a long-term or future trend because the recording period is too short. Arctic temperatures appear to fluctuate on a timescale of 60-80 years, which means that Arctic cooling may occur again soon, rather than warming. Current models also suggest that temperature increases up to about 3ºC may lead to positive mass balance changes at high elevations due to snow accumulations, and negative at low elevations due to snow melts. However, after that threshold is reached (possibly within the next hundred years), the mass balance may begin to reverse.

6. Global Precipitation Trends

A Global Precipitation Climatology Project (GPCP) established by the World Climate Research Programme (WCRP) in 1986 has mapped patterns of precipitation around the world using data collected from satellite and in-situ measurements since 1979. The new Monthly V3.1 analysis, based on 22 years (1998-2019) of TRMM and GPM precipitation data, shows that global precipitation remains stable, with significant positive trends in the deep tropics, particularly along the Pacific ITCZ.

The ESA Climate Change Initiative Water Vapour project (ESA TCWV-COMBI) has produced a tropospheric water vapor data record for assessing interannual variability in global climate models and reanalysis data. Both models and observations predict that boundary layer water vapor will increase by up to 7% per degree Celsius of global warming.

7. Seasonal CO_2 Concentration at Mauna Loa Observatory, Hawaii

Measurements of atmospheric CO_2 concentration measured monthly over past decades at the Mauna Loa Observatory in Hawaii, a fairly remote and non-polluted area, show overall exponential increases with peaks and troughs that appear to correspond with plant growth influences. Peaks correlate with winter seasons (less photosynthesis and rising concentrations), and troughs occur during spring and summer plant cycles that remove CO_2. Anomalies in atmospheric circulation flow patterns associated with the El Niño/Southern Oscillation (ENSO) events also showed significant, though seasonally-variable roles, and it has been suggested that such atmospheric transport variations influencing CO_2 measurements warrant further study. This may account for indications of a decadal shift in the origin of air masses arriving at the observatory during the April-June period that favors lower CO_2 concentrations.

8. Atmospheric CO_2 Influences/Models (NASA Goddard/NASA Facts On Line)

Based upon satellite measurements, it appears that atmospheric sensitivity to greenhouse gases is less than most climate models predict. Some scientists also believe that even if small warming increases are occurring, for example 1ºF (0.5ºC) over 140 years, they may not be outside the range of natural climate variability. It is broadly recognized that more comprehensive and sophisticated climate models are needed, although scientists have little choice but use present ones until better ones are developed.

One of the most challenging needs in creating more reliable climate models is to understand detailed physical processes associated with clouds to determine the extent and nature of water vapor feedback into Earth's temperature. As stated by the IPCC: "*Feedback from the redistribution of water vapor remains a substantial uncertainty in climate models... Much of the current debate has been addressing feedback from the tropical upper troposphere, where the feedback appears likely to be positive. However, this is not yet convincingly established; much further evaluation of climate models with regard to observed processes is needed.*" (Climate Change 1995, IPCC Second Assessment).

9. Aerosols
(Real Climate: Guest Commentary, Ron Miller & Dorothy Koch, NASA GISS)

Estimates of effects of human-made aerosols vary considerably, often with conflicting model results and conclusions. Better understanding of this topic is of particular interest because their concentrations are enhanced by the same fossil combustion processes that create atmospheric greenhouse gases.

According to the IPCC, human-induced radiative forcing (RFR) from aerosols in 2019 contributed to cooling the climate system. However, the trend in aerosol RFR has shifted from negative to positive, indicating that the cooling effect of anthropogenic aerosols has diminished.
. The effects at the top of the atmosphere are even less clear, partly because they fall out or are washed out by rainfall, lasting only a few days or weeks, which does not allow enough time to be mixed uniformly throughout the globe.
Aerosol levels are anticipated to drop faster during the 21st century than greenhouse gases due to more aggressive emission regulations. Although they tend to oppose rather than cause global warming, other problems such as respiratory health problems and acid rain have been of great public concern.

10. Land Use Influences upon Weather/Climate

Agriculture has significantly transformed the face of the planet. Croplands have now replaced natural vegetation over about 18 million square kilometers of land surface with possible feedbacks that have altered the suitability of many regions for growing crops.

Although global changes in suitability are not yet established, regional effects may have occurred, particularly in Canada, Eastern Europe, the former Soviet Union, India and China.

The Great Plains of North America experienced significant overturning of land from natural grassland to irrigated land use during the 20th century. Research results show notably cooler temperatures during the growing season over irrigated areas, with more than 1ºC more cooling since 1945.

Interactions between various land use affects upon climate are complex and are currently poorly understood. This is particularly true on a global level.

11. Global Warming/Cooling Trends

Although there are not enough records available to confidently reconstruct global, or even hemispheric temperature records prior to about 600 years ago, records that do exist do not reveal any multi-century periods when global or hemispheric periods were quite as warm as or warmer than the 20th century. Temperatures during the Medieval Warm Period and the Little Ice Age that followed have been estimated based upon a variety of measurement types and models, including analyses of ice core oxygen isotopes and CO_2 in air bubbles dating from these times. These studies indicate that CO_2 levels (reflecting warm and cold conditions) between AD1530-1810 remained relatively constant (about 280 ppm). After this period the concentrations began to rise abruptly and smoothly.

12. CO_2 Contributions to Global Warming

Atmospheric CO_2 has increased about 25% since the early 1800s, and climatologists at NASA's Goddard Space Flight Center, Green Belt, MD estimate that the increase since 1958 has been about 10%. The level is currently increasing at a rate of about 0.4% per year. Humans add CO_2 to the atmosphere mainly by burning fossil fuels, although other activities such as deforestation and burning/decaying felled timber are also contributors.

Predictions of actual CO_2 impacts are complicated, and most take other greenhouse gases into account such as methane and nitrous oxide. Scientists recognize the importance of understanding these concerns, but cannot yet produce good estimates of future regional climate changes because forecast tools are not sufficiently sophisticated.

The fact that there has been little observed increase in atmospheric CO_2 during the past 50 years raises questions about whether we are really experiencing significant effects from this, or whether other more important factors may be influencing climate. Some believe that a 1°F (0.5°C) change over 140 years may not be out of the range of natural climate variability. Others point out that the weight of carbon emissions by humans differs from that of carbon dioxide, because CO_2 is 3.7 times heavier than carbon alone.

Scientists estimate that about half of the present human carbon emissions are absorbed by the environment, and the half absorbed account for only half of where that goes (the "mystery of the missing carbon") amounting to about 1.8G ton/year. Based upon projections, the 270G tons of human carbon emissions over the past 200 years would have increased CO_2 concentrations from 280 to 415ppm if all had stayed in the atmosphere. But since present CO_2 is about 360ppm, 41% of cumulative emissions have been absorbed by the environment if the carbon source was entirely human, differing from a present 50% estimate.

13. Abrupt Climate Change

Up until the 1990s the general view of climate change was one of the gradual responses to natural and human-induced influences. Observations of more rapid changes during the past few decades suggest that developments can occur much more abruptly as records of past events reveal. Such climate shifts may persist for years or longer, involving marked average temperature changes and altered storm, flood and drought patterns over an entire country or continent. In the context of past abrupt events these changes can occur within a single decade.

Severe droughts and other abrupt climate changes in the past have brought hardships upon societies which have had little time to adapt. The Natural Research Council emphasizes that while it is important not to be fatalistic about these threats, downplaying their significance can be costly. They conclude that "*Increased knowledge is the best way to improve the effectiveness of response; research into cases, patterns, and likelihood of abrupt climate change can help reduce vulnerabilities and increase our abilities to adapt.*"

ENERGY RESOURCES

General

Energy Information Administration. Electric Power Industry 2005: Year in Review. 2005.

Furchtgott-Roth, Diana. "Running on Empty." Missile Defense April/May 2001: 48-49.

Lovins, Amory and E. Kyle Datta. Winning the Oil End Game. Colorado: Rocky Mountain Institute, 2007.

McDonough, William and Michael Braungart. Cradle to Cradle: Remaking the Way We Make Things. New York: North Point Press, 2002.

Nielson, Ron. The Little Green Handbook. New York: Picador, 2006.

Transportation: Engines and Motors:

Compare hybrid cars. www.fueleconomy.gov. <http://www.fueleconomy.gov/feg/hybrid_sbs_cars.shtml>.

Diesel electric hybrid Ricardo i-MoGen Integrated Motor Generator. Ricardo. <http://www.ricardo.com/iMoGen>.

Ford unveils fleet of hybrid NY taxis. CNN. 10 Nov. 2005 <http://english.epochtimes.com/news/5-11-11/34504.html>.

FreedomCAR: Getting New Technology into the Marketplace. U.S. House of Representatives Charters: Committee on Science, Subcommittee on Energy. <http://www.house.gov/science/hearings/energy02/jun26/charter.htm>.

Georgano, N. Beaulieu Encyclopedia of the Automobile. London: HMSO, 2000.

Godfather of the Hybrid. California Institute of Technology: Engineering & Science. 11 Jan. 2006 <http://pr.caltech.edu/periodicals/EandS/articles/LXVII3/wouk.html>.

Helm, Burt. Hybrids: More Power, Less Fuel. Beth Belton. 20 Sep. 2005. Business Week. <http:www.businessweek.com/technology/content/sep2005/tc20050920_8040_tc_217.htm>.

Hybrid Cars Losing Efficiency, Adding Oomph. National Geographic. 8 Aug. 2005 <http://new.nationalgeographic.com/news/2005/08/0808_050808_hybrid_cars.html>.

Hybrid Diesel-Electric Buses. TriMet. http://www.trimet.org/environment/hybridbus.htm.

Hybrid Vehicle Powertrains. Folsom Technologies, Inc. <http://www.fti-cvt.com/hybrid_vehicles_.htm>.

ISE ThunderVolt® Electric and Hybrid-Electric Buses and Trucks Drive Systems also using Ultracapacitators. ISE Corporation. <http://www.isecorp.com/hybrid_information_center/ise-presenations.php>.

Lombardini ECOMOVE Hybrid Power Engine. Lombardini Group. <http://www.lombardinigroup.it/modules/prodotti/innovation/ecomove.php?IDS=2>

Lyons, Pete. "Inside Bruce Crowner's Six-Stroke Engine." Autoweek. 23 Feb. 2006. <http://www.autoweek.com/apps/pbcs.dll/article?AID=/20060227/FREE/302270007/1023/THISWEEKISSUE>.

Marshall, Robert W. An Amazing 75-MPG Hybrid Electric Car. July 1979. Mother Earth News. <http://www.motherearthnews.com/Alternative-Energy/1979-07-01/An-Amazing-75-MPG-Hybrid-Electic-Car.aspx>.

Mercedes Hybrid Sprinter. Daimler Chrysler. <http://www.daimlerchrysler.com/dccom/0,,0-5-7165-49-456546-1-0-0-0-0-0-243-7165-0-0-0-0-0-0-0,00.html>.

Neff, John. "BMW Turbosteamer gets hot and goes." Autoblog. 9 Dec 2005 Weblogs, Inc. Network. <http://www.autoblog.com/2005/12/09/bmw-turbosteamer-gets-hot-and-goes/>.

National Research Council. Review of the Research Program of the Partnership for a New Generation of Vehicles: Seventh Report. Washington, D.C.: National Academy Press, 2001.

Real Hybrid Mileage Database. Green Hybrid. <http://www.greenhybrid.com/compare/mileage/>.

Series, Parallel, and series-parallel systems animation. Hybrid Center. <http://www.hybridcenter.org/hybrid-center-how-hybrid-cars-work-under-the-hood-2.html>.

Siemens Integrated Powertrain Management. Siemens VDO. <http://www.siemensvdo.com/products_solutions/powertrain/hybrid/hybrid.htm>.

Toyota May License Hybrid System to More Carmakers. Fuel Cell Work. 14 Jan. 2005 <http://www.fuelcellsworks.com/Supppage1855.html>.

U.S. Federal Income Tax Deduction for Hybrids Placed in Service. www.fueleconomy.gov. <http://www.fueleconomy.gov/feg/tax_hybrid_old.shtml>.

Use of energy for transportation - U.S. Energy Information Administration (EIA)

Biofuels

Biofuels for Transportation: Global Potential and Implications for Sustainable Agriculture and Energy in the 21st Century. World Watch Institute. <http://www.worldwatch.org/taxonomoy/term/445>.

Cellulosic Ethanol vs. Biomass Gasification. 22 Oct. 2006. R-squared Energy Blog. 19 Nov. 2006 <http://i-r-squared.blogspot.com/2006/10/cellulosic-ethanol-vs-biomass.html>

Dar, William D. "Biofuel Crops: Power to the Poor." What ICRISAT Thinks. Sept 2006 <http://www.icrisat.org/Investors/wit_5/wit_5.htm>.

Ekless, Simon. Biomass Technical Brief. Intermediate Technology Development Group.

PEFI Alcohol Process Development & Demonstration. Ecalene™. <http://www.westernresearch.org/content/technology_areas/altnernative_fuels/alcohols.shtml>
Robertson, Andrew J. "Biodiesel Heating Oil: Sustainable Heating for the Future." The Journal of the Institute of Domestic Heating & Environmental Engineers. Aug 2006. Southampton, 2006.
United States Department of Energy and Department of Agriculture. Biomass as Feedstock for a Bioenergy and Bioproducts Industry. 2005 <http://feedstockreview.ornl.gov/pdf/billion_ton_vision.pdf>.
What is Biodiesel. Michael Happe. 23 Jan. 2006. BIOfuel.be. < http:// www.biofuel.be/whatisbiodiesel.html>
Wood Energy Organization. Holz-Energie-Zentrum Olsberg GmbH. <http://www.holzpellet.com/index_en.htm>.

Fuel

Coal - IEA
Coal and the environment - U.S. Energy Information Administration (EIA)Coal Factsheet_4.18.24.pdf
Factors affecting gasoline prices - U.S. Energy Information Administration (EIA)
Global Flex-Fuel Vehicle Market 2024 by Manufacturers, Regions, Type and Application, Forecast to 2030--GlobalInfoResearch
Natural gas imports and exports - U.S. Energy Information Administration (EIA)
Natural gas intrastate pipeline capacity additions outpaced interstate additions in 2023 - U.S. Energy Information Administration (EIA)
Oil imports and exports - U.S. Energy Information Administration (EIA)
Oil and natural gas production is growing in Caspian Sea region - U.S. Energy Information Administration (EIA)
PE-backed LNG project marred with setbacks and community opposition
Proved Reserves of Crude Oil and Natural Gas in the United States, Year-End 2022
The State of Coal Energy in the U.S. in 2024
U.S. Coal Reserves - U.S. Energy Information Administration (EIA)
Use of natural gas - U.S. Energy Information Administration (EIA)
U.S. energy facts explained - consumption and production - U.S. Energy Information Administration (EIA)
U.S. energy facts - data and statistics - U.S. Energy Information Administration (EIA)
U.S. Natural Gas Imports by Country
2024 World Oil Transit Chokepoints

Geothermal

All About Geothermal Energy. Geothermal Energy Association – Washington, DC. <http://www.geo-energy.org/aboutGE/potentialUse.asp#world>.
Aqua Genesis Testing at San Diego. Aqua Genesis Ltd. 2006 <http://aquagenesis.us/testing.html>.
Enhanced Geothermal Systems. 2006 <http://egs.egi.utah.edu/>.
Geothermal Research Program. Energy and Geoscience Institute at the University of Utah. 2006 < http://geothermal.egi.utah.edu>.
Hillis, R. Hand, M. Mildren, S., Morton, J., Reid, P. and S. Reynolds. "Hot Dry Rock Geothermal Exploration in Australia." Eastern Australian Basins Symposium. 2 (2004).
Introduction to Geothermal Power Generation. Bassfel. 2006 <http://www.bassfel.ch/Energy/assets/Geothermal_Power.pdf>.
McClaren, S. N., Sandiford, M., Hand, M., Newmann, N. L., Wyborn, L. A. and I. Bastrakova. "The Hot Southern Continent: Heat flow and Heat Production in Australian Proterozoic Terranes." Geological Society of America Special Paper. 22 (2003): 157-167.
Neumann, N., Sandiford, M., and J. Foden. "Regional Geochemistry and Continental Heat Flow." Earth and Planetary Science Letter 183 (2000):107-120.
Reynolds, S. D., Mildren, S. D., Hillis R. R., and J. Meyer. "Application of the In Situ Stress Field to Hot Dry Rock Geothermal Energy in the Cooper Basin." Eastern Australian Basins Symposium. 2 (2004).
Somerville, M., Wyborn, D., Chopra, P., Rahman S., Estrella, D., and T Van Der Meulen. "Hot Dry Rocks Feasibility Study." Energy Research and Development Corporation (ERDC) Report. 243 (2004):133.
The Future of Geothermal Energy: Impact of Enhanced Geothermal Systems (EGS) on the United States in the 21st Century. MIT Report. 2006 <http://geothermal.inel.gov/publications/future_of_geothermal_energy.pdf>.
Wyborn, D., DeGraaf, L., Davidson S., and S. Hann. "Development of Australia's first hot fractured rock (HFR) Underground Heat Exchanger, Cooper Basin, South Australia." Eastern Australian Basins Symposium. 2 (2004).

Hydrogen

"A Leading Contributor to Understanding Hydrogen." Hydrogen Pathways Program. Institute of Transportation Studies-UC Davis. 2006 <http://hydrogen.its.ucdavis.edu/>."A Power Grid for the Hydrogen Economy." Scientific American Magazine. July 2006. 2006 <http://sciam.com/article.cfm?chanID=sa006&collD=1&articleID=00003872-159C-1498-959C83414B7F0000>.

Brian, Marshall. "How the Hydrogen Economy Works." American Hydrogen Association. Marie Davis and Roy McAlister. 2006 <http://www.goh2.org/>.

Buswell, Clause, Cohen, Louie, Watkins. Hydrocarbon Fueled Solid Polymer Fuel Cell Electric Power Generation System. Ballard US Patent 5,360,679. 1994.

Crabtree, George W., Dresselhaus, Mildred S., Buchanan, Michelle V. "The Hydrogen Economy." Physicstoday.org. 58 Feb. 2007: v57i12p39. 2006 <http://www.physicstoday.org/vol-57/iss-12/p39.html>.

Hydrogen Economy-Power for Cars, Homes, Businesses and Travel. Hydrogen Power. Hydrogen Economy Resource, 2005. 2006 <http://www.h2-economy.com/>.

Hydrogen Hopes. PBS. 2006 <http://www.pbs.org/saf/1506/>.

Hydrogen Use. Hydrogenuse.info. 2006 <http://www.hydrogenuse.info/copyright>.

Kordesch, K., Simander, G. Fuel Cells and their Applications. New York: VCH Press, 1996.

Lovins, Amory B. Twenty Hydrogen Myths. Snowmass: Rocky Mountain Institute, 2005.

Kreith, F. "Fallacies of a Hydrogen Economy: A Critical Analysis of Hydrogen Production and Utilization." Journal of Energy Resources Technology. 126 (2004):249-257.

McAlister. The Solar Hydrogen Civilization. American Hydrogen Association, 2003.

Nakicenovic, et al. Global Energy Perspectives. Cambridge University Press, 1998.

Pinkerton, Frederick E., Wicke, Brian G. "Botting the Hydrogen Genie." The Industrial Physicist. vol10iss1pg20. 2006 <http://www.aip.org/tip/INPHFA/vol-10/iss-1/p20.html>. Rifkin, Jeremy. The Hydrogen Economy. Tarcher, 2002.

Romm, Joseph J. The Hype about Hydrogen, Fact and Fiction in the Race to Save the Climate. Island Press, 2004.

Wang, M. "Fuel Choices for Fuel Cell Vehicles: Well-to-Wheels Energy and Emissions Impact." Journal of Power Sources. 112 (2002): 307-321.

Wise, Jeff. "The Truth About Hydrogen." Popular Mechanics. Nov. 2006 <http://www.popularmechanics.com/technology/industry/4199381.html>.

Zyga, Lisa. "Why a Hydrogen Economy Doesn't Make Sense." Physorg.Com. 11 Dec. 2006 <http://www.physorg.com/news85074285.html>.

Hydropower

Harvey, Adam. Micro-hydro Power. Intermediate Technology Development Group. 2004. 1 Jan. 2005 <http://wwww.itdg.org/docs/technical_information_service/micro_hydro_power.pdf >.

Hydropower | Department of Energy

U.S. Department of Energy. Microhydropower Systems. 2005 <http://www.eere.energy.gov/consumer/your_home/electricity/index.cfm/mytopc=11050>.

Renewable

Clean Air Act Standards and Guidelines for Electric Utilities | US EPA

Nuclear vs Renewables - which is cheaper? - The Centre for Independent Studies

Renewable energy statistics 2024

Renewable energy explained - incentives - U.S. Energy Information Administration (EIA)

Renewables generated 24.2% of US electricity in 2024 – EIA data | Electrek

Solar Power

An_Overview_of_Heliostats_and_Concentrating_Solar_Power_Tower_Plants.pdf

Concentrating Solar Power | Electricity | 2024 | ATB | NREL

"CSP Technologies Overview." CSP Overview. Concentrating Solar Power and Sun Lab. 2006 <http://www.energylan.sandia.gov/sunlab/overview.htm#trough>.

"Daylighting." Building Technologies Program. U.S. Department of Energy. 2006 <http://www.eere.energy.gov/buildings/info/design/integratedbuilding/passivedaylighting.html>.

"Direct Lighting." Lighting Design Knowledgebase. 2006 <http://www.schorsch.com/kbase/glossary/d_lighting.html>.

Earth Radiation Budget. NASA Langley Research Center. 17 Oct. 2006. <http://marine.rutgers.edu/mrs/education/class/yuri/erb.html>.

Gartner, John. Silicon Shortage Stalls Solar. 28 Mar. 2005. Wired News. 2006 < http://www.wired.com/news/planet/0,2782,67013,00.html>.

"Hydrogen Production and Delivery." Hydrogen and Fuel Cells Research. 13 Feb. 2007. National Renewable Energy Laboratory. 2007 <http://www.nrel.gov/hydrogen/proj_production_delivery.html#solar>.

How Much Do Solar Panels Cost? (Breakdown Guide)- Forbes Home Report Grey

Liepert, B.G. Observed Reductions in Surface Solar Radiation in the United States and Worldwide from 1961 to 1990.

Linear-focusing Concentrator Facilities: DCS, DISS, EUROTROUGH and LS3. Ministerio De Educacion Y Ciencia. 2006 <http://www.psa.es/webeng/instalaciones/parabolicos.html>.

New World Record Achieved in Solar Cell Technology. 5 Dec. 2006. U.S. Department of Energy. 2007 < http://www.energy.gov/news/4503.htm>.

Pichel, Jesse W. and Ming Yang. 2005 Solar Year-end Review & 2006 Solar Industry Forecast. 11 Jan. 2006. Renewable Energy Access. 2006 < http://www.renewableenergyaccess.com/rea/news/story?id=41508>.

Quaschning, Volker. "Technology Fundamentals: Solar Thermal Power Plants." Renewable Energy World. (2003):109-113. 2006 <http://www.volker-quaschning.de/articles/fundamentals2/index_e.html>.

Solar Buildings. National Renewable Energy Laboratory. 2006 <http://www.nrel.gov/docs/fy06osti/29913.pdf>.

"Solar FAQS-Photovoltaics-The Basics." Solar Energy Technologies Program. 8 Feb. 2007. U.S. Department of Energy. 2006 <http://www.eere.energy.gov/solar/cfm/faqs/third_level.cfm/name=Photovoltaics/cat=The%20Basics#Q43>.

Solar Generation System Ready for Production. Energy Services. 2006 < http://www.wapa.gov/es/pubs/esb/1998/98Aug/at_solargen.htm>.

"Solar Hot Water." Learning about Renewable Energy. National Renewable Energy Laboratory. 2006 <http://www.nrel.gov/learning/re_solar_hot_water.html>.

"Solar Maps." Dynamic Maps, GIS Data, & Analysis Tools. National Renewable Energy Laboratory. 2006 <http://www.nrel.gov/gis/solar.html>.

"Solar Timeline." Solar Energy Technologies Program. 5 Jan. 2006. U.S. Department of Energy. 2007 <http://www1.eere.energy.gov/solar/solar_timeline.html>.

Solar Tres Project. 15 July 2005. SolarPaces. 2006 <http://solarpaces.org/SOLARTRES.HTM>.

Special Programs. Solar Energy International – Renewable Energy for a Sustainable Future. 2006 <http://www.solarenergy.org/>.

The Solar Radiation Data Manual for Flat-Plane and Concentrating Collectors. U.S. Solar Radiation. 2006 <http://rredc.nrel.gov/solar/old_data/nsrdb/redbook/atlas/Table.html>.

Zinc Powder Will Drive Your Hydrogen Car. 9 Sep. 2005. Israeli Technology News. 2007 <http://www.isracast.com/Articles/Article.aspx?ID=51>.

Wind Energy

Archer, Cristina L., Mark Z. Jacobson. Evaluation of Global Wind Power. 21 Apr. 2006. http://www.stanford.edu/group/efmh/winds/global_winds.html>.

Burton, T., Sharpe, D., Jenkins, N., and Ervin Bossanyi. Wind Energy Handbook. England: John Wiley & Sons 1st ed. 2001.

Carey, John. "Alternative Power- A Change is in the Wind." BusinessWeekOnline. 4 July 2005. 2006 http://www.businessweek.com/magazine/content/05_27/b3941036_mz011.html

Cost of Wind Energy Review: 2024 Edition

Eggleston, Eric. What are Vertical-Axis Wind Turbines (VAWTs)? American Wind Energy Association. 2006 <http://www.awea.org/faq/vawt.html>.

"Global Wind Energy Markets Continue to Boom – 2006 Another Record Year." Global Wind Energy Council. 2006 <http://www.gwec.net/uploads/media/07-02_PR_Global_Statistics_2006.pdf>.

Global Wind Map Shows Best Wind Farm Locations. 17 May 2005. Environment News Service. 21 Apr. 2006 <http://ens-newswire.com/ens/may2005/2005-05-17-09.asp#anchor6>.

Offshore Wind Research and Development | Department of Energy

Offshore Wind Market Report | Department of Energy

How Wind Turbines Work. U.S. Department of Energy. 2006 <http://eereweb.ee.doe.gov/windandhydro/wind_history.html>.

Renewable Energy- Wind Power's Contribution to Electric Power Generation and Impact on Farms and Rural Communities. Sep. 2004. United States Government Accountability Office. 21 Apr. 2006 <http://www.gao.gov/new.items/d04756.pdf>.

Sizes of Utility-Scale Wind Turbines

U.S. Wind Turbine Database

U.S. Department of Energy Projects Strong Growth in U.S. Wind Power Sector | Department of Energy

What is Wind Energy Conversion System (WECS)? - Energy Theory

WWEA_WPR2023

What are the Basic Wind Turbine Configurations? American Wind. 2006 <http://eereweb.ee.doe.gov/windandhydro/wind_how.html>.

Wind Market Reports: 2024 Edition | Department of Energy

"Wind Energy Frequently Asked Questions." Delivering the UK's Wind, Wave and Tidal Energy. BWEA. 2006 <http://www.bwea.com/ref/faq.html>.

Is Nuclear the Answer? 2006. Sustainable Development Commission. 22 Dec. 2006 <http://www.sd-commission.org.uk/pages/060306.html>.

Jopf, James. World Uranium Reserves. 2004. American Energy Independence. 10 Nov. 2006 <http://www.americanenergyindepeendence.com/uranium.html>.

Levins, Amory; Data, Kyle; Koomey, Jonathan; and Nathan Glasglow. Winning the Endgame. Rocky Mountain Institute, 2004.

McCarthy, John. "Fact from Cohen and Others." Progress and its Sustainability. 2006 <http://www.formal.standford.edu/jmc/progress/cohen.html>.

Mi, Xu. Chinese Fast Reactor Technology Development. 1999. China Institute of Atomic Energy. 10 Nov. 2006 <http://tauon.nuc.berkeley.edu/asia/2000/XuMi.pdf>.

"Nuclear Power: Both Sides." Wilson Quarterly. 2006 <http://ww.wilsoncenter.org/index.cfm?fuseaction=wq.essay&essay_id=203041>.

"Nuclear Reactors Found to be Leaking Radioactive Water." TruthOut. 17 Mar. 2006 <http://www.truthout.org/cgi-bin/artman/exec/view.cgi/58/18461>.

Radioactive Waste Management. 2002. Uranium & Nuclear Power Information Centre. 9 Nov 2006 <http://www.uic.com.au/waste.htm>.

Strengths of Nuclear Plants Against Terrorist Attacks. 2004. World Nuclear Association. 10 Nov. 2006 <http://www.world-nuclear.org/news/resistance.htm>.

The Future of Nuclear Power. 2003. Massachusetts Institute of Technology. 10 Nov. 2006 <http://web.mit.edu/nuclearpower/>.

"The Nuclear Fuel Cycle." Information and Issue Briefs. 2006. World Nuclear Association. 2006 <http://www.world-nuclear.org/info/inf03.htm>.

"The Political Economy of Nuclear Energy in the United States." Social Policy. 2004. The Brookings Institution. 2006 <http://www.brookings.edu/comm/policybriefs/pb138.htm>.

Thomas, Steve. The Economics of Nuclear Power: Analysis of Recent Studies. Sep. 2005. University of Greenwich. 2006 http://www.psiru.org/reports/2005-09-E-Nuclear.pdf>.

Till, Charles. Nuclear Reaction: Why Do Americans Fear Nuclear Power? Public Broadcasting Service. 2006 <http://www.pbs.org/wgbh/pages/frontline/shows/reaction/interviews/till.html>.

"Waste Management in the Nuclear Fuel Cycle." Information and Issue Briefs. 2006. World Nuclear Association. 2006 <http://www.world-nuclear.org/info/inf04.htm>.

"Wind Farms 'Must Take Root in UK.'" BBC News. 19 May 2005. 2006 <http://news.bbc.co.uk/2/hi/science/nature/4560139.stm>.

"Wind Power Capacity in U.S. Increased 27% in 2006 and is Expected to Grow an Additional 26& in 2007." Wind Energy Works for America. American Wind Energy Association. 2007 <http://www.awea.org/newsroom/releases/Wind_Power_Capacity_012307.html>.

Wyatt, Alan. Electric Power: Challenges and Choices. Toronto: Book Press Ltd., 1986.

Nuclear

50 Years of Nuclear Energy. International Atomic Energy Agency. 9 Nov. 2006 <http://www.iaea.org/About/Policy/GC/GC48/Documents/gc48inf-4_ftn3.pdf>.

"Advanced Nuclear Power Reactors." Information and Issue Briefs. 2006. World Nuclear Association. 2006 <http://www.world-nuclear.org/info/inf08.htm>.

Baker, Peter and Dafna Linzer. "Nuclear Energy Plan Would Use Spent Fuel." Washington Post. 26 Jan. 2007 <http://www.washingtonpost.com/wp-dyn/content/article/2006/01/25/AR2006012502229.html>.

Beardsley, Eleanor. "France Presses Ahead with Nuclear Power." NPR. 1 May 2006. 2006 <http://www.npr.org/templates/story/story.php?storyId=5369610>.

China Will Generate More Nuclear Power Than Both France and the United States by 2030 – The Diplomat

<p>China's demonstration HTR-PM enters commercial operation</p> - World Nuclear News

https://www.worldnuclearreport.org/A-Closer-Look-at-Two-Operational-Small-Modular-Reactor-Designs

https://world-nuclear.org/information-library/current-and-future-generation/fast-neutron-reactors

Cohen, Bernard L. The Nuclear Energy Option. 2006 <http://www.phyast.pitt.edu/~blc/book/BOOK.html>.

Domestic Uranium Production Report - Annual - U.S. Energy Information Administration (EIA)

"Environmental Effects of Nuclear Power." The Virtual Nuclear Tourist. 2005. 2006 <http://www.nucleartourist.com/basics/environ1.html>.

Fact Sheet on the Accident at Three Mile Island. U.S. Nuclear Regulatory Commission. 10 Nov. 2006 <http://www.nrc.gov/reading-rm/doc-collections/fact-sheets/3mile-isle.html>.

Fritsche, Uwe R. Comparing Greenhouse-Gas Emissions and Abatement Costs of Nuclear and Alternative Energy Options from a Life-Cycle Perspective. 1997. Oko-Institut. 2006 <http://www.oeko.de/service/gemis/files/info/nuke_co2_en.pdf>.

Habib, Babur, et al. Stemming the Spread of Enrichment Technology. 2006. Woodrow Wilson School of Public and International Affairs. 10 Nov. 2006 <http://www.princeton.edu/~rskemp/Stemming_the_Spreach_of_Enrichment_Plants.pdf>.

IAEA Outlook for Nuclear Power Increases for Fourth Straight Year, Adding to Global Momentum for Nuclear Expansion | IAEA

"Information Digest, 2006-2007." Nuclear Regulatory Commission. 2007 <http://www.nrc.gov/reading-rm/doc-collections/nuregs/staff/sr1350/>.

Is Nuclear the Answer? 2006. Sustainable Development Commission. 22 Dec. 2006 <http://www.sd-commission.org.uk/pages/060306.html>.

Jopf, James. World Uranium Reserves. 2004. American Energy Independence. 10 Nov. 2006 <http://www.americanenergyindepeendence.com/uranium.html>.

Levins, Amory; Data, Kyle; Koomey, Jonathan; and Nathan Glasglow. Winning the Endgame. Rocky Mountain Institute, 2004.

McCarthy, John. "Fact from Cohen and Others." Progress and its Sustainability. 2006 <http://www.formal.standford.edu/jmc/progress/cohen.html>.

Mi, Xu. Chinese Fast Reactor Technology Development. 1999. China Institute of Atomic Energy. 10 Nov. 2006 <http://tauon.nuc.berkeley.edu/asia/2000/XuMi.pdf>.

"Nuclear Power: Both Sides." Wilson Quarterly. 2006 <http://ww.wilsoncenter.org/index.cfm?fuseaction=wq.essay&essay_id=203041>.

Nuclear power plants generated 68% of France's electricity in 2021 - U.S. Energy Information Administration (EIA)

Nuclear explained - data and statistics - U.S. Energy Information Administration (EIA)

"Nuclear Reactors Found to be Leaking Radioactive Water." TruthOut. 17 Mar. 2006 <http://www.truthout.org/cgi-bin/artman/exec/view.cgi/58/18461>.

Power Reactors | NRC.gov

Radioactive Waste Management. 2002. Uranium & Nuclear Power Information Centre. 9 Nov 2006 <http://www.uic.com.au/waste.htm>.

Strengths of Nuclear Plants Against Terrorist Attacks. 2004. World Nuclear Association. 10 Nov. 2006 <http://www.world-nuclear.org/news/resistance.htm>.

The Future of Nuclear Power. 2003. Massachusetts Institute of Technology. 10 Nov. 2006 <http://web.mit.edu/nuclearpower/>.

"The Nuclear Fuel Cycle." Information and Issue Briefs. 2006. World Nuclear Association. 2006 <http://www.world-nuclear.org/info/inf03.htm>.

"The Political Economy of Nuclear Energy in the United States." Social Policy. 2004. The Brookings Institution. 2006 <http://www.brookings.edu/comm/policybriefs/pb138.htm>.

Thomas, Steve. The Economics of Nuclear Power: Analysis of Recent Studies. Sep. 2005. University of Greenwich. 2006 http://www.psiru.org/reports/2005-09-E-Nuclear.pdf>.

Till, Charles. Nuclear Reaction: Why Do Americans Fear Nuclear Power? Public Broadcasting Service. 2006 <http://www.pbs.org/wgbh/pages/frontline/shows/reaction/interviews/till.html>.

"Waste Management in the Nuclear Fuel Cycle." Information and Issue Briefs. 2006. World Nuclear Association. 2006 <http://www.world-nuclear.org/info/inf04.htm>.

World Nuclear Industry Status Report 2024

ENVIRONMENT AND CLIMATE

General

Benton, Michael J. and Richard J. Twitchett. "How to Kill (Almost) All Life: The End-Permian Extinction Event." Trends in Ecology and Evolution.18 (2003).

Bond, Nick; Overland, Jim and Nancy Soreide. "Why and How Do Scientists Study Climate Change in the Arctic?" Arctic Theme Page. National Oceanic and Atmospheric Administration. 2006 <http://www.artice.noaa.gov/essay_bond.html>.

"Earth Radiation Budget." Environmental Science. 23 Apr. 1998. The University of New Jersey. 2007 <http://marine.rutgers.edu/mrs/education/class/yuri/erb.html>.

Fleagle, RG and JA Businger. An Introduction to Atmospheric Physics. 2 (1980).

Fraser, Alistair B. Bad Greenhouse. 2006 <http://www.ems.psu.edu/~fraser/Bad/BadGreenhouse.html>.

Giacomelli, Gene A. and William J. Roberts. Greenhouse Covering Systems. Rutgers University. 30 Mar. 2005 <http://ag.arizona.edu/ceac/research/archive/HortGlazing.pdf>.

Henderson-Sellers, Ann and Kendal McGuffie. A Climate Modelling Primer. John Wiley and Sons Ltd, 2005.

Idso, SB. Carbon Dioxide: Friend or Foe? Arizona: IBR Press, 1982.

Kiehl, J.T. and K. Trenberth. "Earth's Annual Mean Global Energy Budget." Bulletin of the American Meteorological Society. 2 (1997):197-208.
Piexoto, JP and AH Oort. Physics of Climate. American Institute of Physics. 1992.
Ruddiman, WF. "The Anthropogenic Greenhouse Era Began Thousands of Years Ago." Climatic Change. 3 (2003): 261-293.
Wood, R.W. "Note on the Theory of the Greenhouse." Philosophical Magazine 17 (1909):319-320.

Climate Change

"2005 Temperatures." Real Climate. 15 Dec. 2005. 17 Jan. 2007 <http://realclimate.org/index.php?p=231>.
Aber, James S. Birth of the Glacial Theory. Emporia State University. 4 Aug 2006 <http://academia.emporia.edu/aberjame/histgeol/agassiz/glacial.htm>.
"Abrupt Climate Change, Inevitable Surprises." National Research Council (NRC), 2002. Washington, DC: National Academy Press, 2002.
"Activities-Assessment Reports." IPCC. 20 Dec. 2006 <http://www.ipcc.ch/activity/ar.htm#ar4>.
"Activities-Workshop & Expert Meetings." IPCC. 20 Dec. 2006 <http://www.ipcc.ch/activity/workshops.htm>.
"A Guide to Facts and Fictions About Climate Change." The Royal Society. 19 Dec. 2006 <http://www.royalsoc.ac.uk/downloaddoc.asp?id=1630>.
"Antarctic Temperatures Disagree With Climate Model Predictions." Science Daily. 17 Feb. 2007. Ohio State University. 2007 <http://www.sciencedaily.com/releases/2007/02/070215144314.htm>.
"Article 1." United Nations Framework Convention on Climate Change. 2007 <http://unfccc.int/essential_background/convention/background/items/2536.php>.
"Atmospheric Chemistry and Greenhouse Gases." Climate Change 2001. Intergovernmental Panel on Climate Change. 2007 <http://www.grida.no/climate/ipcc_tar/wg1/134.htm#tab42>.
"False Claims by McIntyre and McKitrick regarding the Mann et al. (1998) reconstruction." RealClimate-Climate Science from Climate Scientists. 4 Dec. 2006. 2006 <http://www.realclimate.org/index.php?p=8>.
Baliunas, Sallie and Willie Soon. A Scientific Discussion of Climate Change. 1 Nov. 1997. Harvard-Smithsonian Center for Astrophysics.
Calder, Nigel. The Carbon Dioxide Thermometer and the Cause of Global Warming. Science and Technology Policy Research. England: University of Sussex, 1998.
"Detection of Climate Change and Attribution of Causes." Climate Change 2001. Intergovernmental Panel on Climate Change. 2007 <http://www.grida.no/climate/ipcc_tar/wg1/462.htm>.
Earth System Research Laboratory. U.S. Department of Commerce. 2007 <http://www.esrl.noaa.gov/gmd/>.
"Eight Glacial Cycles from an Antarctice Ice Core." Nature. 429-6992. 10 June 2004 <http://www.nature.com/cgi-taf/DynaPage.taf?file=/nature/journal/v429/n6992/abs/nature02599_fs.html>.
"False Claims by McIntyre and McKitrick regarding the Mann et al. (1998) reconstruction." RealClimate-Climate Science from Climate Scientists. 4 Dec. 2006. 2006 <http://www.realclimate.org/index.php?p=8>.
First International Conference on Global Warming and the Next Ice Age. Canadian Meteorological and Oceanographic Society and the American Meteorological Society. 21 Aug. 2001.
"Global Warming." National Oceanic and Atmospheric Administration. 3 Feb. 2006. US Department of Commerce. 2007 <http://www.ucar.edu/research/climate/>.
Global Warming Site. 20 Feb. 2007. U.S. Environmental Protection Agency. 2007 <http://epa.gov/climatechange/index.html>.
Goddard Institute for Space Studies. Robert B. Schmunk. 29 Dec. 2006. NASA. 17 Jan. 2007 <http://data.giss.nasa.gov/gistemp/2005/>.
Gray, Richard. "Cosmic Rays Blamed For Global Warming." Telegraph. 11 Feb. 2007. 2007 <http://www.telegraph.co.uk/news/main.jhtml?xml=/news/2007/02/11/warm11.xml>.
Gregory, J. M. et al. "An Observationally Based Estimate of the Climate Sensitivity." Journal of Climate. 15 (2002): 3117-21. 18 Jan. 2007 <http://www.gfdl.gov/reference/bibliography/2002/jmgregory0201.pdf>.
Hansen, James E. and Patrick J. Michaels. A Public Debate on the Science of Global Warming. 20 Nov. 1998.
Hansen, James. "The Global Warming Debate." Goddard Institute for Space Studies. Jan. 1999. NASA. 2006 <http://www.giss.nasa.gov/edu/gwdebate/>.
Hinrichs, K.U., Hmelo L. and Sylvia S. "Molecular Fossil Records of Elevated Methane Levels in Late Pleistocene Coastal Waters." Science. 299-5610 (2003); 1214-1217.
Houghton, David D. Introduction to Climate Change: Lecture Notes for Meteorologists. World Meteorological Organization. 2007 <http://www.wmo.ch/web/etr/pdf_web/926E.pdf>.

Houghton, J.T. et al., IPCC, 2001: Climate Change 2001: The Scientific Basis. New York: Cambridge University Press, 2001.
Hoyt, D.V. and K.H. Schatten. "A Discussion of Plausible Solar Irradiance Variations 1700-1992." J. Geophys. Res. 98 (1993): 18895-18906.
Imbrie, John and Katherine Palmer Imbrie. Ice Ages: Solving the Mystery. Cambridge: Harvard University Press, 1979, 1986.
Intergovernmental Panel on Climate Change. TERI. UNEP. 2 Mar. 2007. 2007 <http://www.ipcc.ch/>.
"IPCC WGI Fourth Assessment Report." Climate Change 2007: They Physical Science Basis. 5 Feb. 2007. Intergovernmental Panel on Climate Change. 2007 <http://www.ipcc.ch/SPM2feb07.pdf>.
Jancovici, Jean-Marc. "What is IPCC?" Climate Change. Sept 2003. 2006 <http://www.manicore.com/anglais/documentation_a/greenhouse/IPCC.html>.
Jaworowski, Zbigniew. "Climate Change: Incorrect Information on Pre-Industrial CO2." Environmental Science and Pollution Research. 1997.
Jaworowski, Zbigniew. "The Ice Age is Coming." 21st Century Science and Technology Magazine. 2003-2004.
Kahn, Jeffery. "Astronomical Theory Offers New Explanation for Ice Age." Berkeley Lab, Research News. 1997.
Lean, J.L., Wang, Y.M., and N.R. Sheeley. "The Effect of Increasing Solar Activity on the Sun's Total and Open Magnetic Flux During Multiple Cycles: Implications for Solar Forcing of Climate." Geophys. Res. Lett. 29 (2002): 2224
Lemley, Brad. "The Next Ice Age." Discover. 23-9 (2002).
Lerner, K. Lee and Brenda Wilmoth Lerner (eds). Environmental Issues: Essential Primary Sources. Thomas Gale, 2006.
Lindzen, Richard S. Statement Concerning Global Warming. 10 June 1997. Massachusetts Institute of Technology.
"Mandate and Membership of IPCC." IPCC. 20 Dec. 2006 <http://www.ipcc.ch/about/about.htm>.
"Maps and Graphics." United Nations Environment Programme/GRID-Arendal. 2006. 2007 <http://maps.grida.no/go/collections>.
Marsh, Nigel and Henrik Svensmark. Cosmic Rays, Clouds and Climate. 10 May 2000. Danish Space Research Institute. 11 Feb. 2007 <http://www.dsri.dk/~hsv/SSR_Paper.pdf>.
McIntyre, Steve. "Statistical Criticism of "Hockey Stick" Climate History Reconstructions." Climate Audit. 2006 <http://climateaudit.org/>
Meehl, Gerald A. et al. "How Much More Global Warming and Sea Level Rise?" Science. 307 (2005): 1769-1772. AAAS. 18 Jan. 2007 <http://www.sciencemag.org/cgi/content/full/307/5716/1769>.
Muscheler, Raimund. "How Unusual is Today's Solar Activity?" Nature. 436 (2005): E3-E4. 11 Feb. 2007 <http://www.cgd.ucar.edu/ccr/raimund/publications/Muscheler_et_al_Nature2005.pdf>.
Muscheler, Raimund; Joos, Fortunat; Muller, Simon A. and Ian Snowball. "Climate: How Unusual is Today's Solar Activity?" Nature. 436 (2005): E3-E4.
Oerlemans, J. "Extracting a Climate Signal from 169 Glacier Records." Science. 308 (2005): 675-677.
"Official Documents." IPCC. Dec 2006 <http://www.ipcc.ch/meet/meet.htm>.
Oreskes, Naomi. "Beyond the Ivory Tower: The Scientific Consensus on Climate Change." The Scientific Consensus on Climate Change." Science. 306 (2004): 1686. 2006 <http://www.sciencemag.org/cgi/content/full/306/5702/1686>.
Past Climate Change, the Last 2,000 Years. 2006. US Environmental Agency.
"Physical Climate Processes and Feedbacks." Climate Change 2001. Intergovernmental Panel on Climate Change. 2007 <http://www.grida.no/climate/ipcc_tar/wg1/295.htm>.
Pielke, Roger, Jr., and Daniel Sarewtiz. "Wanted: Scientific Leadership on Climate." Issues in Science and Technology. 19-2 (2002): 27-30.
"Principles Governing IPCC Work." IPCC. 19 Dec. 2006 <http://www.ipcc.ch/about/princ/pdf>.
"Radiated Forcing of Climate Change." National Research Council (NRC), 2005. Washington, DC: National Academy Press, 2005.
"Report on the Twentieth Session of the IPCC." IPCC. 19 Feb. 2006. 20 Dec. 2006 <http://www.ipcc.ch/meet/session20/finalreport20.pdf>"Researcher: Cosmic Rays Cause Warming." Science Daily. 11 Feb. 2007. 2007 <http://www.sciencedaily.com/upi/index.php?feed=Science&article=UPI-1-20070211-16012900-bc-britain-globalwarming.xml>.
Rowland, F. Sherwood. "Climate Change (Global Warming and Ozone Depletion." The Vega Science Trust. 2001. Florida State University. 2007 <http://vega.org.uk/video/programme/119>.
Ruddiman, William F. Earth's Climate Past and Future. New York: Princeton University Press, 2001.
"Scientists Baffled." RealClimate-Climate Science from Climate Scientists. 11 Jan. 2006. 2006 <http://www.realclimate.org/index.php?p=236>.
Shaviv, Nir J. "Carbon Dioxide or Solar Forcing?" Science Bits. 2006 <http://www.sciencebits.com/CO2orSolar>.

Smith, Thomas M. and Richard W. Reynolds. A Global Merged Land Air and Sea Surface Temperature Reconstruction Based on Historical Observations (1880-1997). National Climatic Data Center. 2006 <http://www.ncdc.noaa.gov/gcag/merge.pdf>.

Solanki, S.K. et al., "Unusual Activity of the Sun During Recent Decades Compared to the Previous 11,000 Years." Nature. 431 (2004): 1084-1087.

Spencer, Roy. "Let Them Confess Their Faith." TCS Daily. 6 Feb. 2004. 2006 <http://www.techcentralstation.com/020604C.html>.

"Summary for Policymakers." Climate Change 2001. Intergovernmental Panel on Climate Change. 2007 <http://www.grida.no/climate/ipcc_tar/wg1/006.htm>.

"Sun's Direct Role in Global Warming May Be Underestimated, Duke Physicists Report." Duke University. 30 Sep. 2005. 11 Feb. 2007 <http://www.dukenews.duke.edu/2005/09/sunwarm.html>.

"Surface Temperature Reconstructions for the Last 2,000 Years." National Research Council (NRC), 2006. Washington, DC: National Academy Press, 2006.

Svensmark, H. and E Friis-Christiansen. "Variation in Cosmic Ray Flux and Global Cloud Coverage: A Missing Link in Solar-Climate Relationships." Journal of Atmospheric and Solar-Terrestrial Physics. 59 (1997): 1225-1232.

"The Carbon Cycle and Atmospheric Carbon Dioxide." Climate Change 2001. Intergovernmental Panel on Climate Change. 2007 <http://www.grida.no/climate/ipcc_tar/wg1/123.htm>.

"The IPCC Controversy." Science & Environmental Policy Project. 2006 <http://www.sepp.org/ipcccont/ipcccont.html>.

The National Center for Atmospheric Research & the UCAR Office of Programs. 2006. UCAR. 2007 <http://www.ucar.edu/research/climate/>.

"The Science of Climate Change." The Royal Society. 19 Dec. 2006 <http://www.royalsoc.ac.uk/displaypagedoc.asp?id=11509>.

"Twentieth Session." IPCC. 19 Feb. 2006. 20 Dec. 2006 <http://www.ipcc.ch/meet/drepipcc20att-a.pdf>.

U.S. Environmental Protection Agency. 14 Dec. 2006. USEPA. 9 Feb. 2007 <http://epa.gov/climatechange/basicinfo.html>.

"U.S. Has Cooler September After Near Record Warm Summer, Global September Temperature Fourth Warmest On Record." National Oceanic and Atmospheric Administration. 16 Oct. 2006. U.S. Department of Commerce. 2006 <http://www.noaanews.noaa.gov/stories2006/s2720.htm>.

Wirth, Timothy E. Our Global Future: Climate Change. U.S. Government Printing Office, 1997.

Cover: Surabhi Sharma, SICSA

Part I:

1. Bernd Untiedt: www.commons.wikimedia.org
2. Osvaldo Gago: www.geocities.com
3. NATO
4. The Green Car Company. New Zealand: www.cleangreencar.co.nz
5. NASA
6. Michael Jastremski: www.friendster.com
7. Walter Siegmond: www.apo.nmsu.edu
8. J. Stein Carter, University of Cincinnati, Clermont College: www.biology.clc.uc.edu
9. Kyle R. Burton, Stanford University Info Lab: www.infolab.standford.edu
10. Filip Majikovic: www.papeline.no/photos/moelen
11. Turbo Squid: www.turbosquid.com
12. Stan Shebs, IV Design: www.ivtechservices.
13. Larry Bell
14. Larry Bell
15. Eindhoven Technical University: www.w3.tue.nl
16. The Cavalier Dailey: www.cavalierdailey.com
17. Simon Ho: www.simonho.org
18. Israel Export & International Cooperation Institute: www.export.gov.il
19. Filipe Fortes: www.fortes.com
20. Trek Design: www.fortes.com
21. Trek Design: www.fortes.com
22. Langfeld Designs: www.langfeldesigns.com
23. Space Daily: www.spacedaily.com
24. NASA
25. NASA: Spitzer Space Telescope image
26. NASA: Spitzer Space Telescope image
27. NASA
28. NASA
29. Washington Native Plant Society: www.wnps.org
30. Associated Content: scienceclarified.com
31. Spirit Alembic: www.spirit-alembic.
32. Microsopy UK Magazine: www.microsopy-uk.org
33. JW Adams Combined School: www.education.jlab.org
34. Society for Industrial Archaeology: www.lloyedminsterheavyoil.com
35. NASA
36. Video Weather & Weatherstreet: www.weatherquestions.com
37. Carleton College Science Education Research Center: www.serc.carleton.edu
38. NASA
39. Case Western Reserve University: www.filercase.edu
40. NOAA- www.noaa.gov.com
41. Science TREK- https://sciencetrek.org/topics/food-web
42. Purves, et al., Life: the Science of Biology: www.emc.marcopa.edu
43. Oregon State University- https://open.oregonstate.education/cellbiology/chapter/mitochondria-and-chloroplasts/
44. Larry Bell
45. Surabhi Sharma, SICSA; Bigelow Laboratory for Ocean Sciences
46. Bigelow Laboratory for Ocean Sciences: www.bigelow.org
47. Britannica- https://www.britannica.com/science/boundary-ecosystem
48. Its About Time: www.its-about-time.com
49. Larry Bell
50. Larry Bell/Candy Feuer
51. Windows to the Universe, University Corp. for Atmospheric Research: windows.ucar.edu
52. US Geological Survey: www.geolor.com/geoteach
53. Google Answers: www.content.answers.com
54. Estrella Mountain Community College, Online Biology Books: www.maricopa.edu
55. Northern Arizona University: jan.ucc.nau.edu
56. Northern Arizona University: www.usca.org
57. Global Warming Art: www.globalwarmingart.com
58. Hannes Grobe, Alfred Wegener Institute for Polar and Marine Research, Bremerhaven, Germany
59. NASA: www.gsfc.nasa.gov
60. NASA: www.gsfc.nasa.gov
61. Larry Bell
62. NASA: www.gsfc.nasa.gov
63. Larry Bell/Candy Feuer
64. Larry Bell/Candy Feuer
65. Larry Bell/Candy Feuer
66. NASA
67. Candy Feuer
68. Monash University Australia
69. Tony Boom
70. Looking Glass Productions, Ltd: www.looking.glass.productison.com
71. Chapel Electrics: www.newchapel.co.uk/nice
72. Oak Ridge National Laboratory: www.ornl.gov
73. World Energy Council: www.worldenergy.org
74. BBC Company, UK: www.newsring.bbc.uk
75. US Government: www.whitehouse.gov
76. Balanced Energy: www.balancedenergy.org
77. Power Technology: www.power-technology.com
78. NASA
79. Theresa Knott, Chemistry Daily: www.chemistrydaily.com
80. EIA- https://www.eia.gov/todayinenergy/detail.php?id=9991
81. US DOE-EIA- https://www.iea.org/countries/united-states/energy-mix
82. Larry Bell
83. Larry Bell
84. Ontario, Canada Ministry of Natural Resources
85. Precast Consulting Services: www.precastconsulting.com
86. Kanagawa Prefectural Education Center, Japan
87. Mikko Mikkola-Helsinki University of Technology

88. Kanagawa Prefectural Education Center, Japan
89. Anna Brown and Keith Jupp-Loughborough University
90. Natural Resources, Canada, Keely Net, Chemistryland, U of Michigan, Voice of America and US Geological Services
91. Larry Bell
92. Environmental Science and Engineering Magazine, March 2004
93. Agency for Toxic Substances and Disease Registry (ATSDR): www.atsdr.cdc.gov
94. Larry Bell
95. Larry Bell
96. National Renewable Energy Laboratory: www.nrel.gov
97. USDOE-EIA: www.eia.doe.gov
98. USDOE-Energy Efficiency and Renewable Energy Program: www.eere.energy.gov
99. Larry Bell
100. University of Minnesota: www.umn.edu
101. SOS-Arsenic: www.sos-arsenic.net
102. PT Pupuk Kujang: www.pupuk-kujang.co.id
103. KNOWASTE: www.knowaste.com
104. NASA
105. USDOE, Office of Fossil Energy: www.fossil.energy.gov
106. Larry Bell
107. National Renewable Energy Laboratory: www.nrel.gov
108. National Renewable Energy Laboratory: www.nrel.gov
109. Sandia National Laboratories: www.sandia.gov
110. US EIA- https://www.eia.gov/energyexplained/us-energy-facts/
111. USDOE-Energy Efficiency and Renewable Energy Program: www.eere.energy.gov
112. USDOE-Energy Efficiency and Renewable Energy Program: www.eere.energy.gov
113. Treehugger: www.adrants.com
114. Enviro-Energy Technologies: www.enviro-energytech.com
115. Cages by Design
116. Water History: www.waterhistory.org
117. Water History: www.waterhistory.org
118. Computer Science at St. Andrews: www.dcs.st-andrews.ac.uk
119. Lienhard Schulz: www.wikipedia..org/wiki/Benutzer
120. Philip Greenspun: www.philip.greenspun.com
121. Atlantisstrom Germany: www.atlantisstrom.de
122. Marine Current Turbines, Ltd: www.marineturbines.com
123. Wave Dragon: www.wavedragon.net
124. Evgeni Sergeev: www.meta.wikipedia.org
125. Andy Darvill: www.home.clara.net/darvill
126. ePrairie, Inc.: www.midwestbusiness.com/privacyasp
127. Genersys Ireland: www.genersys-ireland.com
128. National Renewal Energy Laboratory: www.nrel.gov
129. USDOE- Energy Efficiency and Renewable Energy Program: www.eere.energy.gov
130. Pacific Environment: www.pacificenvironment.org
131. Michigan Solar Solutions
132. Sacred Power Corp. www.sacredpowercorp.com
133. German Energy Agency
134. Fraunhofer Institute for Solar Energy Systems: www.ise.fhg.de/english
135. William M. Connolley. British Antarctic Survey: en.wikipedia.org
136. James Fraser: www.thefraserdomain.typepad.com
137. US Department of Labor- USHA: www.osha.gov
138. Henrik Storch-Methanol production via solar reforming of methane- Researchgate
139. USDOE-Energy Efficiency and Renewable Energy Program: www.eere.energy.gov
140. Energy- https://www.energy.gov
141. USDOE-Energy Efficiency and Renewable Energy Program: www.eere.energy.gov
142. Geothermal Energy Association: www.geo-energy.org
143. Renova- https://www.renovainc.com/en/business/geothermal/
144. Geothermal Energy Association: www.geo-energy.org
145. Geothermal Energy Association: www.geo-energy.org
146. Picture Newsletter: www.picture-newsletter.com
147. NASA
148. Bill and Vicki Tracey: www.flickr.com
149. Nuclear Energy Institute: www.nel.org
150. US EIA: https://www.eia.gov/uranium/
151. IAEA
152. IAEA- https://www.arcgis.com/
153. IEA- https://www.iea.org/data-and-statistics/charts/nuclear-power-capacity-by-region-in-the-announced-pledges-scenario-2010-2050
154. Tungsten: Public domain/Wikipedia
155. NASA/STAIF, Space Technology and Applications International Forum
156. Nuclear Power- https://www.sciencedirect.com/topics/
157. NASA/STAIF, Space Technology and Applications International Forum
158. Astronomy- www.astronomy.com
159. NASA
160. NASA
161. USDOE
162. Finland Regional Government
163. Elena's Puerto Rico Travel Guide, USA: www.elyunque.com
164. Farm credit east

165. Larry Bell
166. Ben Swihart: www.flickr.com/photos/graylight
167. Tyler Hamilton Foundation: www.tylerhamilton.com
168. USDOE-Energy Efficiency and Renewable Energy Program: www.eere.energy.gov
169. USDOE-Office of Civilian Radioactive Waste Management: www.ocrwm.doe.gov
170. GeoCities: www.geocities.com
171. National Renewable Energy Laboratories: www.nrel.gov
172. USDOE-Energy Efficiency and Renewable Energy Program: www.eere.energy.gov
173. US Bureau of Reclamation: www.usbr.gov
174. Berkeley Lab
175. USDOE-EIA: https://www.eia.gov/electricity/issuestrends/
176. Derek Jensen, Desert Morning News: www.desnews.com
177. Enoch Lau, Wikipedia: en.wikipedia.org
178. US Coast Guard: www.uscg.mil
179. Dennis Mojado: photography.mojado.com
180. Texas A&M University, TAMMI: www.tammi.tamu.edu
181. New York State Government Office: www.gorr.state.ny.us
182. USDOE-Energy Efficiency and Renewable Energy Program: www.eere.energy.gov
183. USDOE: ocrwm.doe.gov
184. Carplus: www.carplus.org.uk
185. National Railway Historical Society: www.nrhs.com
186. Phoenix International: www.phoenixint.com
187. Mobile Review: www.mobile-review.com
188. USDOE-EIA: www.eia.doe.gov
189. Larry Bell

Part II:

190. Imagesource: www.magesource.allposters.com
191. Amemco: www.amemco.net
192. University of California Museum of Paleontology, Berkeley: www.ucmp.berkeley.edu
193. NASA: www.earthobservatory.nasa.gov
194. Pennsylvania State University: www.rps.psu.edu
195. ABC Body Building: www.abcbodybuilding.com
196. RAD eCORE: www.radcore.org
197. PBS: www.pbs.org
198. University of Strathclyde, Institute for Energy and Environment: www.estru.strath.ac.uk
199. Aero Ecology, yair Zyndik et al.,: en.wikipedia.org
200. FCIPT/Institute for Plasma Research, India: www.plasmaindia.com
201. Sowerproject: www.sowerproject.org
202. Ymparisto: www.ymparisto.fi
203. Larry Bell
204. Larry Bell
205. Larry Bell
206. Larry Bell
207. University of Texas Environmental Sciences Institute: www.esi.utexas.edu
208. Larry Bell
209. Sierra Pacific Innovations (SPI): www.ir55.com
210. Silberwolf
211. Silberwolf
212. ConocoPhillips: www.conocolubricants.com
213. Keyosk: www.keyosk.co.uk
214. The Way Things Work, Simon & Shuster (1967)
215. JLC Electromet Pvt. Ld: www.jlcelectromet.com (India)
216. Jean-Pierre Daniel: www.commons.wikipedia.org
217. Ben Ostrowski: www.flicker.com
218. US Bureau of Reclamation: www.usbr.gov
219. Adrian Pingstone: commons.wikimedia.org
220. Popular Mechanics: www.popularmechanics.com
221. NASA
222. Larry Bell
223. Larry Bell
224. Enviro-Energy Technologies: www.enviro-energytech.com
225. Purdue University: www.purdue.edu
226. E-Marine, Inc.: www.e-marineinc.com
227. Micro-vett: www.micro-vett.it
228. University of Birmingham: www.aacg.bham.ac.uk
229. Larry Bell
230. Larry Bell
231. Pacific Summit Energy: www.pacificsummitenergy.com
232. Parker Construction, Inc: www.parkerconstruction.ca
233. Enter Stage Right: www.enterstageright.com
234. Kleenheat Gas Australia: www.kleenheat.com.au
235. Rawhideleasing- https://www.rawhideleasing.com/equipment/
236. Western Natural Gas: www.westernnaturalgas.com
237. SGC Engineering, LLC: www.sgcen.com
238. USDOE-EIA: www.eia.doe.gov
239. Yucca Mountain Information Office: www.yuccamountain.org
240. Colorado State University, Engineering: www.engr.colostate.edu
241. Alan D. www.flickr.com
242. Milica Sekulic: www.flickr.com
243. USDOE-Energy Efficiency and Renewable Energy Program: www.eere.energy.gov
244. Larry Bell
245. Chevron: www.chevron.com
246. Larry Bell
247. Larry Bell
248. Larry Bell
249. Larry Bell
250. The Pasty: www.thepasty.com
251. Subiaco Sustainable Demonstration Home: www.subiacosustainable.au.com
252. University of Warwick: www.ww2.warwick.ac.uk

253. MSNBC: www.msnbc.media.msn.com
254. Larry Bell
255. Larry Bell
256. Colorado State University, Warner College of Natural Resources: www.warnercnr.colostate.edu
257. ICL- icl-sf.com
258. Larry Bell
259. Agri Teach: www.agriteach.com
260. Energy Encyclopedia
261. Norbert Lechner, Heating, Cooling, Lighting (2001), Wiley and Sons
262. Norbert Lechner, Heating, Cooling, Lighting (2001), Wiley and Sons
263. Norbert Lechner, Heating, Cooling, Lighting (2001), Wiley and Sons
264. Norbert Lechner, Heating, Cooling, Lighting (2001), Wiley and Sons
265. Norbert Lechner, Heating, Cooling, Lighting (2001), Wiley and Sons
266. DIMEC, University of Salerno: www.dimec.unisa.it
267. DIMEC, University of Salerno: www.dimec.unisa.it
268. Theade- https://kojenturk.org/en/fuels-for-cogeneration-1285
269. Kelecraft E&M Consulting Engineers: www.iceloroft.com.hk
270. Larry Bell
271. Larry Bell
272. Larry Bell
273. Larry Bell
274. Larry Bell
275. SKF Evolution: www.evolution-skf.com
276. US Government
277. Tynagh Energy, Ltd.: www.tynaghenergy.ie.ndramart.com
278. Auto Junction, Indian Automobile Industry: www.auto.unc
279. Executive Branch, Student Government: www.studentorgs.edu
280. Oglata Lakota College: www.olc.edu
281. Larry Bell
282. Natural Resources Canada: www.nrcan.gc.ca
283. Natural Resources Canada: www.nrcan.gc.ca
284. Derek Jensen, Desert News: www.desertnews.com
285. SatCon Technology: www.satcon.com
286. American Wind Energy Association: www.awea.org
287. MIT https://energy.mit.edu/news/a-healthy-wind/
288. USDOE Energy Efficiency and Renewable Energy Program: www.eere.energy.gov
289. Dragonfly Energy
290. Iowa Energy Center: www.energy.iastate.edu
291. International Network for Sustainable Energy: inforseeinforse.org
292. USDOE Energy Efficiency and Renewable Energy Program: www.eere.energy.gov
293. Larry Bell
294. Water History: www.waterhistory.org
295. Energy Encyclopedia
296. US Army Corps of Engineers: www.usace.mil
297. Voith Siemens: www.voithsiemens.com
298. Forcefield: www.otherpower.com
299. World Ocean Review- https://worldoceanreview.com/en/wor-1/energy/renewable-energies/
300. Engineering Business: www.engb.com
301. SMD Hydrovision: www.smhydrovision.com
302. Dicander & C. O.: www.dicander.com
303. Larry Bell
304. BWEA Marine: www.bwea.com
305. Wavegen: www.wavegen.co.uk
306. Ocean Power Delivery: www.oceanpd.com
307. Wave Dragon: www.wavedragon.net
308. University of Oregon: www.uoregon.edu
309. C.R.Nave, Georgia State University: www.phy-astr.gsu.edu
310. TXU Energy: www.txu.com
311. Consumer Energy Council of America: www.cecarf.org
312. Helmut Hutten Motoren: www.hbraru.thinkquest.org
313. Siemens: www.siemens.com
314. Dal Engineering Group: www.dal.com
315. Global Power Resources: www.globalpowersales.com
316. Toshiba International Corporation: www.toshiba.com
317. Wikipedia: www.en.wikipedia.org , Public Domain
318. Wikipedia: www.en.wikipedia.org , Public Domain
319. Austrian Energy Agency: www.energytech.at
320. EEA- https://chp.ecatalog.ornl.gov/resources/faq
321. National Maritime Research Institute of Japan
322. PES Network, Inc: www.pureenergysystems.com
323. Karis Ruhe University: www.uni-karisruhe.de
324. Larry Bell
325. Matthia Sebulke:www.msebulke.tripod.com
326. Madgasters Photos: www.flickr.com
327. Popular Mechanics: www.popularmechanics.com
328. National Council of Educational Research and Training: www.ncert.nic.in
329. Eric Pierce: www.ericpierce.com
330. Hawaii Community College: www.hawaii.hawaii.edu
331. Watt Power Canada: wattpower.com
332. EIA- https://www.eia.gov/environment/emissions/carbon/
333. Honda
334. Honda
335. Romary: fr.wikipedia.org

336. US DOE Energy Efficiency and Renewable Energy Program: www.eere.energy.gov
337. Staxon GmbH: www.staxonide.com
338. European Commission: www.ec.europe.eu
339. JM Ogden et al., Journal of Power Sources (1999)
340. JM Ogden et al., Journal of Power Sources (1999)
341. JM Ogden et al., Journal of Power Sources (1999)
342. JM Ogden et al., Journal of Power Sources (1999)
343. Wisconsin Distributed Resources Collaborative: www.wisconsindr.org
344. Whole Building Design Guide: www.wbdg.org
345. US DOE Energy Efficiency and Renewable Energy Program: www.eere.energy.gov
346. NJ Solar Power: www.njsolarpower.com
347. US DOE Energy Efficiency and Renewable Energy Program: www.eere.energy.gov
348. US DOE Energy Efficiency and Renewable Energy Program: www.eere.energy.gov
349. US DOE Energy Efficiency and Renewable Energy Program: www.eere.energy.gov
350. NJ Solar: www.njsolarpower.com
351. US DOE Energy Efficiency and Renewable Energy Program: www.eere.energy.gov
352. US DOE Energy Efficiency and Renewable Energy Program: www.eere.energy.gov
353. Kelvin Husley Illustration, Inc.: www.khusley.com
354. Effpower, Sweden: www.effpower.com
355. Effpower, Sweden: www.effpower.com
356. Larry Bell
357. Axeon Power: www.axeonpower.com
358. NEDO Japan: www.nedo.go.jp
359. ITI Aturo Maligani: www.maligani.ud.it
360. Ian Smith: blog.joesgoals.com
361. Minnesota Legislature: www.leg.state.mn.us
362. Environmental and Energy Study Group: www.eesi.org
363. Marcello Moura: www.sxc.hu
364. Seeds of Change: www.seedsofchange.com
365. Fairfax County Public Schools: www.fcps.edu
366. www.flagstaffotos.com
367. US Fish and Wildlife Service: www.fws.gov
368. USDA- Agricultural Research Service: www.ars.usda.gov
369. Stan Shebs: en.wikipedia.org
370. Council of Industrial Boiler Owners: www.cibo.org
371. Biomass Energy: www.biomassenergy.org.uk
372. Larry Bell
373. INRA Montpellier: www.montpellier.inra.fr
374. Larry Bell
375. Clarke Energy: www.clarke-energy.co.uk
376. Larry Bell
377. Nebraska Corn Processing
378. Danish Ministry of Research and Information Technology: www.risoe.dk
379. Fleet Challenge Canada: www.fleetchallenge.ca
380. Larry Bell
381. Larry Bell
382. Claus Hindsgaul: en.wikipedia.org public domain
383. Alcyon Engineering: www.alcyon.cn
384. Madur: www.madur.com
385. European Commission, DG Research: www.combio-project.com
386. Larry Bell
387. University of North Dakota
388. US DOE Fossil Fuel Energy Office: www.fossil.energy.gov
389. Missoula News: www.missoulanews.com
390. Larry Bell
391. Larry Bell
392. SKF Group: www.skf.com
393. EV World: www.evworld.com
394. How Stuff Works: www.howstuffworks.com
395. Australian Cooperative Research Center for Renewable Energy: www.phys.murdoch.edu.au
396. Larry Bell
397. Larry Bell
398. Larry Bell
399. Larry Bell
400. Larry Bell
401. Greenpeace: www.greenpeace.org
402. University of Georgia: www.uga.edu
403. Minnesota Legislature: www.house.leg.state.mn
404. Sci/Tech: www.scitechinfo.com
405. About Spas: www.spas.about.com
406. Green Car Journal: www.greencar.com
407. Ministry of Tech: www.ministryoftech.com
408. The Tomorrow Bulletin: www.tomorrowproject.net
409. US DOE Energy Efficiency and Renewable Energy Program: www.eere.energy.gov
410. Umpqua Watersheds: www.umpqua-watersheds.org
411. Research Institute for Sustainable Energy: www.rise.org.au
412. USDOE-National Renewable Energy Laboratory: www.nrel.gov

Part III

413. Texas A&M University: www.cheweb.tamu.edu
414. USDOE Energy Efficiency and Renewable Energy (EERE): www.eere.energy.gov
415. CA Fuel Cell Partnership: www.cafcp.org
416. Celilo Group Media: www.media.nwcurrent.com
417. G-Man: Wikipedia, Public Domain Release
418. Touchstone Home Energy Systems: www.dixieec.apogee.net
419. Montana Department of Environmental Quality: www.deq.state.mt.us
420. Gunder-friend Productions: www.gunderfriend.com
421. Eau Claire, Wisconsin Economic Development: www.eauclaire-wi.com

422. Energy Xchange: www.energyxchange.org
423. Project GHB: www.ghb.org.uk
424. Tewkesbury Borough Council: www.tewkesburybc.gov.uk
425. Cismac Electronique: www.cismac.com
426. The Worlds of David Darling: www.daviddarling.info
427. Yelm Earthworms & Castings: www.yelmworms.com
428. How Stuff Works: static.hwostuffworks.com
429. US EIA- https://www.eia.gov/outlooks/ieo/consumption/sub-topic-01.php
430. University of Rochester: www.troi.rochester.edu
431. Las Vegas Review Journal: www.reviewjournal.com
432. Clipmarks: www.images.encarta.msn.com
433. Age Concern Cymru: www.accymru.org.uk
434. EIA - https://www.eia.gov/totalenergy/data/monthly/
435. EIA- https://www.eia.gov/energyexplained/us-energy-facts/
436. USDOE- Energy Efficiency and Renewable Energy Program
437. New Mexico Passenger Rail Action Group
438. USDOE-EIA Commercial Buildings Energy Conservation Survey, 1995
439. US EIA
440. US EIA- https://www.eia.gov/international/analysis/regions-of-interest/Caspian_Sea
441. USDOE-EIA https://www.eia.gov/todayinenergy/detail.php
442. College of Computer and Information Science: www.ccs.neu.edu
443. OEC- https://oec.world/en/profile/hs/crude-petroleum
444. Permianpartnership- https://permianpartnership.org/
445. US EIA- https://www.eia.gov/todayinenergy/detail.php?id=62444
446. MSCI- institute; EIA- https://www.msci-institute.com/charts/global-primary-energy-consumption-generation-in-2023/
447. USDOE-EIA World Energy Outlook, 2024: www.eere.energy.gov
448. US Government, National Energy Policy, National Energy Policy Group, May 2001
449. Independent Petroleum Association of America: www.ipaa.org
450. USDOE-EIA, CAPP, NRCan: www2.nrcan.ge.ca
451. US EIA - https://www.eia.gov/maps/
452. US EIA- https://www.eia.gov/energyexplained/natural-gas/natural-gas-pipelines.php
453. LNGallies- https://lngallies.com/charts/
454. FERC- https://www.ferc.gov/libraries/pdf.js/
455. USDOE-EIA: www.eia.doe.gov
456. US EIA- https://atlas.eia.gov/datasets/eia::petroleum-refineries-1/about
457. The Kentucky Network: www.ket.org
458. USDOE-EIA: https://www.eia.gov/coal/data/browser/#/topic
459. USDOE- EIA- https://wolfstreet.com/2024/
460. State of South Dakota Dept of Environment and Natural Resources: www.state.sd.us
461. NEI- https://www.nei.org/resources/
462. USDOE Grand Junction Project Office (1979): www.eia.doe.gov
463. DOE- https://www.kunc.org/environment/2019-10-21/
464. The Global Change Program, University of Michigan: www.globalchange.umich.edu
465. EIA- https://www.eia.gov/petroleum/ethanolcapacity/
466. Texas State Energy Conservation Office: www.seco.cap.state.tx.us
467. Arizona Solar Center: www.azsolarcenter.com
468. Iowa Energy Center: www.energy.iastate.edu
469. Bruce Center for Energy Research and Information: www.friendsofbruce.ca
470. Research Institute for Sustainable Energy: www.rise.org.au
471. Greenpeace: www.greenpeace.org.uk
472. Oak Ridge National Laboratory- https://storymaps.arcgis.com/
473. Oak Ridge National Laboratory- https://storymaps.arcgis.com/
474. US Geological Service: www.wfrc.usgs.gov
475. The Atlas of North America: www.nationalatlas.gov
476. North Carolina Department of Administration: www.doa.nc.us
477. WisconSUN: www.wisconsun.org
478. National Energy Research Laboratory: www.nrel.gov
479. The Global Change Program, University of Michigan: www.globalchange.umich.edu
480. Los Alamos National Laboratory: www.lanl.gov
481. Oak Ridge National Laboratory: www.ornl.gov
482. Western Libraries, University of Western Ontario: www.lib.uwo.ca
483. MYAA Site: www.filelibrary.myaasite.com
484. The Livingroom: www.livingroom.org.au
485. Larry Bell
486. Larry Bell
487. Larry Bell
488. Larry Bell
489. Larry Bell
490. Weber State University- https://physics.weber.edu/schroeder/energy/PowerPlantsMap.html?
491. Wikimedia- https://commons.wikimedia.org/
492. Larry Bell, compiled from USDOE- Energy Efficiency and Renewable Energy Program data
493. USDOE- Energy Efficiency and Renewable Energy Program

494. Larry Bell, compiled from American Coal Foundation, US Census and USDOE- Energy Efficiency and Renewable Energy Program data
495. USDOE/Static How Stuff Works: static.howstuffworks.com
496. Weber State University- https://physics.weber.edu/schroeder/energy/PowerPlantsMap.html?
497. Larry Bell, compiled from USDOE- Energy Efficiency and Renewable Energy Program data
498. CNBC- https://www.cnbc.com/2022/
499. Larry Bell compiled from GEC, National Renewable Energy Laboratory and Env. Systems Research Inst. Data
500. Larry Bell, compiled from USDOE-EIA data
501. Larry Bell, compiled from Governors' Ethanol Coalition (GEC), National Renewable Energy Laboratory and USDOE data
502. EIA- https://www.eia.gov/todayinenergy/detail.php?id=53539
503. Larry Bell, compiled from National Renewable Energy Laboratory GEC and American Coal Foundation (ACF) data
504. Larry Bell, compiled from National Renewable Energy Laboratory, GEC and ACF data
505. Larry Bell, compiled from US Census (2000), LLNL, ACF and GEC data
506. Larry Bell
507. Alachua County, Florida: www.alachuacounty.us
508. SEIA- https://seia.org/research-resources/solar-market-insight-report-q2-2023/
509. National Rural Electric Cooperative Association: www.nreca.org
510. Larry Bell
511. Wiregrass Coop Greenpower: www.wiregrass.coop/greenpower.html
512. Department of Sewage, Tokyo Metropolitan Government: www.gesui.metro.tokyo.jp
513. Practical Action: www.itdg.org
514. Larry Bell
515. Larry Bell
516. Lewis County Rural Electric Cooperative: www.lewiscountryrec.org
517. Virginia Tech Department of Mines, Minerals and Energy: www.geo4va.vt.edu
518. Water Wheel Factory: www.waterwheelfactory.com
519. Otherpower.com
520. ABS Energy
521. Canadian Renewable Energy Network: www.canren.gc.ca
522. Britannica
523. Britannica
524. Iowa Energy Center
525. Iowa Energy Center
526. Iowa Energy Center
527. NASA
528. IEA Photovoltaic Power Systems Program: www.iea-pvps.org
529. European Solar Engineering School: www.eses.org
530. Energy Supermarket: www.solardirect.com
531. Fraunhofer Institute for Solar Energy Systems: www.ise.fhg.de
532. Fraunhofer Institute for Solar Energy Systems
533. Provision Technologies: www.provisiontechnologies.com
534. SkyBuiltPower: www.skybuilt.com
535. National Photovoltaics: www.nationalphotovoltaics.com
536. Lawrence Berkeley National Lab: www.gaia.lbl.gov
537. University of Cambridge: www.building.act.cam.ac.uk
538. Walt Pyle et al., HomePower #39, February/March 1994
539. NJ Tech- https://njtechweekly.com/hopewell-hydrogen-house-dedication-draws-hundreds/
540. Panasonic: www.panasonic.co.jp
541. Larry Bell
542. Larry Bell
543. Larry Bell
544. The Apache Trail Archives: www.hmarc.com
545. National Renewable Energy Laboratory (NREL): www.nrel.gov
546. Powered Living: www.poweredliving.co.nz
547. European Solar Thermal Industry Federation: www.estif.org
548. Larry Bell/Olga Bannova
549. Green energy: www.greenenergy.org.uk
550. Larry Bell/Olga Bannova
551. USDOE- Energy Efficiency and Renewable Energy Program
552. Norbert Lechner, Heating, Cooling, Lighting (2001) Wiley and Sons
553. USDOE- Energy Efficiency and Renewable Energy Program
554. USDOE- Energy Efficiency and Renewable Energy Program
555. Norbert Lechner, Heating, Cooling, Lighting (2001) Wiley and Sons
556. USDOE- Energy Efficiency and Renewable Energy Program
557. USDOE- Energy Efficiency and Renewable Energy Program
558. Norbert Lechner, Heating, Cooling, Lighting (2001) Wiley and Sons
559. USDOE- Energy Efficiency and Renewable Energy Program
560. Larry Bell
561. Norbert Lechner, Heating, Cooling, Lighting (2001) Wiley and Sons
562. Larry Bell
563. Norbert Lechner, Heating, Cooling, Lighting (2001) Wiley and Sons

564. Accurate Building Inspectors: www.accuratebuilding.com
565. Larry Bell
566. US Department of Housing and Urban Development
567. The Worlds of David Darling: www.daviddarling.info
568. Norbert Lechner, Heating, Cooling, Lighting (2001) Wiley and Sons
569. Kbv Research- https://www.kbvresearch.com/
570. Energy Efficiency Guide for Utah Businesses: www.utahefficiencyguide.com
571. Oak Ridge National Laboratory: www.ornl.gov
572. Larry Bell
573. Georgetown County School District: www.gcsd.k12.sc.us
574. University of Illinois College of Veterinary Medicine: www.cvm.uiuc.edu
575. Direct Vehicle Leasing UK: www.directvehicleleasing.co.uk
576. Phronk: www.phronk.com
577. USDOE- National Renewable Energy Laboratory
578. All 4 Engineers: www.all4engineers.com
579. National Alternative Fuels Training Consortium: www.naftc.wvu.edu
580. Fair-PR, Hannover Messe 2007 Exhibit: www.fair.pr.com
581. Mitsui Bussan: www.mitsui-tr.co.jp
582. American Public Transportation Association: www.apta.com
583. Pictures, Direct News: www.pictures.directnews.org.uk
584. National Sustainable Agriculture Information Service: www.ncat.org
585. G.E. Dolbear & Associates, Inc.: www.coalscience.com
586. USDOE- Energy Efficiency and Renewable Energy Program
587. Intertek Petroleum Industry Affiliations: www.intertek-cb.com
588. USDOE
589. Larry Bell/Candy Feuer
590. Pyron Solar: www.pyronsolar.com
591. ESI Africa: www.esi-africa.com
592. Wikipedia Commons: en.wikipedia.org, Public domain
593- Energy
594- IAEA- https://www.iaea.org/newscenter/news/what-is-nuclear-fusion
595- Green Car Congress- https://www.greencarcongress.com/2014/
596- Energy Matters
597- Larry Bell
598- Larry Bell
599- Larry Bell
600- Chevron Corporation: www.chevron.com
601. Auckland City Council: www.aucklandcity.gov.nz
602. Earthworks Recycling, Inc.: www.earthworksrecycling.com
603. Truman State University: www.trumantoday.truman.edu
604. Rochester Environment: www.rochesterenvironment.com
605. State of South Dakota: state.sd.us

Part IV

606. Science Daily: www.sciencedaily.com
607. Jewish Community Center of Southern Nevada: wwww.jccsn.org
608. Marina del Ray, California: www.visitthemarina.com
609. Carla Perrotti: www.carlaperrotti.com
610. The Nature Conservancy: www.nature.org
611. Fox Island: www.foxisland.net
612. Rec Travel USA-Canada/Google: www.travel-destination-pictures.com
613. Nature Products Network: www.natureproducts.net
614. Dalhousie University: www.flame.cs.dal.ca
615. Windstar Wildlife Institute: www.windstar.org
616. Brad DeLong: www.myric.typepad.com
617. Larry Bell
618. Association for the Study of Peak Oil: www.aspo-usa.com
619. Larry Bell
620. NOAA- Trends in CO2 - NOAA Global Monitoring Laboratory
621. WFLA- https://www.wfla.com/weather/
622. Hendrick Avercamp, Dutch Painter (1585-1634)
623. NASA Goddard: www.rst.gsfc.nasa.gov
624. Content Answers: www.content.answers.com
625. Archaeology Info: www.archaeologyinfo.com
626. Michigan State University: www.msu.edu
627. Marian Koshland Museum of the National Academy of Sciences: www.koshland-science-museum.org
628. University Corporation for Atmospheric Research: www.car.edu
629. NASA: www.helios.gsfc.nasa.gov
630. NASA: www.svs.gsfc.nasa.gov
631. Isaac Newton Group: www.ing.iac.es
632. Orbit@Home: www.seti.dyndns.org
633. Petit et al., Nature, 1999: www.nature.com/nature/Journalv399
634. W. Broecker, Columbia University: www.anl.gov/Media_Center
635. Larry Bell
636. Robert A. Rohde, Global Warming Art: www.globalwarmingart.com

637. MDPI- https://www.mdpi.com/2072-4292/16/20/3776
638. NSIDS
639. ESIRO Marine and Atmospheric Research: www.cmar.au
640. NASA www.gsfc.nasa.gov
641. US Geological Survey: www.geochange.er.usgs.gov
642. US Climate Change Science Program: www.climatescience.gov
643. University Corporation for Atmospheric Research: ucar.edu
644. College of Arts and Science, University of Missouri: www.rcp.missouri.edu
645. Executive Action Team, Water Resources: www.exact-me.org
646. Julian Pye's Photodiary: www.photodiary.org
647. Solcomhouse: www.solcomhouse.com
648. Australian Academy of Science: www.science.org.au
649. Ljungqvist (2010)
650. NOAA - https://research.noaa.gov/no-sign-of-greenhouse-gases-increases-slowing-in-2023/
651. Larry Bell
652. Larry Bell
653. Larry Bell
654. Larry Bell
655. NOAA/NASA
656. Asel Electronic: www.asel.com/tc
657. Reaching Teachers to Teach Technology: www.rst2.eedu
658. NOAA Center for Coastal Env. Health and Biomolecular Research: www.chbr.noaa.gov
659. Welfare Service Club: www.clubs.ntu.edu.sg
660. Larry Bell
661. Larry Bell
662. Larry Bell
663. Children's Research Center: www.childrens-research-center.open.ac.uk
664. National Park Service: www.nature.nps.gov
665. US Department of State, Larisa Dubriansky: www.usinfo.state.gov
666. Salem County AVA Center: www.salemcountyava.org
667. Chemistryland: www.chemistryland.com
668. Chemistryland: www.chemistryland.com
669. NASA
670. NASA
671. Scientific American: www.sciam.com
672. European Space Agency: www.esa.int
673. United Nations Magazine: www.ourplanet.com
674. Vestal Design Blog: www.vestaldesign.com
675. Regional Municipality of Wood Buffalo: www.ab.ca
676. NASA: www.history.nasa.gov
677. Sydney Morning Herald: www.smh.com.au
678. The Dynamic Unity of Reality: www.spaceandmotion.com
679. Planetsave: www.watersecretblog.com
680. NASA
681. Our World- https://ourworldindata.org/co2-emissions
682- Larry Bell
683. Missouri Secretary of State: www.sos.mo.gov
684. URFAN: www.urfan.com
685. Hawaiian Airlines Star Bulletin: www.starbulletin.com
686. City of Cambridge: Cambridge.gov
687. NASA
688. Particle Physics and Astronomy Research Council: www.pparc.ac.uk

ILLUSTRATIONS

www.ingramcontent.com/pod-product-compliance
Lightning Source LLC
Chambersburg PA
CBHW061354240926
51846CB00007B/1400
9781960405531